Data Requirements for Integrated Urban Water Management

Urban Water Series – UNESCO-IHP

ISSN 1749-0790

Series Editors:

Čedo Maksimović
Department of Civil and Environmental Engineering
Imperial College
London, United Kingdom

Alberto Téjada-Guibert
International Hydrological Programme (IHP)
United Nations Educational, Scientific and Cultural Organization (UNESCO)
Paris, France

Data Requirements for Integrated Urban Water Management

Edited by

Tim D. Fletcher and Ana Deletić

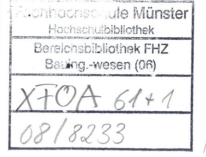
Cover illustration

New York City, NASA/GSFC/MITI/ERSDAC/JAROS, and U.S./Japan ASTER Science Team, with kind permission from NASA.

Published jointly by

The United Nations Educational, Scientific and Cultural Organization (UNESCO)
7, place de Fontenoy
75007 Paris, France
www.unesco.org/publishing

and

Taylor & Francis The Netherlands
P.O. Box 447
2300 AK Leiden, The Netherlands
www.taylorandfrancis.com – www.balkema.nl – www.crcpress.com
Taylor & Francis is an imprint of the Taylor & Francis Group, an informa business, London, United Kingdom.

Typeset by Charon Tec Ltd (A Macmillan company), Chennai, India
Printed and bound in Hungary by Uniprint International (a member of the Giethoorn Media-group), Székesfehévár.

ISBN UNESCO, paperback: 978-92-3-104059-7
ISBN Taylor & Francis, hardback: 978-0-415-45344-8
ISBN Taylor & Francis, paperback: 978-0-415-45345-5
ISBN Taylor & Francis e-book: 978-0-203-93247-6

Urban Water Series: ISSN 1749-0790

Volume 1

The designations employed and the presentation of material throughout this publication do not imply the expression of any opinion whatsoever on the part of UNESCO or Taylor & Francis concerning the legal status of any country, territory, city or area or of its authorities, or the delimitation of its frontiers or boundaries.
The authors are responsible for the choice and the presentation of the facts contained in this book and for the opinions expressed therein, which are not necessarily those of UNESCO nor those of Taylor & Francis and do not commit the Organization.

British Library Cataloguing in Publication Data
A catalogue record for this book is available from the British Library

Library of Congress Cataloging-in-Publication Data
Data requirements for integrated urban water management / Edited by Tim D. Fletcher and Ana Deletic.
        p. cm.
    Includes bibliographical references.
    ISBN 978-0-415-45344-8 (hardcover : alk. paper) – ISBN 978-0-415-45345-5 (pbk. : alk. paper) – ISBN 978-0-203-93247-6 (e-book)
    1. Sewerage. 2. Municipal water supply. I. Fletcher, Tim D. II. Deletic, Ana.

TD655.D38 2007
628.109173'2–dc22

                                                                2007032488

Foreword

The complexity and acuteness of urban water issues is increasing worldwide. Critical challenges to livability and sustainability include increasing demand and decreasing quality of available water, deteriorating infrastructure in cities in industrialized countries, and the water and sanitation needs of sprawling megacities in the developing world. These problems can no longer be addressed piecemeal; an overall holistic approach is required to formulate effective solutions.

The concept of Integrated Urban Water Management (IUWM) has been advanced to confront this situation. IUWM gives due consideration to all aspects and interactions of the urban water cycle, including water supply, wastewater, stormwater, groundwater, aquatic ecosystems and human health, with the aim of achieving sound and joint management. IUWM also considers interactions between the bio-physical and social components and their impact on the urban water cycle.

A fundamental prerequisite to IUWM is the availability of appropriate data. This allows for examination of the behaviour of individual water cycle components, thereby, enabling their interactions to be understood and predicted.

This book has been written to address the pertinent issues related to data requirements and management within the IUWM context. This publication is a part of a series of urban water management books produced within the framework of the Sixth Phase of UNESCO's International Hydrological Programme (2002–2007) and is the main output of the IHP project 'Data requirements for integrated urban water management'.

This book provides guidance to those involved in urban water management, focusing in particular on the collection, validation, storage, assessment and utilization of data to improve the integrated management of urban water resources. It presents a number of guiding principles that should be considered in developing any monitoring programme to support IUWM, as amply described in the Introduction.

In order to address the broad range of approaches for data gathering, management and quality control in urban water management, UNESCO established a working group combining representatives from various scientific backgrounds and experience. Their deliberations and joint efforts, under the capable leadership of the lead editors Tim Fletcher and Ana Deletić of Monash University, Australia resulted in this book. The full listing of contributing authors, encompassing a remarkable range of international affiliations is given in the List of Contributors and Acknowledgements.

The book has been produced under the responsibility of J. Alberto Téjada-Guibert, Deputy Secretary of IHP and officer responsible for the Urban Water Management

Programme of IHP, assisted by the Programme Specialist Sarantuyaa Zandaryaa and consultants Biljana Radojević, Wilfried Gilbrich and David McDonald.

We are grateful to all the contributors and the editors for their sustained efforts in the production of this publication, and are certain that the concepts and methodologies put forward will prove valuable for urban water managers, researchers, users and stakeholders in general across the globe.

András Szöllösi-Nagy
Secretary of UNESCO's International Hydrological Programme (IHP)
Director of UNESCO's Division of Water Sciences
Deputy Assistant Director-General for the Natural Sciences Sector of UNESCO

Preface

Pressure on urban water resources is increasing throughout the world, with problems of declining water quality, insufficient water to meet human demands, and degraded aquatic ecosystems. The concept of Integrated Urban Water Management (IUWM) has evolved from the recognition that the traditional approaches have not been adequate. IUWM attempts to manage all aspects of the urban water cycle (e.g. water supply, wastewater, stormwater, groundwater, aquatic ecosystems, and human health) together. This places a much greater emphasis on the interactions between various components (see Chaper 1, Figure 1.1), including interactions between the bio-physical and social components.

Often, these interactions are unintended, and their consequences are therefore rarely predicted. For example, discharge of polluted stormwater through an infiltration basin may result in polluted groundwater. In turn, rising groundwater levels may act as a conduit for leaking wastewater systems to transmit contaminated wastewater into leaking water supply systems.

A fundamental prerequisite to IUWM is *the availability of appropriate data to allow the behaviour of the individual water cycle components, and their interactions, to be examined, understood, and predicted.* This book provides guidance to those involved in urban water management, on the collection, validation, storage, assessment and utilization of data *to improve the integrated management of urban water resources.* Under IUWM, the paradigm changes from one of understanding the functioning of individual systems well, to understanding the overall system well. Such a paradigm shift brings substantial challenges, both technical and institutional.

There are a number of guiding principles that should be considered in developing any monitoring programme to support IUWM. The first overarching principle is that when collecting data on an individual water cycle component (e.g. water supply or wastewater), the monitoring should be designed to be useful for understanding other system components. Generally, this requires people with expertise in other system components (e.g. aquatic ecosystem health, groundwater quality) to be *involved in the design* of the monitoring programme from its inception.

Another critical principle is to ensure that the objectives are clearly identified *before* the monitoring commences. The objectives should be based on a conceptual model of the system(s) to be monitored and explicit identification of the application of the data. All relevant stakeholders should be involved in the definition of objectives. Objectives should be at first developed without consideration for financial constraints, so that the impacts of any financial constraints can be clearly identified. Once these constraints are identified, it is possible to identify the 'gap' between what *needs* to be done and what *can* be done. Such a transparent process will facilitate the decisions that then need to be made about either (i) how to get extra resources or (ii) which objectives are to be cut or reduced to meet the budget.

Based on the objectives, appropriate variables to measure will be selected, covering aspects such as water flow, water quality, infrastructure characteristics, ecological condition, economics, as well as social aspects (e.g. community preferences). The temporal and spatial scale aspects of each variable to be measured needs to be considered. For example, the *duration* and *frequency* of sampling needs to be determined, as does the *spatial extent* and *intensity* of measurements. A study of drinking water demand, for example, might need only annual data collected over a very wide area to identify broad trends in demand (for future infrastructure planning) or data collected every 15 minutes over a few months to identify periods of peak demand and optimize water distribution. These decisions can only be made based upon prior knowledge (e.g. a pilot study) of the system.

Before determining how measurements are to be taken, consideration needs to be given to the acceptable level of *uncertainty*, based on the objectives. Uncertainty consists of two types of error – *random* (caused, for example, by variations in measurement equipment and in natural processes) and *bias* (systematic errors, due to erroneous sensor calibration, for example). Processes for quantifying uncertainty derive from relevant European and US standards. It is important that decisions be made with explicit recognition of the uncertainty in predictions, so that risk can be considered in the decision-making process.

Monitoring equipment will be selected according to the required level of accuracy and precision and also the context of application (e.g. the available funds, the expertise of personnel, the ability to service and maintain equipment, etc.). The ability to be able to reliably calibrate equipment is critical. *Data validation* should be applied to *all data* collected, both static and dynamic, using a sequence of error detection and diagnosis. The project of the *Observatoire de Terrain en Hydrologie Urbaine* [Field Observatory for Urban Hydrology] (OTHU) in France provides a good example, using an automated seven-criteria validation process. Careful recording and maintenance of *metadata*, which informs future data users of the quality of the data and any modifications made to it is critical.

The process of turning data into useful information in order to generate knowledge involves data cleaning, integration, extraction, transformation, as well as pattern evaluation and finally representation of the new information and knowledge. Statistical analysis, simulation models, self-organizing maps and GIS can help in these tasks. However, data will only be useful if they are *available to those who need them*. This means that attention needs to be paid to data handling and storage to ensure security, durability and efficient sharing of data. For IUWM, data-sharing networks provide great opportunities, but also risks, especially with regard to version control, data compatibility and potential corruption. Protocols to manage these risks are outlined.

Organizations involved in IUWM need to demonstrate *leadership and commitment* to integrated data collection, to encourage *public participation* in the process (in order to use local knowledge and to understand community attitudes and values) and to use the data to ensure *transparency and accountability*. Appropriately collected data can also serve action-based learning, or adaptive management, both fundamental elements of IUWM. *Efficient data sharing* between agencies and other stakeholders is critical, since each agency will manage different parts of the urban water cycle, which may interact. Monitoring each component of the urban water system has specific requirements, especially given the need to understand each component's interactions with the other parts of the urban water system.

Urban meteorology is a fundamental driver for all aspects of the urban water cycle, affecting water demand, stormwater runoff, leakage into wastewater, groundwater recharge, and aquatic ecosystems. Monitoring should consider spatial and temporal variability and aim to characterize various cycles (e.g. diurnal variations in temperature, or seasonal variations in rainfall). Appropriate scales will depend on the monitoring objectives. Generally, a long time series of data on climate (e.g. two to three times the design life of proposed water infrastructure) are required in order to understand the influences of climate variability and change.

The *water supply* system affects wastewater production, aquatic ecosystems (e.g. through water extraction) and human health and is in turn potentially influenced by aspects such as groundwater quantity and quality. Advice is given on monitoring the performance of the water supply system, its inputs, storage, distribution network and consumers. Monitoring will be used to check water transfers, operate water quality treatment and diagnose problems and leaks.

Similarly, the *wastewater* system must be monitored for its interactions with components such as water supply (e.g. potential leakages into drinking water) and aquatic ecosystems (e.g. spills into streams). Measurements of wastewater quality usually involve measures of biochemical oxygen demand (BOD) and nutrient loads (nitrogen and phosphorus). Sewer leakage can be detected using indirect measures (e.g. water balance calculations or groundwater modelling) or direct measures (e.g. pressure testing, or tracer studies). Detection of cross-connections is also critical, but can be very difficult. Effects on receiving water must be measured (e.g. BOD, herbicides and endocrine disruptors), if these impacts are to be mitigated.

Stormwater may also have significant impacts on aquatic ecosystems, groundwater, wastewater systems and even drinking water (where stormwater forms part of the water source, for example). Measurement of catchment characteristics (e.g. proportion of impervious surfaces) along with the monitoring of climate, water flows and water quality are all required, matched to data on, for example, the condition of streams and aquatic ecosystems. Monitoring at short time-intervals is often required to capture the variability in stormwater quantity and quality, particularly during wet weather. *Combined sewers* carry wastewater and stormwater in a unified network; their behaviour is stochastic, and events, such as combined sewer overflows (CSOs), require specific monitoring approaches.

Groundwater is often forgotten in the urban water cycle, but it can be both a source of and sink for water and pollutants, and its behaviour (groundwater level and water quality) needs to be monitored in parallel with many of the surface water components. Similarly *aquatic ecosystems* express the impacts of water supply extractions and changes to water quality and flow regimes from stormwater, wastewater or groundwater. Aquatic ecosystem monitoring must simultaneously measure stressors (e.g. water quality) along with measures of ecosystem composition (e.g. species diversity) and function (e.g. ratio of photosynthesis to respiration).

There are, unfortunately, very few examples of fully integrated monitoring to support IUWM in the world. Two case studies are outlined in this book. While neither can claim to represent a fully integrated urban water monitoring system, covering all aspects discussed in this book, both illustrate some important principles and advances towards integration. The first case study is an integrated programme of monitoring climate, stormwater and groundwater in Lyon, France, namely, the OTHU Project. The second case study describes a wireless monitoring network applied to urban water supply and wastewater infrastructure in Boston, USA.

Tim D. Fletcher and Ana Deletić

Contents

List of Figures

List of Tables

Acronyms

AD	Anomaly detection strategy
ADAM	Anomaly detection and mitigation strategy
AET	Actual evapotranspiration
Ag/AgCl	Silver/silver chloride electrode
AIAA	American Institute of Aeronautics and Astronautics
APUSS	Assessing Performance of Urban Sewer Systems Project (EC project)
AQC	Analytical quality control
ARI	Average recurrence interval
ASIC	Application Specific Integrated Circuit
ASME	American Society of Mechanical Engineers
BMP	Best management practices
BOD	Biochemical oxygen demand
BOD_5	Biochemical oxygen demand (over a 5-day period)
CBA	Cost–benefit analysis
CFD	Computational fluid dynamics
Chl_a	Chlorophyll *a*
CIPM	*Comité International des Poids et Mesures*
CIS	Customer information systems
COD	Chemical oxygen demand
CSAIL	Computer Science and Artificial Intelligence
CSO	Combined sewer overflow
CSS	Combined sewer systems
Cu	Copper
DAVE	Data validation engine (software)
DBL	Daily BOD load
DBMS	Database management systems
DMA	District metering areas
DOC	Dissolved organic carbon
DRIPS	Data-rich information-poor syndrome
DSS	Decision support system
DTM	Digital terrain model
DWF	Dry weather flow
EC	European Community; electrical conductivity
EFD	European Framework Directive (EU)
EMC	Event mean consumption
EPA	Environmental Protection Agency (USA)

EPANET	computer model
ESML	Earth science markup language
ET	Evapotranspiration
EU	European Union
FRP	Filterable reactive phosphorous
FS	Full scale
GDP	Gross domestic product
GIS	Geographic information system
GLP	Good laboratory practice
GML	Geographic markup language
GPRS	General Packer Radio Service
GPS	Global satellite positioning system
GW	Groundwater
HEC-RAS	Computer model
HYSTEM-EXTRAN	Computer model
IBD	*Indice Biologique Diatomées* (Diatom Index)
IHP	International Hydrological Programme, UNESCO
INSA	*Institut National des Sciences Appliquées* (France)
ISO	International Standards Organization
IUWM	Integrated urban water management
IWA	International Water Association
LCC	Life-cycle costing
LCS	Leakage control system
LGCIE	*Laboratoire de Génie Civil et d'Ingénieurie Environnemental* (France)
LID	Low impact design
LIDAR	Light detection and ranging (dataset)
LIMS	Laboratory information management system
MCA	Multi-criteria analysis
MID	Measuring Instruments Directive (EU)
MMS	Maintenance management system
MOUSE	(stormwater model)
MVC	Model view controller
NATO	North Atlantic Treaty Organization
NGO	Non-government organization
NH_3	Ammonia
NO_{3-}	Nitrate
NO_X	Oxides of nitrogen
NRW	Non-revenue water
NTU	Nephelometric turbidity units
OEM	Original equipment manufacturer
OIML	*Organization International de Métrologie Légale*
OO	Object-oriented concept
OPTIMA	Optimisation for Sustainable Water Resources Management
OTHU	*Observatoire de Terrain en Hydrologie Urbaine* [Field Observatory for Urban Hydrology] (France)
PAH	Poly-aromatic hydrocarbon

PCA	Principal component analysis
PCB	Polychlorinated biphenyls
PET	Potential evapotranspiration
PLC	Programmable logic controller
PMP	Probable maximum precipitation
PO_4^{3-}	Phosphate
PON	Particulate organic nitrogen
PRV	Pressure regulating valves
QA	Quality assurance
RSS	Root sum of squares
RTC	Real-time control
SBC	Single-board computer
SCADA	Supervision, control, data acquisition and data analysis system
SEBAL	(remotely sensed data)
SIGNAL	(a benthic macro-invertebrate index)
SMART	Sustainable Management of Scarce Resources in the Coastal Zone
SMC	Site mean concentration
SMS	Short messaging service
SoE	State of environment
SOM	Self-organizing map
SQL	Standard query language
SWMM	Storm Water Management Model
TBL	Triple-bottom-line analysis
TDN	Total dissolved nitrogen
TDR	Time domain reflectometry
TIA	Total impervious service area
TKN	Total Kjeldahl Nitrogen
TN	Total nitrogen
TOC	Total organic carbon
TP	Total phosphorous
TSS	Total suspended solids
UGROW	(computer model)
UNESCO	United Nations Educational, Scientific and Cultural Organization
UV	Ultraviolet
VICAS	*Vitesse de Chute en Assainissement* (laboratory protocol)
WFD	Water Framework Directive (EU)
WHO	UN World Health Organization
WMO	UN World Meteorological Organization
WSN	Wireless sensor network
WSS	Water supply system
WSSS	Water supply and sewer systems
WWRG	Urban Water Research Group
WWTP	Waste water treatment plant
XML	Extensible markup language
YOPI	Young, old, pregnant or immuno-deficient person

Glossary

Ag/AgCl electrode   Standard electrode consisting of a silver (Ag) wire coated with silver chloride (AgCl) and dipping into a solution of chloride ions.

Aquifer   Geological formation capable of storing, transmitting and yielding exploitable quantities of water.

Aquitard; *syn.* **semi-confining bed**   Geological formation of low hydraulic conductivity which transmits water at a very slow rate.

Artificial neural network   A computing system modelled on the neural network in the animal nervous system, which converts one or more input signals to one or more output signals by means of an interconnected set of elementary non-linear signal processors called neurons.

Baseflow; *syn.* **base runoff**   Discharge entering a stream channel mainly from groundwater, but also from lakes and glaciers, during long periods when no precipitation or snowmelt occurs.

Benthic species/communities   Aquatic organisms that spend all or part of their life cycle at the sediment–water interface.

Biocenoses   a set of organisms, plants and animals living in ecological balance based on the chemical, physical and biological conditions in the local environment.

Biochemical oxygen demand (BOD)   The amount of dissolved oxygen consumed by micro-organisms as they decompose organic material in polluted water; a water quality indicator; BOD_5 is the biochemical oxygen demand over a 5-day period, that is, the amount of oxygen consumed by microorganisms over a five-day period as they decompose organic matter in polluted water.

Backwater   A body of water in which the flow is slowed or turned back by an obstruction such as a dam, an opposing current, or the movement of the tide. In this situation, the water level is said to be 'downstream-controlled'.

Chemical oxygen demand (COD)   Water quality indicator characterizing the potential of dissolved oxygen consumption based upon the chemical oxidation (in general by potassium dichromate) of organic and mineral compounds in the water.

Conservative matter; *syn.* **conservative pollutants**   Non-degradable matter or pollutants.

Cost–benefit analysis   Assessment of the direct (and possibly indirect) economic and social costs and benefits of a proposed project or programme.

Data noise   Any large quantifiable fluctuation in individual statistical data results, relative to the prominent data results; fluctuation resulting from the way that the data is recorded, not from the data itself.

Data sonde   A device that can monitor different properties of water, such as depth, temperature, conductivity, pH, dissolved oxygen, photosynthetic light penetration, water transparency, and chlorophyll content, usually used to monitor continuously over extended periods, and capable of storing the collected data internally, or externally via connection to a data-logger.

Δ p transducer; *syn.* **differential pressure transducer**   Instrument designed to receive waves from two independent and simultaneous pressure sources. The output is proportional to the pressure difference between the two sources.

Doppler sensor; *syn.* **radar sensor, sonar sensor**   A spatial type sensor that measures velocity of water from the Doppler frequency shift (ie. the change in the frequency and wavelength of the wave emitted from the sensor, as a function of its reflection from the flowing water).

Dual-reticulation system   A water supply system with two separate sets of pipelines, one for potable water, and the other for non-potable (usually recycled) water.

Event mean concentration   The average concentration of a pollutant concentration during a storm event, calculated by integrating the product of flow volume and pollutant concentration at a number of sampling times over an event, and dividing this value by the total event flow volume.

Exfiltration Flow of water out of a porous medium, such as the soil into the groundwater table below.

Externality   Outside effect, such as social and environmental benefits and costs, not included in the market price of goods and services being produced, or a consequence of an action that affects someone other than the agent undertaking that action and for which the agent is neither compensated nor penalized. Externalities can be positive or negative.

Flume; *syn* measuring flume   Constructed channel with clearly specified shape and dimensions, which may be used for the measurement of discharge.

Fourier transform   A mathematical technique (a set of mathematical formulas) used to convert a time domain signal to a frequency domain signal (Fourier analysis) and vice versa (Fourier synthesis).

Frequency domain analysis   A method of representing a wave form by plotting its amplitude against frequency.

Fuzzy logic   Mathematic technique capable of using qualitative, linguistic, and imprecise information. Relationships are based on a linguistic implication between an antecedent and its corresponding consequent.

Green accounting; *syn.* **environmental accounting**   Physical and monetary accounts of environmental assets and the costs of their depletion and degradation, accompanied by conventional economic accounts (e.g. measures of gross domestic product) with the ultimate objective of providing a comprehensive measure of the environmental consequences of economic activity.

Greywater   Domestic wastewater generated from dish washing, laundry and bathing, excluding blackwater (ie. wastewater from toilets).

Hydric exchange   The exchange of water between local environments, for example, between a stream and its riparian zone, or between the surface and sub-surface of a stream environment.

Hydrograph   A graphical representation of stage, that is, water depths above some datum, or discharge as a function of time.

Hypogean species   Animal species living underground (including the dark zone of caves) such as cave-restricted fishes.

Hyporheic organism   Species that live in underground water ecosystems, such as in the saturated sub-surface of a stream.

Infiltration   Flow of water through the soil surface into a porous medium, such as the soil, or from the soil into a drainage pipe.

Influent   Water (raw or partially treated) or other liquid flowing into a reservoir, basin, treatment process or treatment plant, for example, influent concentration refers to the concentration of a pollutant flowing into such a system.

Interflow   (1) That portion of the precipitation not passed down to the water table, but discharged from an area as subsurface flow into stream channels.
(2) flow of water from ephemeral (temporary) zones of saturation by moving through the upper strata of a geological formation at a rate much in excess of normal base-flow seepage.

Interstitial water; *syn.* **pore water;**   Water contained in the interstices, or voids of porous or fractured media, such as soil

Lentic   Referring to a freshwater habitat characterized by calm or standing water.

Life-cycle costing   The cost of a good or service over its entire life cycle, taking into account its capital costs (e.g. design and construction), ongoing costs (e.g. maintenance, periodic renewal) and decommissioning cost.

Limnimeter   A gauge used to measure water levels in rivers and lakes.

Linearity   In general, linearity describes the situation where there is a linear (straight-line) relationship between two variables. In relation to monitoring equipment, it describes how closely an instrument measures actual values of a variable through its effective range; a measure used to determine the accuracy of an instrument.

Lotic   Referring to fresh water habitat characterized by running water.

Markup language   Any computer language, such as HTML, that uses tags and attributes surrounding text to convey information such as format.

Measurand   Particular quantity subject to measurement.

Metadata   Information describing the type and characteristics of data sets and their location in a data archive, that is, data that defines, summarizes and describes other data.

Metrology   Science of measurement.

Monochloramine (NH_2Cl)   A chemical compound formed by the reaction of free available chlorine with nitrogen compounds (e.g. ammonia) under controlled conditions; a chemical hazard in drinking water.

Multi-criteria analysis (MCA)   A method used to formalize issues for decision making with several objectives being considered simultaneously, using both quantitative and qualitative indicator. MCA is not always intended to yield an optimum solution, and may instead help to clarify differences between options. The method includes techniques for ranking, rating, and pair-wise comparisons.

Nephelometric turbidimeter; *syn.* **nephelometer**   An instrument used to measure turbidity (the scattering of light by particulate matter within water). A nephelometer passes light through a sample and the amount of light deflected (usually at a 90-degree angle) is then measured. Turbidity is often used as a surrogate measure of the amount of suspended sediment in water.

Nowcasting   Short-term (varying from minutes up to a few hours) forecasting techniques used to predict severe weather.

Oligochaetes   Any class or order (Oligochaeta) of hermaphroditic terrestrial or aquatic annelid worms lacking a specialized head, such as earthworms.

Parshall flume   A device used to measure the flow in an open channel. The flume narrows to a throat of fixed dimensions and then expands again. The rate of flow can be calculated by measuring the difference in head (pressure) before and at the throat of the flume.

Perceptron; *syn*. neural net   A feedforward (i.e. information flows only in one direction, from input to output) type artificial neural network used for simple pattern recognition and classification tasks.

Piezometer   A well, totally encased except at its lowest end, used to measure the hydraulic head at that point.

Piezometric flow   Flow based on piezometric gradient and aquifer characterization.

Piezoresistive sensor   A sensor made of a material which resistance changes with stress and used in instruments, pressure-transducers, for measuring depth by pressure.

Pipe invert   The bottom of the inside of a pipe.

Pollutograph   A plot of pollutant concentration over time.

Redox; *syn*. oxidation/reduction reaction   A reaction that involves oxidation and reduction (i.e., transfer of electrons from one substance to another). The redox potential is an index giving a quantitative measure for the oxidation or reduction potential of an environment, which has major effects on water and soil chemistry and biochemistry (including retention or leaching of pollutants, level of microbial activity, etc).

Settling basin   A holding area for wastewater or stormwater, in which suspended solids are removed by the process of sedimentation.

Species of nitrogen   Chemical compounds containing nitrogen. In wastewater, four species (or forms) of nitrogen are commonly determined, namely: organic nitrogen; ammonia nitrogen, including both ammonium ion and free ammonia; nitrites; and nitrates.

Spectrometer   An instrument for measuring the intensity of radiation as a function of wavelength.

Stationarity   A form of homogeneity in a single characteristic (i.e. the property of a time series in which probability distributions involving values of the time series are independent of time translations).

Stormwater   Runoff from buildings and land surfaces resulting from storm precipitation.

TCP/IP protocol   Transmission Control Protocol/Internet Protocol; the most common protocol used for the transmission of data over the Internet.

Tensiometer   An instrument that consists of a porous cup, inserted in a soil and connected by a water-filled tube to a pressure measuring instrument (manometer), which makes it possible to measure the capillary tension or suction.

Throughflow   Shallow, lateral groundwater flow, usually above a perching layer (i.e. a confined layer which restricts one groundwater layer from percolating further below to an underlying groundwater layer).

Time-series analysis   Evaluation and interpretation of a set of hydrological or other measurements usually taken at regular intervals to determine the way in which these data vary over time.

Total suspended solids (TSS)   In a water sample, the total weight of suspended constituents per unit volume or unit weight of the water.

Treatment train   A series of stormwater treatment measures to provide a staged approach to removal of pollutants (most commonly from urban runoff) by changing the flow and characteristics of surface runoff such that the pollution impact on the receiving waters is minimized.

Triple-bottom-line analysis   An evaluation of a company's performance against three criteria, namely, economic, social and environmental impact.

Urban water cycle   A water cycle including all the components of the natural water cycle with the addition of urban flows from water services, such as the provision of potable water and collection and treatment of wastewater and stormwater.

Vadose zone; *syn.* unsaturated zone, zone of aeration   Subsurface zone above the water table in which the spaces between particles are filled with air and water, and the water pressure is less than atmospheric.

Venturi meter   An instrument for measuring flow of water or other fluids through closed conduits or pipes.

Wastewater   Water that has been used for domestic or industrial purposes and contains particulate matter and bacteria in solution or suspension, discharged typically into a sewage system.

Xenobiotic   A chemical compound that is not produced by, and often cannot be degraded by, living organisms; a chemical foreign to a biological system and displaced from its normal environment.

List of Contributors

Editors
Tim Fletcher, Monash University, Australia
Ana Deletić, Monash University, Australia

Principal Authors
Jean-Luc Bertrand-Krajewski, INSA-Lyon, France
Pascal Breil, Cemagref, France
Rebekah R. Brown, Monash University, Australia
Francis Chiew, CSIRO, Australia
François Clemens, Delft University of Technology, The Netherlands
Ana Deletić, Monash University, Australia
Tim Fletcher, Monash University, Australia
Shane Haydon, Melbourne Water, Australia
Čedo Maksimović, Imperial College London, UK
V. Grace Mitchell, Monash University, Australia
Marian Muste, University of Iowa, USA
Dubravka Pokrajac, University of Aberdeen, UK
Dusan Prodanovic, University of Belgrade, Serbia
William Shuster, US Environmental Protection Agency, USA

Authors
Tom Adams, NOAA, USA
Sylvie Barraud, INSA-Lyon, France
David Butler, University of Exeter, UK
Cécile Delolme, ENTPE, France
Janine Gibert, Université Claude Bernard Lyon 1, France
Ralph Kling, Intel Research, USA
Michel Lafont, Cemagref, France
Sam Madden, MIT, USA
Florian Malard, Université Claude Bernard Lyon 1, France
Lama Nachman, Intel Research, USA
Allison Roy, US Environmental Protection Agency, USA
Alan Seed, Bureau of Meteorology, Australia
Ivan Stoianov, Imperial College London, UK
Andrew Whittle, MIT, USA
Thierry Winiarski, ENTPE, France
Yu Zhang, US Environmental Protection Agency, USA

Acknowledgements

This book was developed under the auspices of the UNESCO International Hydrological Programme (IHP-VI). In particular, we wish to express our appreciation for the assistance and guidance provided by José Alberto Téjada-Guibert, Biljana Radojević and Wilf Gilbrich. The Department of Civil Engineering and Institute for Sustainable Water Resources at Monash University (Australia) provided resources to support this project, in particular the editorial support of Belinda Hatt. We appreciate also comments on the manuscript by Marie Keatley (University of Melbourne, Australia). We also express our sincere gratitude to the URGC-HU laboratory of INSA Lyon (France), which provided resources to assist with coordination and editing of this book. Lastly, we acknowledge the contribution of Thomas Einfalt, who provided information on urban meteorology.

Chapter 1

Introduction

T.D. Fletcher[1], V.G. Mitchell[1], A. Deletić[1] and
Č. Maksimović[2]

[1]Department of Civil Engineering (Institute for Sustainable Water Resources), Building 60, Monash
University, Melbourne 3800, Australia
[2]Department of Civil and Environmental Engineering, Imperial College London, London SW7 2AZ, UK

with contributions from

P. Breil[3], D. Butler[4], F. Clemens[5], D. Prodanović[6], and
D. Pokrajac[7]

[3]Unité Hydrologie-Hydraulique, Cemagref, bis Quai Chauveau, Lyon, 69336 CEDEX, France
[4]Centre for Water Systems, School of Engineering, Computer Science and Mathematics, University of
Exeter, Harrison Building, North Park Road, Exeter EXF 4QF, UK
[5]Delft University of Technology, Faculty of Civil Engineering, Sanitary Engineering Section, 2600 GA
Delft, The Netherlands
[6]Institute for Hydraulic and Environmental Engineering, Faculty of Civil Engineering, University of
Belgrade, Bulevar Kralja Aleksandra Belgrade, Serbia
[7]Department of Engineering, Kings College, Aberdeen, Scotland, UK

1.1 INTRODUCTION

1.1.1 Background and context

Increasing pressure on both the quantity and quality of water resources is occurring throughout most, if not all, of the populated world (Maksimović and Tejada-Guibert, 2001). In urban areas, that pressure is evident across all aspects of the water cycle – including water supply, wastewater, stormwater, groundwater and aquatic ecosystems. As a result, there is increased attention to the management of the urban water cycle in a more integrated way with the ultimate aim of increasing the sustainability of water resources and of attaining or maintaining high levels of public health, flood protection and enhancement of aquatic ecological status.

Integrated Urban Water Management (IUWM), as is it often termed, is, however, easier said than done. Integrated management requires the individual components of the urban water cycle to be designed and managed *together*, rather than individually. Consequently, there is also a need to manage the *interactions* between urban water cycle components. For example, if wastewater is recycled, this may have an impact not only on potable water demands, but also on flow regimes in downstream rivers and creeks. *A fundamental prerequisite to IUWM is thus the availability of appropriate data to allow the behaviour of the individual water cycle components, and their interactions, to be examined, understood, and predicted.*

1.1.2   Objectives and scope

This book provides guidance to those involved in urban water management, to allow them to collect, store, validate, assess and utilize data which will improve the integrated management of urban water resources. This book deals with the physical and chemical aspects of IUWM (i.e. the data required for describing the quality and quantity of water and interaction with ecosystems) as well as the social and institutional aspects, both in terms of the type of data required and also how institutional capacity may affect data collection, interpretation and decision-making. The book does not provide detailed description of sensor technologies nor detailed monitoring protocols for specific water components. There are many existing local, national and international guidelines providing this information (refer to Section 1.3). This book aims, therefore, to provide guidance on the considerations necessary when *the ultimate use of data is to manage urban water resources in an integrated manner*.

Specifically, this book provides guidance on:

- defining objectives and applications of monitoring programmes to support IUWM
- monitoring design and selection of variables to measure
- considering spatial and temporal scales
- selecting of monitoring equipment
- data validation, handling, and storage
- understanding and evaluating uncertainty (and its subsequent use in decision-making)
- utilizing data to provide information and knowledge
- considering institutional arrangements necessary to achieve integration
- considering specific components of the urban water cycle and their interactions.

Part I of this book provides a description of general principles, while Part II provides more specific descriptions of each water cycle component often with illustrative examples. Part III then provides two case studies, both demonstrating many of the principles discussed in this book; they illustrate attempts to deliver a more integrated monitoring system to enable more efficient management of the urban water cycle.

The target audience for this book includes the wide range of stakeholders who have responsibility for managing urban water resources, regardless of whether they operate within a government, education, research or private institutions. For example:

- *Water system operators*, who may need to understand how water system components will operate with alternative supplementary water resources (e.g. stormwater harvesting or recycling of wastewater), and who may need to use data for real-time control of water systems (e.g. flood management systems).
- *Water system planners*, who may design integrated urban water systems for new developments, or for retrofitting existing service areas and who may wish to develop programmes to assess the performance of their designs once implemented.
- *Ecosystem managers*, who need to understand the impacts of the urban water system on aquatic ecosystems and how these impacts may change under alternative management scenarios.

- *Researchers, model developers and users,* who rely on (often complex) models to predict the performance of water systems, and who need data suitable for developing, calibrating, validating and applying such models.

1.1.3 Why is an integrated data collection approach needed?

There are many examples throughout the world of where a non-integrated approach to urban water management has led to problems, because interactions between various components were not well understood and thus not foreseen. For example:

- Discharge of polluted stormwater through infiltration and subsurface treatment may result in pollution of groundwater.
- Rising groundwater levels may act as a conduit for leakage from wastewater systems, transmiting contaminated water into susceptible water supply systems.
- Increasing water use by consumers may result in increased wastewater discharge, which may have consequences for stormwater and natural ecosystems (through leaks and spills), potentially also impacting human health and food resources.

Unless integrated monitoring of individual elements of urban water systems, and their interactions, is undertaken, such effects will typically be identified only once they have already caused major impacts, rather than early on when they can be mitigated more readily.

1.2 OVERVIEW OF INTEGRATED URBAN WATER SYSTEMS

1.2.1 IUWM concepts

Traditionally urban water management has treated water supply, wastewater, stormwater and groundwater as largely separate elements, with planning, installing and operating these components of the urban water cycle would making little reference to one another. Despite this approach, the reality is that these systems have substantial interaction, whether intended or not. For example, as an urban area grows, the carrying capacity of the water infrastructure such as a sewer system can be exceeded, leading to problems such as uncontrolled overflows of poor quality water into rivers. Similarly, leaking water supply systems may result in infiltration into wastewater systems.

Integrated Urban Water Management (IUWM) takes a comprehensive approach to urban water services, viewing water supply, stormwater, groundwater and wastewater as components of an integrated physical system and recognizing that the physical system sits within an organizational framework and a broader natural landscape (Mitchell, 2006). As Niemczynowicz (1999) stated 'we are beginning to talk not only about some new isolated technologies but instead about new *total system solutions*'. Such an integrated approach is needed in order to identify opportunities that are not apparent when separate planning is conducted for each individual component (Anderson and Iyaduri, 2003). This integrated approach will enable synergies to be identified and realized, providing opportunities for more sustainable, more cost-effective solutions to the challenges of managing the urban water cycle in the twenty-first century. Therefore, to capture the multiple benefits that IUWM seeks to achieve, planning and management should be integrated across traditional infrastructure domains and institutions.

The key concept of IUWM is that urban water systems should be seen as a whole, placing greater emphasis on how components 'interact'. This focus has arisen from the realization that urban water systems are complex with feedback loops that must be considered in planning and management to ensure that unintended negative outcomes of operations can be minimized or averted. This represents a substantial change of paradigm and requires changes in the way in which urban water services are designed and operated. This also leads to subsequent changes in data needs, the focus of this book. For further description and analysis of the emerging trends in IUWM, refer to Butler and Parkinson (1997), Chocat et al., (2001), Harremoes (1997), Maksimović and Tejada-Guibert (2001), Mitchell (2006) and Niemczynowicz (1999).

Currently, data are collected generally with a narrow scope, usually serving a single purpose only. Such data are insufficient to capture the complexity inherent in urban water systems. For example, a residential water efficiency programme will not only reduce the water usage of individual households, it will reduce the amount of water harvested from supply catchments, change the way in which the bulk transfer system of trunk mains and balancing reservoirs are operated, alter the peak flows in the local water reticulation infrastructure, reduce both the volume and concentration of wastewater discharged to the local wastewater reticulation infrastructure, reduce the amount of wastewater processed at the sewage treatment plant and reduce the amount of water ultimately released back into the environment, possibly affecting surface and groundwaters and their ecosystems. All of these effects may be assessed as positive, but there may be circumstances where one or more could have negative side effects, such as the reduction in water usage could reduce the flow within sections of the local wastewater network to below the self-cleansing velocity, thus creating clogging problems. This is an illustration of an unintended negative outcome that could occur if the potential connection between water supply services and wastewater services is not considered.

Considering how individual components interact with surrounding systems raises the question of system boundaries. For example, the provision of water supply often requires the harvesting of water from surface or groundwater catchments which are located a considerable distance from the urban area in question. This book provides guidance primarily on processes occurring within the urban boundary, although it is important to consider the impacts that planning and operational decisions have on the broader system. Sustainable urban development requires the identification, quantification and often preservation of the dynamic of natural systems on which the urban development depends.

IUWM also challenges the view of urban water service provision being a purely technical function, resting within the realm of engineers and scientists alone. The change from traditional practice to IUWM has considerable social and economic implications for urban communities and the range of public and private institutions providing urban water services to that community. Increasingly, integration is being implemented in managerial functions, in organizational structures and even in legislation, for example, the European Water Framework Directive (European Community, 2000). Integrated urban water management can thus only be achieved if the organizations responsible for different parts of the urban water cycle are prepared to work together and to share data and knowledge. All stakeholders need to be aware of the data available and to have ready access to them.

We have argued the case for a shift in paradigm towards *integrated* urban water management, which in turn will require integrated datasets. This is a challenge in its own right requiring standardization of system component descriptions, data format and data handling frameworks. The details of these requirements and guidance on how to achieve them will be provided in subsequent Chapters.

1.2.2   Urban water system components

Urban water systems contain a myriad of components (Figure 1.1), which collectively provide water supply to urban populations as well as wastewater (sanitation), flood protection, surface and groundwater management. Urban water systems also contain components which serve the environment through ecosystem maintenance and protection, upstream and downstream of an urban area, as well as within its boundaries. These urban water system components are far from discrete, but rather, interact significantly. Consideration of these often complex interactions is needed, in order to maximize the positive and minimize the often unintended, sometimes negative outcomes.

This section aims to provide an overview of the components that comprise the urban water system, discuss how these components interact and explain the implications for data collection and management.

Expanding on the view of an urban water system provided in Figure 1.1 and further exploring how urban water systems interact within an urban area, Figure 1.2 highlights a few of the ways in which water supply, wastewater and stormwater systems interact

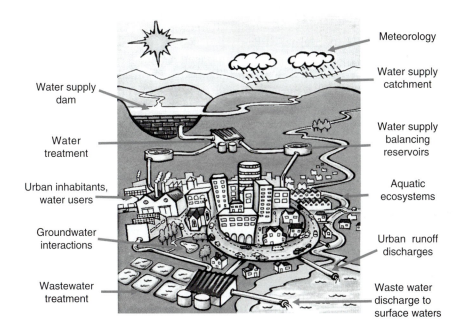

Figure 1.1   An urban water system site within the broader landscape from which it draws water for supply and into which it discharges stormwater and wastewater (see also colour Plate 1)

Source: Adapted from Parliamentary Commissioner for the Environment, 2000.

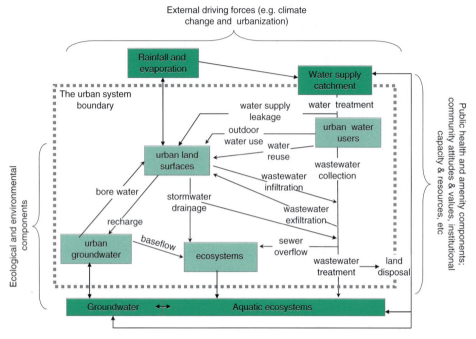

Figure 1.2   Interactions within an urban water system

with each other and with urban aquatic and groundwater ecosystems. It also highlights
the external driving forces on the urban water system as well as the public health,
amenity, ecological and environmental components of the system. In addition, the
whole urban water system sits within a social and institutional framework. For example, operation of the urban water system will reflect community values and expectations. The capability, attitudes and resources of organizations responsible for managing
different parts of the urban water cycle, will also reflect these values and expectations.

1.2.2.1   Water supply

An urban water supply system typically consists of a number of components: water
withdrawal, conveyance of untreated water, supply reservoirs, water treatment,
conveyance of treated water, distribution network and finally, the consumers (or water
customers). The water supply system is intrinsically linked with the rest of the urban
water system. For example, the majority of water flowing through the wastewater
system comes from the water supply system. In situations where there are leaks in the
water supply reticulation system, wastewater can be sucked into the supply pipe if its
pressure drops below the ambient atmospheric pressure, potentially compromising
public health. Stormwater, groundwater and treated wastewater are all potential
sources of water supply within an urban area, be it for potable or non-potable applications. Furthermore, leakage from the water supply reticulation system can impact
on urban groundwater. A well-designed water supply system monitoring programme
should take account of these interactions and collect sufficient data to evaluate them.

Each component of the water supply system, such as dams, reservoirs, treatment plants and distribution network, has specific monitoring requirements in order to manage and optimize its performance. However, the overarching monitoring requirement is the quantification of the water balance for each of the components in the water supply system and the system as a whole. When different components or sections of the system are operated by different organizations, it is also important that the water balance at the point of exchange is accurately monitored. Data sharing between agencies then becomes a critical requirement.

Maintenance of adequate supply pressure and water quality throughout the conveyance and distribution system is an important performance criterion. Given the importance of pipe leakage minimization on both the conservation of water supply and avoiding the raising of urban groundwater levels, it is also a significant focus of the monitoring system. Therefore, the continuous monitoring of water flow rate, pressure or level and water quality parameters should occur at key locations throughout the system. Increasingly the monitoring of water supply systems is quite sophisticated requiring the use of informatics.

1.2.2.2 Wastewater management

The provision of wastewater sanitation systems has led to a significant decrease in the occurrence of water-borne disease in urban areas. Since the mid-nineteenth century urban sanitation in the developed world has taken the form of infrastructure to collect and transport wastewater out of urban areas, in conjunction with the provision of reliable drinking water services. Surprisingly, the effectiveness of this approach to sanitation in terms of increase in life expectancy relative to the financial investment is not well quantified. Many countries in the developing world remain without reliable water supply or wastewater treatment.

In many countries, wastewater systems are integrated with the urban drainage system through the use of combined sewer systems. These systems collect both wastewater and stormwater, transporting them in a single system of channels and/or underground pipes. In most cases the collected wastewater and stormwater is treated and then discharged to receiving surface water bodies, usually downstream of the urban area. Combined sewer systems are the predominant type in the developed world due to their relative robustness, low investment cost and simple operation and maintenance in comparison to separated wastewater and stormwater systems (there are notable exceptions though, for example throughout Australia where urban areas use separate wastewater and stormwater systems). The combination of an urban sewer system (be it separate stormwater and wastewater systems or a combined system) and wastewater treatment plant is referred to as the wastewater system.

The presence of effective wastewater systems helps to protect water quality in receiving waters, thus reducing impacts on aquatic ecosystem. However, despite substantial efforts to reduce wastewater discharges to receiving waters, wastewater water systems and particularly combined sewer systems, frequently spill diluted wastewater during rainfall events when the sewer system storage and discharge capacity is exceeded, leading to discharges via 'emergency outlets'. While separate wastewater and stormwater systems reduce the probability of such occurrences, infiltration of stormwater into the wastewater network can still result in occasional spills. These instances show the importance of monitoring the performance of wastewater systems – in their various forms – to ensure their efficient operation and maintenance. This may involve monitoring the stormwater system, wastewater system, as well as the health and functioning of downstream aquatic ecosystems.

Experience also shows that leakage from sewers into groundwater and vice versa, can increase significantly as the sewer pipes age. It is possible for up to half of the dry weather flow in a sewer to be due to groundwater infiltration. These enhanced wastewater flows during dry weather adversely effect the operation of treatment plants, increasing energy and chemical consumption, and decreasing treatment effectiveness due to dilution of the wastewater. Monitoring of dry weather wastewater flows in the pipe collection system during periods of low water usage, such as the middle of the night, is one method to detect the infiltration of groundwater. In comparison, the leakage of wastewater out of sewer pipes into urban groundwater bodies poses an even greater challenge. This is because its occurrence is very hard to detect and often goes unnoticed. One way in which it is sometimes detected is through the quality control system implemented when the groundwater is being used as a source of water for drinking water production.

Effective wastewater management involves detailed monitoring to understand aspects of the operation and behaviour of wastewater, such as:

- the effect of rainfall and subsequent discharges into receiving water bodies
- the effects of wastewater discharges (both dry weather discharges and combine sewer overflows) on receiving water bodies
- the presence of endocrine disruptors in wastewater and receiving waters
- the incidence of water-borne diseases
- the effectiveness of separated systems (and other alternatives to combined systems such as dry sanitation) compared with combined systems in relation to the resources and investment required.

1.2.2.3   Stormwater

Stormwater describes the water that originates from precipitation on the land surface, which then runs off (predominantly from impervious surfaces, in the urban context), often through a constructed drainage network before entering a receiving water body, such as a lake, stream, river or estuary. The quantity and quality of stormwater will depend on climate, land use, particularly land surface cover, and the nature of the drainage network. As described above, stormwater may interact directly with wastewater, through combined sewer systems. To manage urban stormwater and its impacts, we therefore need to know something about its generation, transport and effect on receiving waters and other parts of the urban water cycle.

Data are therefore required on the characteristics of the catchment, such as area, proportion of impervious surfaces (specifically those directly connected to the drainage network and thus deemed to be hydrologically effective [Hatt et al., 2004]), average slope and proportion and location of different types of land use. However, we also need to know something of the meteorology, especially the amount and distribution (spatial and temporal) of rainfall, as well as evapotranspiration. Combined, the catchment and meteorological data will allow us to estimate both the *quantity* and *quality* of data.

To understand the transport of stormwater, we need data on the drainage network, whether separate from sanitary sewers or in a combined system, and details of the slopes, dimensions and condition of the network. We will then make measurements of the quantity and quality of stormwater within the network, in order to calibrate predictions based on the catchment and drainage network characteristics.

Lastly, we wish to understand the impacts of stormwater on receiving waters and other urban water cycle components, such as streamflow, groundwater and wastewater. We must therefore measure fluxes of both the water and the constituent pollutants which it may be carrying. This may involve sampling at points within the catchment and drainage network and also at discharge points. In particular, monitoring is likely to be required upstream and downstream of stormwater treatment systems, such as detention basins or constructed wetlands, to determine changes in flow and quality.

Ultimately, however, to understand impacts of stormwater on other urban water cycle components, integrated monitoring of both the stormwater component and the other(s), will be required. For example, in catchments with separate stormwater and sanitary sewer systems, stormwater may infiltrate into the wastewater system, exceeding the capacity of wastewater treatment plants. Conversely, sewer overflows may result in contamination of stormwater systems with major potential human and ecosystem health impacts. Integrated monitoring of stormwater, wastewater and aquatic ecosystems will thus be required.

A key consideration in the monitoring of stormwater, its impacts and its interactions with other urban water system components, is the short temporal scale at which stormwater often operates. Particularly in small, urban catchments with high levels of imperviousness and thus very rapid production of runoff, data collected at short intervals throughout the drainage network is needed to understand the implications of storm events.

1.2.2.4   Groundwater

Urban groundwater interacts with all other urban water systems, acting either as a source or sink of water. This interaction may be deliberate, such as groundwater abstraction for water supply or groundwater recharge from infiltration wells. More often the interaction is neither planned nor deliberate and occurs because of the proximity of groundwater to a network of leaking pipes. In this way water supply pipes usually recharge groundwater, except during the pipe maintenance, when the pipes may also act as drains if groundwater levels are high. In the former case the artificially raised groundwater levels may create a potential problem, while in the latter case there is a potential for the pollution of the water supply system.

The interaction of groundwater with sewers occurs in both directions, depending on the groundwater level and the water level or pressure in the sewer. If groundwater level around a sewer is higher than the sewer itself, groundwater will feed the sewer, decrease its capacity to receive the stormwater/wastewater and increase the load on the treatment plant. If groundwater level around a sewer is deep, water from the sewer will exfiltrate through the cracks in the pipe and recharge groundwater. The main problem this can create is groundwater pollution, although the resultant raising of groundwater levels could also be undesirable.

Since there is significant interaction between groundwater and other urban water system components, particularly water supply and sewer pipes, the management of urban aquifers is closely related to the management of other urban water systems. As their management requirements are often conflicting, an integrated approach is needed, so that all of these systems are considered simultaneously.

1.2.2.5   Urban aquatic ecosystems

As illustrated in Figure 1.2, urban aquatic ecosystems are linked to both urban surface waters and urban groundwater. Aquatic ecosystems can be degraded by the changes in flow and water quality, a result of urbanization and its related impacts on the urban water cycle. For example, the long-term load of pollutants accumulating in large water bodies and the occurrence of an acute pollution event that releases toxic concentrations of pollutant into a waterway can severely impact the condition of the ecosystem. In smaller aquatic ecosystems, such as ponds, wetlands and small streams, ecological conditions are very sensitive to the degree to which the natural flow regime has been disturbed, for example, from stormwater inflows or conversely from extractions for water supply, as well as to variations in pollutant concentrations.

Whatever the physical habitat features, water quality is a limiting condition for aquatic ecosystem function. Any toxic or persistent pollution source can seriously impair the aquatic ecosystem. Because the ecosystems of lentic waterbodies, such as lakes and ponds, do not instantaneously react to the pollution input and develop their own response dynamics to physical or chemical changes, continuous water quality monitoring is needed. Continuous monitoring of water quality parameters relevant to aquatic ecosystems, such as dissolved oxygen, pH and water temperature, can be easily implemented using current technology with low ongoing maintenance costs.

However, it is also important that water flows (or fluxes) are monitored in conjunction with any water quality parameters that are monitored. This enhances the interpretability of the water quality data and supports more robust management decisions. More importantly, degradation of an aquatic ecosystem is often a product of *both* water quality and hydrologic changes and their interactions. Thus, in order to preserve aquatic ecosystem function, environmental flows should be provided or maintained along with the management or protection of water quality. The interaction between surface water bodies and the urban environment should be managed to provide an effective water flow control strategy and enable the bio-assimilation function an aquatic ecosystem to develop. For example, a primary concern is often the amount of surface runoff diverted away from urban aquatic ecosystems by combined sewer systems and the seepage of groundwater into sewer and stormwater pipes.

Improvements to the physical habitat characteristics of urban aquatic ecosystems will not succeed if the pollution input exceeds the local bio-assimilation capacity of an aquatic ecosystem. Data on phenomena such as groundwater table levels, groundwater temperature, combined sewer system overflows to receiving waters (measured or modelled), as well as other water and wastewater discharges, should be integrated into a common database, in an attempt to explain the relationships between various aspects of the urban water cycle and the resultant ecosystem condition and function. Such a database could then be used to develop a comprehensive urban aquatic ecosystem management plan.

1.2.2.6   Social and institutional components

In addition to the biophysical data, understanding the social and institutional parameters is critical to IUWM. Of all components, it is perhaps the social and institutional that have the broadest interactions with all other components of the urban water system; they act as over-arching parameters. For example, social attitudes will affect preferences regarding water supply, management of wastewater and groundwater and the values placed on aquatic ecosystems. Understanding community receptivity to IUWM is critical knowledge for the urban water manager. Similarly, the implementation of IUWM depends on the capacity of organizations which are responsible for managing the urban water cycle. Assessing this capacity will help to identify opportunities and constraints, so that appropriate management strategies can be developed.

A social profile of the relevant catchment will give important information about land use, demographics, as well as community attitudes, preferences and behaviours. Similarly, organizational profiling can be used to determine the incentives and constraints for IUWM within and between organizations and the level of knowledge, skills available to individual organizations, or across multiple organizations. Although the collection of social and institutional data is not as well developed as that of the bio-physical aspects of urban water management, it is critical that the social and institutional components are assessed *at the same time* as bio-physical data are being collected if true integration is to be achieved.

1.2.3   Variations in urban water systems worldwide

It is important to recognize that the way in which water services are provided to urban communities varies significantly around the world. For example:

- There is variation in use of separate or combined sewer systems, where the former conveys stormwater and wastewater in separate infrastructure networks, aiming to minimize the amount of urban runoff that enters the wastewater system, while the latter uses a single set of infrastructure, intentionally mixing stormwater and wastewater flows.
- The availability of water supply in a reticulation system may be either on-demand (i.e. 24 hours a day) or scheduled for a set period of each day. There are towns and cities which manage scarcity by limiting the period when users can access water in the reticulation system to, say, two hours each day, with the result of users often storing water themselves to ensure continuity of supply through the rest of the day.
- The collection of roof runoff has been developed in several countries and regions of the world. In some less developed countries this approach to water supply was taken because there was a lack of or limited accessibility to surface and groundwater resources, or there was a lack of water supply networks. There has also been resurgence in the use of rainwater tanks in water stressed urban areas in developed countries such as Australia.

As a result of the variations in urban water systems, care should be taken in translating data and data collection approaches from one location to another without consideration of any local variation in management practices.

1.2.4 Different data applications

Data is employed to meet many different objectives during the planning, design and operation of an urban water system and, therefore, different types of data can be more appropriate than others depending on the question at hand and the objectives of the persons posing the question. Also, the way in which the data will be used should shape and guide the data collection process. This is because the data required for, say, a detailed modelling investigation of the behaviour of the urban stormwater network under a range of conditions would be quite different from a high-level performance assessment of the health of urban waterways. This detailed modelling investigation will have quite specific data requirements dictated by the selection of modelling tool, for instance, a continuous time series of specific variables at a certain spatial and temporal resolution. In comparison, the high level performance assessment may have less rigid data requirements and be limited to a single point in time, for example, to provide a snapshot of water quality in urban streams.

1.3 RELATIONSHIP TO OTHER GUIDELINES

There are many local, national and international guidelines on data requirements for urban water management, but to date, little attention has been given to the specific considerations necessary to facilitate more integrated urban water management (IUWM).

This book should preferably be read in conjunction with, rather than instead of, local or component-specific guidelines. We do not provide in this book the necessary level of detail, for instance, on the appropriate sensor technologies to measure flow in a specific application, but instead provide guidance on the principles of sensor selection deployment and application within an IUWM context. For example, urban water managers should be aware of local requirements that might be imposed by their municipality or local environment protection authority, while also considering component-specific guidelines, such as meteorological guidelines and relevant International Standards Organization (ISO) recommendations (Figure 1.3). Specific information will also be needed to understand the capabilities and limitations relating to precision, accuracy, reliability and suitability of relevant sensor technologies and of techniques for the collection, storage, management and analysis of data. This book, touching on all of these subjects, places primary emphasis on the *integration of these considerations* in order to enhance integrated urban water management.

- In particular, the reader is referred to guidelines provided by:
- *World Meteorological Organization*: The WMO provides guidelines on the collection, analysis and presentation of meteorological data, including technical regulations, available at www.wmo.ch/web/www/QMF-Web/Meetings/ICTT_geneva07/Intro-WMO49.pdf (Accessed 03 July 2007.)
  WMO also published guidelines on gauging rainfall and runoff (quality and quantity) in urban areas entitled 'Rainfall and Floods in our Cities'.
- *World Health Organization*: The WHO is concerned with achieving and maintaining human health. Among its activities, WHO develops guidelines relating to water management and sanitation, and the potential impact on human health. For example, guidelines have been developed on drinking water quality, bathing waters, water

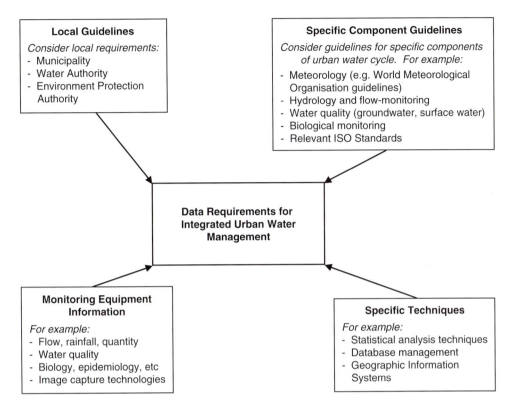

Local Guidelines

Consider local requirements:
- Municipality
- Water Authority
- Environment Protection
  Authority

Specific Component Guidelines

*Consider guidelines for specific components
of urban water cycle. For example:*
- Meteorology (e.g. World Meteorological
  Organisation guidelines)
- Hydrology and flow-monitoring
- Water quality (groundwater, surface water)
- Biological monitoring
- Relevant ISO Standards

**Data Requirements for
Integrated Urban Water
Management**

**Monitoring Equipment
Information**

For example:
- Flow, rainfall, quantity
- Water quality
- Biology, epidemiology, etc
- Image capture technologies

Specific Techniques

For example:
- Statistical analysis techniques
- Database management
- Geographic Information
  Systems

Figure 1.3   Guidelines and information useful for determining data requirements for an integrated
urban water management programme

resources quality, monitoring of water supply and sanitation, wastewater use and the relationships between water health and economics. Further information is available at: http://www.who.int/topics/water/en/ (Accessed 02 July 2007).

- *International Water Association:* The IWA (www.iwahq.org.uk/ [Accessed 02 July 2007]) is a global network of water professionals, spanning the continuum between research and practice and covering all facets of the water cycle. It contains many specialist groups of great relevance to IUWM. For example, the Hydroinformatics Group publishes the Journal of Hydroinformatics, while the Group on Systems Analysis and Integrated Assessment provides guidance on decision-making and optimization of water management (see: http://www.ensic.inpl-nancy.fr/iwa-saia/ [Accessed 02 July 2007]). The Specialist Group on Standards and Monitoring considers the current needs for water quality standards and which water quality parameters should be monitored or measured to meet the regulations.
- *International Standards Organization:* ISO are responsible for the development of internationally accepted standards on a wide range of issues. It is a non-government network of national standards institutes. Standards are available for issues such as uncertainty analysis and calibration of flow measurements, the design and implementation of water quality sampling programmes. Further information is available at: http://www.iso.org/iso/en/ISOOnline.frontpage (Accessed 02 July 2007).

- *European Union Commission*: The EU Framework has made directives on integrated river basin management for Europe, as well on urban wastewater, groundwater management, drinking water and a range of other relevant topics (see: http://europa.eu.int/comm/index_en.htm [Accessed 02 July 2007]).
- *US Environmental Protection Agency (US EPA)*: In addition to publishing various technical documents on methodologies it also provides standards, for example, for selection of instrumentation and measuring ranges for water quality monitoring (US EPA, 1983) (see: www.epa.gov/ [Accessed 02 July 2007]).

Notes on a selection of helpful guidelines, papers, conference proceedings and other documents are provided in the list of references concluding this chapter. Additional references to other guidelines and documents on individual components of the urban water cycle are provided in all subsequent chapters.

1.4   RELATIONSHIP WITH OTHER UNESCO IHP PROJECTS

The UNESCO International Hydrological Programme (IHP) was established in 1975 to study the hydrological cycle and to develop strategies and policies for sustainable management of water resources. The IHP work programme is organized in six-year phases with each phase developed in response to the identified current needs. IHP-VI (2002–2007) focuses on 'Water Interactions: Systems at Risk and Social Challenges', with six themes:

Theme 1 – Global Changes and Water Resources focuses on global estimates of water resource supply and quality, as well as demand and consumption. It considers the impacts on these of global changes in land-use and climate.

Theme 2 – Integrated Watershed and Aquifer Dynamics examines the impact of extreme events on land and water management, and considers techniques for integrated river management, particularly for river basins and aquifers which cross international boundaries.

Theme 3 – Land Habitat Hydrology deals with drylands, wetlands and mountains, as well as small islands and coastal zones. It also includes a focus on urban areas and rural settlements (Focus Area 3.5), for which this book was produced.

Theme 4 – Water and Society explores the links between water, civilization and ethics, including the value of water, and how to prevent and resolve water conflicts. Water security is examined, particularly in relation to water-related disasters and degrading environments.

Theme 5 – Water Education and Training identifies techniques for education professionals and the community about water, with a focus on the use of new technologies, and international networks.

This book represents one of the outputs from the first of nine projects conducted under the theme of Land Habitat Hydrology (Theme 3) and together with Project 9 has an overarching role (Figure 1.4), in guiding other projects addressing specific components of the urban water cycle and with their own outputs. More information regarding the UNESCO International Hydrologic Programme is available at www.unesco.org/water/ihp (Accessed 02 July 2007).

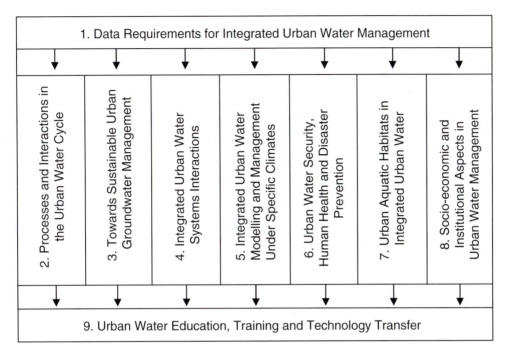

Figure 1.4   Projects undertaken within Theme 3 (Land Habitat Hydrology) of the UNESCO International Hydrological Programme IV (IHP-VI). Project 1, Data Requirements for Integrated Urban Water Management, provides guidance across the range of urban water issues (Projects 2-8) with all contributing to training materials for urban water education, training and technology transfer (Project 9)

REFERENCES

Anderson, J. and Iyaduri, R. 2003. Integrated urban water planning: big picture planning is good for the wallet and the environment. *Water Science and Technology*, Vol. 47, No. 7–8, pp. 19–23.

Arsov, R., Marsalek, J., Watt, W.E. and Zeman, E. (eds) 2003. *Urban Water Management Science Technology and Service Delivery*. Proceedings of the NATO Advanced Research Workshop, Borovetz, Bulgaria, 16–20 October 2002, Dordrecht, Kluwer Academic Press. (NATO Science Series: IV: Earth and Environmental Sciences, Vol. 25).
Covers a range of urban water management issues, and includes information on technologies in urban water management (including some case studies on GIS).

Berne, A., Delrieu, G., Creutin, J-D. and Obled, C. 2004. Temporal and spatial resolution of rainfall measurements required for urban hydrology. *Journal of Hydrology*, Vol. 299, No. 3, pp. 166–79.
Provides recommendations on the temporal and spatial resolution of rainfall measurements required for urban hydrological applications, based on quantitative investigations of the space-time scales of urban catchments and rainfall.

Bertrand-Krajewski, J.-L., Barraud, S. and Chocat, B. 2000. Need for improved methodologies and measurements for sustainable management of urban water systems. *Environmental Impact Assessment Review*, Vol. 20, No. 3, pp. 323–31.

Rationale for better understanding, monitoring of urban water systems, and outlines the OTHU (*Observatoire de Terrain en Hydrologie Urbaine*, i.e. Field Observatory for Urban Hydrology).

Butler, D. and Parkinson, J. 1997. Towards sustainable urban drainage. *Water Science and Technology*, Vol. 35, No. 9, pp. 53–63.

Chocat, B., Krebs, P., Marsalek, J., Rauch, W. and Schilling, W. 2001. Urban drainage redefined: from stormwater removal to integrated management. *Water Science and Technology*, Vol. 43, No. 5, pp. 61–68.

Church, P.E., Granato, G.E. and Owens, D.W. 1999. *Basic Requirements for Collecting, Documenting, and Reporting Precipitation and Stormwater-Flow Measurements.* Washington DC, US Department of Interior and US Geological Survey (Open-File Report 99-255). http://ma.water.usgs.gov/fhwa/products/ofr99-255.pdf (Accessed 02 July 2007.)

European Community. 2000. *Directive Establishing a Framework for Community Action in the Field of Water Policy*, 2000/60/EC. Available at http://europa.eu.int/eur-lex/pri/en/oj/dat/2000/l_327/l_32720001222en00010072.pdf (Accessed 02 July 2007.).
Monitoring requirements of the EU Water Framework Directive. Objectives of the three broad monitoring types under the Water Framework Directive (WFD): Surveillance, Operational and Investigative.

European Union. 2003. *Guidance on Monitoring for the Water Framework Directive.* Copenhagen, Denmark. Water Framework Directive Common Implementation Strategy Working Group 2.7 (Monitoring).

Haramancioglu, N.B., Ozkul, S.D., Fistikoglu, O. and Geerders, P. (eds) 2003 *Integrated Technologies for Environmental Monitoring and Information Production.* Proceedings of the NATO Advanced Research Workshop, 16–20 October 2002, Borovetz, Bulgaria. Dordrecht, Kluwer Academic Press. (NATO Science Series: IV. Earth and Environmental Sciences, Vol. 23) Outcomes of a conference which focus on how integration should be accomplished to meet the prevailing demand for reliable information systems.

Harremoes, P. 1997. Integrated water and waste management. *Water Science and Technology*, Vol. 35, No. 9, pp. 11–20.

Hatt, B.E., Fletcher, T.D., Walsh, C.J. and Taylor, S.L. 2004. The influence of urban density and drainage infrastructure on the concentrations and loads of pollutants in small streams. *Environmental Management*, Vol. 34, No. 1, pp. 112–24.

IHP/OHP German National Committee. 2003. *Hydrological Networks for Integrated and Sustainable Water Resources Management.* Report of International Workshop, Koblenz, Germany, 22–23 October, 2003. Koblenz, Bundesanstalt für Gewässerkunde.
Outlines hydrologic (including water quality) networks throughout the world, including discussion of uncertainty, optimization, etc.

Maksimovic, C. and Tejada-Guibert, J.A. (eds) 2001. *Frontiers in Urban Water Management: Deadlock or Hope.* London, IWA Publishing.
Description of the challenges of urban water management, including issues of integration, economic, social, institutional and regulatory issues, and problems of developing countries.

Mitchell, V.G. 2006. Applying integrated urban water resource management concepts: a review of Australian experience. *Environmental Management*, Vol. 37, No. 5, pp. 589–605.

Niemczynowicz, J. 1999. Urban hydrology and water management – present and future challenges. *Urban Water*, Vol. 1, pp. 1–14.

Parliamentary Commissioner for the Environment. 2000. *Aging Pipes and Murky Waters, Urban Water System Issues for the 21st Century.* Wellington, Parliamentary Commissioner for the Environment.

Savic, D., Marino, M., Savennije, H.G. and Bertoni, J.C. (eds) 2005. *Sustainable Water Management Solutions for Large Cities.* Wallingford, UK, International Association of Hydrological Sciences. (IAHS Publication 293).

Outcomes of a conference on urban water sustainability and includes information on water quality monitoring and data.

Szollosi-Nagy, A. and Zevenbergen, C. 2004. *Urban Flood Management*. New York, A.A. Balkema Publishers.
Survey of recent developments in urban flood management, including spatial analysis, flood planning, and risk management.

US EPA. 1983. *Methods for Chemical Analysis of Water and Wastes*. Washington DC, United States Environment Protection Agency. (US-EPA 600/4-79-020).

USGS. 2005. *National Field Manual for the Collection of Water-Quality Data*. Washington DC, United States Geological Survey (USGS). Available at: http://water.usgs.gov/owq/Field Manual/ (Accessed 02 July 2007.)
Comprehensive overview of equipment and techniques for water quality sampling in wide variety of catchment conditions.

Vignolles, M., Woloszyn, E., Niemczynowicz, J., Maksimović, Č. and Marsalek, J. 1986. *Rain and Floods in Our Cities, Gauging the Problem*. Geneva, World Meteorological Organization.
Detailed methodologies for measurement and data processing of quantity and quality of rainfall and runoff for urban areas.

Part I

Guiding principles for data acquisition, management and use

Chapter 2

Overview of guiding principles

A. Deletić and T.D. Fletcher

Department of Civil Engineering (Institute for Sustainable Water Resources), Building 60, Monash, Monash University, Melbourne 3800, Australia

2.1 INTRODUCTION

This chapter describes the main issues that should be addressed in developing a monitoring programme to meet the needs of Integrated Urban Water Management (IUWM). It is not a step-by-step guide to setting up an IUWM monitoring programme, nor does it provide detailed specification of the design of monitoring components, such as installation of monitoring equipment. For advice on these issues, local guidelines (reflecting local availability of technology) should be consulted. Rather, this chapter provides an in-depth discussion of the main principles and issues that should be considered *before* the monitoring system is established.

The guiding principles are listed below, along with a brief description, while an in-depth discussion of each of these principles is presented in the following sections. Some of the principles are simply universal good practice when setting up a monitoring programme for any water systems, while others are specific to the needs of delivering a monitoring system to enable urban water resources to be managed in a more integrated manner.

2.2 PRINCIPLES

2.2.1 Integration as an overarching principle

An IUWM monitoring programme must aim to develop an understanding of *the whole of the urban water system, rather than simply collect separate sets of information on individual urban water components*. Data collected on a particular component (e.g. precipitation, water supply, wastewater, stormwater drainage, urban waterways, etc.) should be collected in a manner which, whenever possible, is useful for understanding the functioning of other components and interactions among them. In other words, when designing the monitoring programme for a particular component, the data needed for understanding other components should be considered from the outset.

To achieve integration across all streams of the urban water system, it is important also to achieve *integration of the team of people involved in the project*. People with expertise in the different urban water system components, which may potentially interact, should all be involved in the monitoring system design, to ensure that interactions between system components are monitored, not just the performance of the individual components themselves. Finally, as the following sections highlight, all the listed principles are highly interconnected (and therefore integrated) with one leading to another. *Integration is the key principle that is interwoven through all other principles listed below.*

2.2.2 Defining the objectives of monitoring

As in all other monitoring schemes, it is important to clearly identify the aims and objectives of the IUWM monitoring programme prior to detailed planning and programme design. To maximize the use of the data and minimize the costs it is necessary to understand the issue to be addressed, the information needed to address it. Being familiar with the existing and potential users of the data is also important, as is a conceptual model of the system(s) and their interactions. Examples of monitoring programmes that have failed because objectives were not *explicitly* identified are numerous. It is also at this initial stage that integration should be clearly identified. In other words, one objective will be to collect data in a way that facilitates an understanding of the system as a whole, rather than just individual components. At this stage involvement of all potential users of the data is critical.

2.2.3 Identifying the variables to monitor

Selection of the variables to monitor will be governed by the previously defined objectives. Variables may include those related to characteristics and conditions of natural environments and infrastructure. Water quantity parameters (such as flow rate, level, pressure, consumption), water quality parameters (such as pollutant concentrations) are also commonly monitored as are climate characteristics. Environmental monitoring, for example, on aquatic ecosystem health and function, may be required to understand the impact of the urban water system on the environment. As IUWM has social and economic aspects as well as physical, there may also be a need to measure a range of non-physical variables, such as community attitudes towards IUWM technologies. Together with economic data (from lifecycle cost analysis, which account for both capital costs and operating and maintenance costs), these data can contribute to some form of triple-bottom-line (TBL) or multi-criteria analysis (MCA).

2.2.4 Considering spatial and temporal scale effects

Depending on the aims, objectives and the variability and reliability of the measurements, the scale of monitoring has to be appropriately selected. *In other words, the decision has to be made where to monitor (spatial resolution), how often and for how long, or how many episodes, (temporal resolution).* Again, the integration principle is the key; *the data collected should be applicable across a number of components* (e.g. because the data collected for one component of the system are usually used for analysis of the performance of some other component, the frequency of data recording should match the maximum frequency required across the number of system components. The representativeness of the monitored data across the selected time and space scale should be ensured for each variable.

2.2.5 Assessing and managing uncertainty

Any monitoring programme should deliver not just measurements, nor just the (calculated or predicted) results that derive from them, but should also include quantification of the uncertainty of these measurements and resulting predictions. In defining the monitoring objectives, the allowable uncertainty should have already been specified. If the resulting uncertainties are greater than allowable levels, remedial action should be taken, such as replacement of more accurate and/or precise sensors. In some situations, it may even be necessary to cancel a monitoring programme, or at least to cease monitoring some variables where they cannot be measured within an acceptable level of uncertainty. Uncertainty is made up of two errors, being bias errors (fixed, systematic errors, which come from erroneous sensor calibration, for example), and random errors, which are caused, for example, by variations in measurement equipment and natural processes. Processes for quantifying these sources of uncertainty are outlined in Chapter 6 with appropriate implementation examples. There is also philosophical discussion of the role of uncertainty in the decision-making process.

2.2.6 Selecting monitoring equipment

Recent and ongoing advances in monitoring equipment are making it easier to collect more and more complex data, with increasing reliability (but not always increasing accuracy). Unfortunately, the increase in technology and capability has also increased the complexity of selecting appropriate equipment. Users must consider issues of accuracy or uncertainty and repeatability of both the sensors (relative to objectives) and the whole monitoring system, as well as system stability, resolution, linearity, measurement range and dynamic response capability. Equipment which can be readily and reliably calibrated should also be given preference, as should the ability to undertake diagnostics and self-tests. Ultimately, the selection of equipment will be a product of the objectives and available resources.

Considering the potential users of the data is important (including their capacity to the use the equipment and/or the data that comes from it). In some cases it may be cheaper to use ad hoc manual sampling than to use expensive continuous monitoring techniques. If effective integration across the urban water cycle is to be achieved, rigorous time synchronization will need to be enforced (so that data from across a diverse monitoring network can be integrated). As well as technical considerations, users should consider aspects such as after-sales support and the applicability to its local context (environmental conditions, as well as the capacity of local users to operate the equipment). In many cases the most sophisticated equipment may not be the best for a given purpose.

2.2.7 Validating the data

Without appropriate validation procedures, all data can be assumed to be inaccurate. This means that *all* data should be submitted to a rigorous data validation procedure, including not only the *dynamic* data, such as flow or water quality measurements, but also the *static* data, such as information about catchment areas or pipe diameters. For large *in situ* monitoring datasets, there is a need for automatic validation procedures, both for efficiency, but also to maintain transparency and objectivity. Such systems will detect outliers, either coming from the sensor, or from environmental conditions (such as upstream maintenance activities). Following the detection phase, a diagnostic phase is needed, to identify why the outlier occurred, and determine what (if anything) can be done with the data point. Such a staged approach will also help to reduce the occurrence of such outliers in the future.

A critical aspect of data validation is to maintain metadata which informs users of the data about both the quality of the data and any actions which have been applied to it. An example of an automated validation process is given, based on the French OTHU project. This process outlines seven criteria for assessing (and diagnosing) data quality, including (i) status of the sensor (*on*, *off*, or *hold*), (ii) assessment of possible range of the data, (iii) range of most frequently observed values, (iv) maintenance periodicity, (v) signal gradient, (vi) sensor redundancy and (vii) signal redundancy.

2.2.8 Data handling and storage

If data are to facilitate integration across the urban water cycle, then issues relating to data portability, application of data for different models as well as data security and lifetime become very important. An important principle is the use of a well designed central DataBase Management System (DBMS) to handle the flow of data. The DBMS must keep track of data and important metadata providing information about the origin, quality and assumptions regarding the collected data. The metadata are also used to distinguish raw data and synthetic (e.g. interpolated values) or processed data. The DBMS also provides the central hub to facilitate integration. For example, data from Geographic Information Systems (GIS), Supervisory, Control, Data Acquisition and Data Analysis systems (SCADA) and other systems can all be brought together, either within the same organization, or across organizations.

There are some critical considerations for data handling and storage. Firstly, a copy of the raw format data must always be marked as such and maintained, no matter what other processing or analysis occurs. Metadata must also be maintained, preferably according to agreed standards. To avoid the risk of data loss, strict version control must be enforced as should some degree of software and hardware redundancy. Ensuring a common timeline for data will allow easier data sharing between applications and users. Data compression will help to reduce storage requirements in a database, but it should be used with great caution, due to the risk of data loss. Perhaps the most critical considerations relate to data longevity and security. Careful selection of software is necessary to ensure the data do not end up 'trapped' in a redundant software platform; open-source or standardized platforms can help to reduce this risk. Data security through careful control of access and restrictions to key authorized users is necessary, particularly where databases are connected to the internet. There should always be a mirrored version of the database which is not internet-connected.

2.2.9 Using data to develop information and knowledge

Data by itself is not useful. The objective of data is to provide information, which in turn is transformed into knowledge so that decisions can be made. The conversion from data to information is typically a computerized process, while the conversion from information to knowledge is intrinsically a human process. This process can be broken into seven steps:

(1) data cleaning (to remove noise)
(2) data integration (to combine different sources of data)
(3) data selection (to extract relevant data for analysis, often by sampling)

(4) data transformation (to process the data into a form suitable for mining)
(5) data mining (to search for patterns)
(6) pattern evaluation and interpretation
(7) knowledge representation and usage.

Geographical information systems, statistical analysis and simulation models can all help in this process (primarily in Steps 5 and 6), as can pattern-detection software (either unsupervised or supervised). A more recent development is self-organizing maps, which reduce data dimensionality by producing maps showing the similarities of data by grouping.

   The process of turning information into knowledge is the fundamental bottleneck in the overall process, because of the limited capacity of humans. Information overload (a human condition) can be reduced with the assistance of software to standardize the production of information so that it is in a form more readily interpreted by the user and with appropriate aggregation and disaggregation, as required. Ultimately, however, the ability of make decisions from the knowledge determines the success of a data collection process. Decision support systems can help in this regard.

2.2.10 Budgeting and financial considerations

The available resources are crucial in achieving the objectives of a monitoring programme. *Proper budgeting that includes sufficient allowance for contingency* (e.g. for equipment breaks) *has to be done at the very start of a monitoring programme.* The most important step is to ensure that the *objectives are determined first without consideration of potential financial* constraints. This is not to imply that financial constraints will not be considered. However, it is important that the objectives are formed *free* from the influence of financial constraints. Then, the financial resources can be evaluated, and the two 'sides' compared. If there is a gap, a transparent decision-making process can commence, where the impacts of reducing the scope of monitoring are clearly identified and discussed by all stakeholders. In this situation, the potential users of the data may find other sources of funding, for example, through creating partnerships with other relevant agencies. When the cost of monitoring is compared to the potential costs of construction and operation of a water treatment plant, it becomes evident that compared to the cost of retrofitting a malfunctioning plant, a properly-resourced monitoring programme will likely save a substantial amount of money in the long term.

   Where cost savings do need to be made, possible strategies could include (prudent) use of existing data resources, use of new technologies which may reduce monitoring cost, and the sharing of monitoring networks. In some cases an integration of modelling and monitoring will provide the required data reliability at a reduced cost. Such decisions should always be considered against the original objectives of the programme and not made in an ad hoc manner.

2.2.11 Social and institutional considerations

Effective and efficient collection, analysis and use of data depend not only on technical considerations, but also on the role of institutions and the community. Five key principles need to be considered when designing and admininistering data collection and management as part of IUWM:

(1) *Leadership and commitment*: organizations with strong leadership and a clear commitment to IUWM will have a clear *vision*, and will work to collect the necessary data to achieve that vision, using clearly-stated targets as indicators of their performance. The commitment will extend not only to implementation, but to *ongoing collection of data to better understand the urban water system, and changes in it.*

(2) *Public participation*: involving local people recognizes both the value of local (or 'indigenous') knowledge, and the importance of involving the community in determining what information is needed, in order to secure their long-term commitment to its collection (for example, through their preparedness to pay taxes for such purposes).

(3) *Transparency and accountability*: clear and transparent reporting of the performance of organizations responsible for IUWM builds trust with the community, in turn increasing support for IUWM implementation. Consideration needs also to be given to the manner of communication, ensuring that it is appropriate to the target audience(s).

(4) *Coordinated data access and sharing*: the very nature of IUWM means that organizations depend on information often held by others, to be able to manage 'their

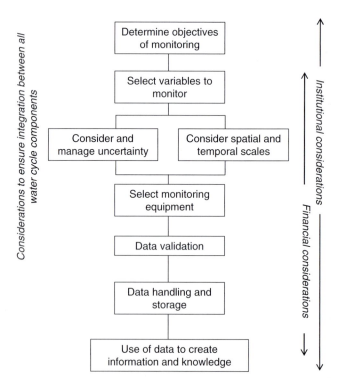

Figure 2.1 Summary of considerations for data collection and use in integrated urban water management

component' of the urban water cycle effectively. A well coordinated and equitable system is needed to ensure that all stakeholders in IUWM have access to the required data, *and are aware* of the availability of that data.

(5) *Evaluation and action learning*: advances in the management of urban water systems depend on ongoing improvement in the understanding of the system, and its interactions. An 'action-learning' philosophy encourages organizations to accept risk, and to commit to collecting data to enhance their capacity to design, operate and maintain water systems.

2.3 SUMMARY OF PRINCIPLES

The main considerations in the planning of monitoring programme for IUWM systems are presented in the diagram in Figure 2.1. They are organized in three layers, according to the order of their application, with the principle of integration going across all layers.

It is important to note also, that we define *monitoring* as an ongoing activity, rather than a short-duration, or 'one-off' investigation. There are often good reasons to undertake specific short-term investigations, however, given that by definition, we are dealing with components of the urban water cycle operating over different spatial and temporal scales, a long-term programme of continuous monitoring is a prerequisite of integrated urban water management.

Chapter 3

Defining objectives and applications of monitoring

T.D. Fletcher[1] and J.-L. Bertrand-Krajewski[2]

[1]*Department of Civil Engineering (Institute for Sustainable Water Resources), Building 60, Monash University, Melbourne 3800, Australia*
[2]*Laboratoire LGCIE, INSA-Lyon, 34 avenue des Arts, Villeurbanne Cedex, France*

3.1 INTRODUCTION

It is commonly argued (e.g. ANZECC and ARMCANZ, 2000; Bertrand-Krajewski et al., 2000) that defining clear objectives is the most important phase of the design of a monitoring programme. Without clear objectives the likelihood of success is limited, and the ability to evaluate the degree of success is even more limited. Given the fundamental importance of monitoring, considerable resources should be devoted to the definition of objectives. Appropriate dedication of personnel, time and gathering of information necessary to inform the process, should be built into the budget for the design and implementation of the monitoring programme.

The design of a monitoring programme, such as the specification of the appropriate measurement techniques and the required level of confidence, depend on having explicit objectives. Collecting data without a clear purpose and application for the data is likely to lead to inappropriate variables being monitored or inappropriate methods being employed with ultimately little use made of the collected data. Thus the objectives of a monitoring programme will be to gain the knowledge and understanding necessary to manage a particular issue or set of issues, not just to gain information about them (refer to Chapter 10). Therefore, the actual objectives should be framed with a clear idea of *what is to be done* with the data, that is, how they are to be used to address the issue(s) of concern.

The objectives should also be linked to the functional responsibilities of the stakeholders undertaking, or benefiting from, the monitoring effort. Therefore, potential stakeholders should be involved right from the start in defining the objectives. They should include those with expertise and/or responsibility for each aspect of the urban water cycle directly or indirectly affected.

3.2 DEFINING THE OBJECTIVES

Monitoring objectives derive from the issue that is of concern and from the *management objectives* that have been determined for that issue. In framing the monitoring objectives, one may ask some key questions based on information needed to meet the management objectives, for example:

- What information and knowledge is currently lacking, that is needed to improve management of the issue, and its interaction with other aspects of the water cycle?
- Who will use the data, and how?

- What level of uncertainty can be accepted?
- What level of resources do we need to undertake the monitoring?

Questions about *how* the monitoring is to be undertaken are not of primary concern at this stage. Once the objectives have been established, design of the monitoring programme may show that the objectives cannot be met, and a review of the objectives and available resources will then be required (refer to Chapter 12). However, it is important that cost not be the primary consideration at the objective-definition stage, so that any gaps between the monitoring objectives, and the ability to meet them, are made explicit. Incorporation of financial constraints into the process of setting objectives will make it more difficult to objectively assess the consequences of reducing or increasing the resources available to meet the objectives.

Developing monitoring objectives should be based on a rigorous process that ensures the issue of concern, and the system within which it exists, are well understood. Figure 3.1 summarizes the process of defining objectives, involving five principal stages:

(1) Definition of the management issues
(2) Identification of information requirements
(3) Collection of existing information
(4) Creation of a conceptual model of the system with an understanding of component interactions
(5) Definition of monitoring objectives

Eventual implementation of the monitoring programme depends very heavily on the clarity and specificity of the objectives. Different objectives will require different monitoring and data analysis techniques. Definition of the objectives is also critical to be able to determine the level of confidence required in the data, and thus affects choice of methodology (refer to Chapter 6).

In order to help define objectives a methodology may be suggested to help set appropriate variables to be measured and time scales to be used. The starting point is the problem or the question to be answered, i.e. *what do we want to know, to understand, to test, to verify?* Then, by using the body of existing knowledge, concepts, hypotheses and theories, we can define the variables to be measured. As an example, let us consider the measurement of rainfall.

Step 1. Problem to be solved

At which time step should rainfall intensity be measured in order to simulate the rainfall-runoff process at the outlet of the catchment?

Step 2. Body of knowledge, theories, and results of previous research to consider

- There is a cause-effect relationship between rainfall and discharge at the outlet, which is very well established from past experience and measurement.
- Models (for example, the linear reservoir model) are able to simulate this process
- Rainfall is a stochastic process whose fluctuations are very significant at very short time scales (typically one minute). Thus it would be recommended to measure rainfall at very short time steps, typically one minute.

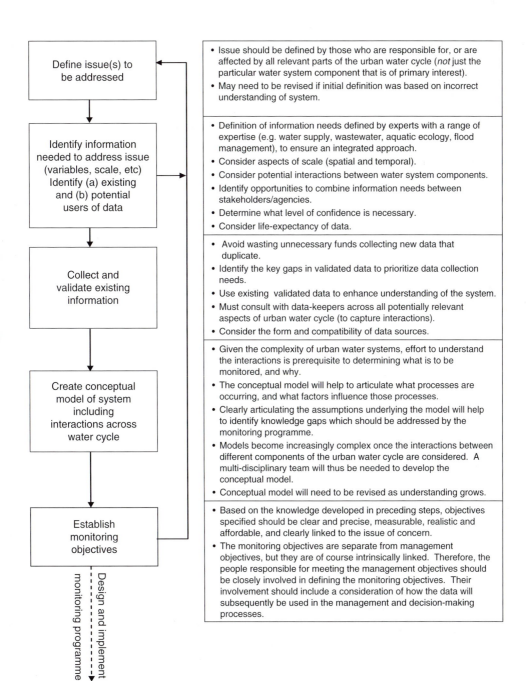

Define issue(s) to be addressed	• Issue should be defined by those who are responsible for, or are affected by all relevant parts of the urban water cycle (*not* just the particular water system component that is of primary interest). • May need to be revised if initial definition was based on incorrect understanding of system.
Identify information needed to address issue (variables, scale, etc) Identify (a) existing and (b) potential users of data	• Definition of information needs defined by experts with a range of expertise (e.g. water supply, wastewater, aquatic ecology, flood management), to ensure an integrated approach. • Consider aspects of scale (spatial and temporal). • Consider potential interactions between water system components. • Identify opportunities to combine information needs between stakeholders/agencies. • Determine what level of confidence is necessary. • Consider life-expectancy of data.
Collect and validate existing information	• Avoid wasting unnecessary funds collecting new data that duplicate. • Identify the key gaps in validated data to prioritize data collection needs. • Use existing validated data to enhance understanding of the system. • Must consult with data-keepers across all potentially relevant aspects of urban water cycle (to capture interactions). • Consider the form and compatibility of data sources.
Create conceptual model of system including interactions across water cycle	• Given the complexity of urban water systems, effort to understand the interactions is prerequisite to determining what is to be monitored, and why. • The conceptual model will help to articulate what processes are occurring, and what factors influence those processes. • Clearly articulating the assumptions underlying the model will help to identify knowledge gaps which should be addressed by the monitoring programme. • Models become increasingly complex once the interactions between different components of the urban water cycle are considered. A multi-disciplinary team will thus be needed to develop the conceptual model. • Conceptual model will need to be revised as understanding grows.
Establish monitoring objectives	• Based on the knowledge developed in preceding steps, objectives specified should be clear and precise, measurable, realistic and affordable, and clearly linked to the issue of concern. • The monitoring objectives are separate from management objectives, but they are of course intrinsically linked. Therefore, the people responsible for meeting the management objectives should be closely involved in defining the monitoring objectives. Their involvement should include a consideration of how the data will subsequently be used in the management and decision-making processes.

Design and implement monitoring programme

Figure 3.1 Defining monitoring objectives

Source: Adapted from ANZECC and ARMCANZ, 2000.

- Catchment surfaces, plus the natural and constructed drainage system, act to transform rainfall into discharge. This process acts to attenuate the discharge hydrograph, relative to the rainfall hyetograph, reducing the frequency of variability. The attenuation increases with catchment area. This process acts as a low-pass filter, only low frequencies explaining the shape of the hydrograph at the outlet. Moreover, the filtering of high frequencies increases with increasing catchment areas and increasing concentration and propagation times in the system.

Step 3. Decision

It is sufficient to measure the rainfall intensity with a frequency (or a time step) equal to twice the highest frequency (or half of the corresponding time step) contained in the discharge time series. A Fourier transform can be used in the analysis of the signal frequencies. As an order of magnitude, it is necessary to measure rainfall intensity with a time step of 1 minute to 2 minutes for a catchment area of less than 100 ha to 200 ha, while a time step of 5 minutes to 10 minutes is sufficient for catchment areas ranging from 200 to 1000 ha.

Of course, each problem has its own appropriate solution. If the problem was instead 'at which time step should we measure rainfall intensities to simulate the sediment erosion process at the surface of the catchment', a different answer would result. For each objective of the monitoring programme, such an analysis is thus required.

3.3 APPLICATIONS OF MONITORING: INTEGRATING ACROSS FUNCTIONAL RESPONSIBILITIES

The definition of monitoring objectives clearly depends on the ultimate application of the data. Table 3.1 outlines some common applications of the data resulting from monitoring programmes, and identifies some of the considerations that should be given to interaction between these applications. Table 3.1 is not meant to be exhaustive, but rather to provide examples that illustrate the considerations relevant to different monitoring applications. The important thing is to consider how the data collected will be used, *and to identify potential interactions between water cycle components*, so that the monitoring objectives encompass the important interactions.

Applications of monitoring will be closely related to the functional responsibilities of the organization which actually collect the data. For example, data may be collected by a water authority making decisions about the need to upgrade a wastewater treatment plant. In that case, the decisions regarding objectives and applications may be considered by a relatively small group of stakeholders, with a narrow range of expertise (design and operation of wastewater treatment plants).

If a monitoring programme is to assist in the integrated management of urban water, a greater range of perspectives and expertise is needed, extending beyond traditional functional objectives and involving several agencies. *The objectives must be defined by a multi-disciplinary team, with the combined expertise and responsibility across the several components of the urban water system.* Similarly, if the objectives for a monitoring programme are focused upon a more integrated application of the data, greater consideration will need to be given to collecting data with consistent protocols, formats and storage systems, so that data may be shared more readily between agencies and disciplines (Vogt and Petersen, 2003).

Table 3.1 Categories of typical monitoring applications, and considerations regarding interactions and integration

Application	Description of data required	Interaction and integration considerations
System and process understanding	Data about the behaviour of water systems – either natural or constructed. Typically require data on rates of processes, factors influencing processes, variability. Initial monitoring programme will often be based on imperfect knowledge of the system (which is why the data are being collected), and so the monitoring objectives and methodology will need regular review.	Interactions between systems are likely to be great, and monitoring programmes will often need to extend beyond the core issue of concern. Monitoring the core issue only may lead to incomplete understanding, and inability to properly manage the issue (and its 'knock-on' effects). For example, if groundwater level and quality is of concern, the monitoring programme may also need to consider: 1. surface water flows (quality, quantity), abstractions and uses, wastewater discharges, 2. catchment physical characteristics (including imperviousness), 3. catchment rainfall and evapotranspiration, 4. functioning and health of groundwater-dependent ecosystems, and 5. effects on drinking water quality and human health outcomes.
Assessment for planning and design	Nature of the planning/design will determine the scope of data required (e.g. from single wastewater treatment plant to entire water supply, waste- and storm-water system). Data on rates, variability, process, as well as long-term trends, socio-economic issues (demand, preferences, etc) needed.	For large, complex systems (e.g. water supply, disposal drainage systems for new development areas) there is a very obvious need to measure wide range of variables, including those which may be impacted (e.g. flow and quality in streams) by the system. For simpler systems (e.g. single wastewater treatment plant), data needs should be considered for all systems on which the plant will depend (e.g. human population, water use, catchment characteristics, etc) and which may be impacted by the plant (e.g. stream flow and quality, aquatic ecosystem health, non-potable water demand, etc).
Planning for maintenance and asset management	Data on the age, condition and performance of water system assets, both natural and anthropogenic. Data on changes in performance with asset age and maintenance status, as well as costs of maintenance, etc. Data that describes how a water system	Consider the impact of water system assets on other parts of the water system. For example, lack of maintenance may result in reduced performance, or system failure (e.g. overflows), impacting on the health of receiving waters, or on the health of humans who depend on these receiving waters. All aspects should be monitored in an integrated way, in order to understand the system, and its future data requirements.

(Continued)

Table 3.1 (Continued)

Application	Description of data required	Interaction and integration considerations
Long-term performance assessment	(e.g. water treatment plant, stormwater quality treatment system) performs over the long-term. With information on average performance, variability and exceptions, factors influencing performance, interactions between systems.	Data on performance may help to identify problems, but will not help in identifying solutions. Data about the factors influencing performance should be an integral part of performance monitoring (e.g. climate data), as should data on other systems which interact with the system of interest (Vogt and Petersen, 2003). A simple example would be that of a wastewater treatment system; its performance will be related to the performance of the water supply system, which in turn will be related to climate and streamflows, etc. Similarly, long-term performance of 'sustainable urban drainage systems' should usually be integrated with data on rainfall, streamflows, and environmental impacts.
Compliance reporting	Data on whether a particular asset or system is performing in accordance with legislated requirements.	Collect data in a form which provides more information than merely compliance. For example, if samples are recorded simply as 'complying' or 'non-complying', then no other information is provided. However, recording the data as concentrations with flows, will allow load estimates, which may be useful for planning and design purposes. Data should also be collected on interacting systems. For example, the compliance or non-compliance of a drinking water treatment system may be a function of rainfall, streamflow and climate. Similarly, data on the impacts of non-compliance (e.g. human health, ecosystem health) will improve understanding of appropriate targets and contingencies.
Operational data and real-time control	Data which provides the information necessary for decisions on systems operations. For example, real-time data may be used to guide the operation of flood control systems, or to determine dosing rates in water treatment plants.	Monitoring should be based on a clear understanding of the operation not only of the specific system, but its interactions with and/or impacts on other systems and components of the urban water cycle. For example, the impact of flood control systems on the ecology of streams can only be understood by collecting data from both systems, demonstrating their interaction. Similarly, understanding variations in water quality within the drinking water supply system will likely require data on consumption levels, groundwater levels, as well as treatment plant performance. A 'source-to-tap' approach is required, including all possible interactions from other components along that path.
Model development, validation and calibration	Data used to make predictions about the behaviour of components of the urban water cycle. Models may be empirical or conceptual, but all depend	Models are, by definition, an imperfect representation of the phenomenon or phenomena that they are attempting to represent. Therefore, data help in the ongoing refinement of models, and so data collection should encompass other potentially related variables, which may be important.

	on data for their calibration, and their validation.	Consideration should also be given to collecting data necessary to develop models of related, interacting phenomenon.
Protection and management of natural environment	Data providing information on the structure, composition, diversity, functioning (e.g. rates) of aquatic ecosystems (e.g. bays, rivers, creeks, lakes) (refer also to Chapter 20).	Ecosystems are influenced by many factors, including climate (and season), along with natural and anthropogenic disturbances. Ecological indicators are thus 'integrators' of many variables. However, to understand how to manage an ecosystem, data are also required about all the potential influences (e.g. changes to flow regimes, such as abstractions or changes in land-use and imperviousness, changes in water quality, changes in groundwater-surface water interactions). In turn, management responses will depend on understanding what are the driving forces (e.g. why is water quality changing – is it because of diffuse sources related to land use, point sources, or hydrologic changes?).
Demand management	Data regarding the amount, timing and variability of demand for water (e.g. agricultural irrigation, domestic or industrial supply), and the factors that influence that demand. Data on responses to demand management strategies.	Water demand will be influenced by water quality, as well as climate, amongst other things. Changes in water demand will influence the operation of the water harvesting, treatment, storage and distribution systems. In turn, these may influence natural water cycle elements (such as surface and groundwater quantity and quality). Monitoring of all these influences and impacts will be prerequisite to making optimal decisions about managing demand.
Pollution load estimates	Data regarding the quantity of pollution in a particular water stream, along with its variability influences, impacts.	Pollution loads may be of interest in almost all components of the urban water cycle, including rainfall, stormwater, streamflow, groundwater, drinking water, wastewater, and greywater. Many of these components will be linked. For example, pollution loads in rainfall will obviously affect streamflow, which may affect the quality of drinking water and of groundwater. Monitoring of pollution loads should thus be based on a conceptual model of the systems and their interactions, and data should be shared between agencies with different functional *responsibilities*.
Multi-criteria decision-making processes	Data to assist making decisions that consider economic, social and environmental outcomes (Roy, 1990).	By definition, data requirements for multi-criteria decision making are diverse, highly interdisciplinary, including economics and the social sciences, and span many aspects of the urban water cycle.
Diagnostic measurements	Data to identify the cause of variations in performance, or of failures (such as leakages, water quality variations).	Often the causes of performance failures in urban water systems will be complex, resulting from not just one individual factor, but from unintended and unpredicted interactions between several components.

Although current technology offers a wide range of facilities to achieve integration of environmental data, there are several important *institutional issues* (see also Chapter 11) that need attention for successful implementation (Geerders, 2003):

- strong and long-term institutional commitment to the principle of integrated data management
- unambiguous definitions of terminology and methodology for data acquisition, which are compatible across the range of potential applications for the data
- development and consistent application of criteria for the evaluation of environmental indicators in accordance with the established objectives of environmental management in each specific sub-domain
- development of data policy that has considered the requirements of, and thus satisfies, all stakeholders.

REFERENCES

ANZECC and ARMCANZ. 2000. *Australian Guidelines for Water Quality Monitoring and Reporting*. Canberra, Australian and New Zealand Environment and Conservation Council (ANZECC) and the Agriculture and Resource Management Council of Australia and New Zealand (ARMCANZ).

Bertrand-Krajewski, J.-L., Barraud, S. and Chocat, B. 2000. Need for improved methodologies and measurements for sustainable management of urban water systems. *Environmental Impact Assessment Review*, Vol. 20, pp. 323–31.

Geerders, P. 2003. Information-integration-inspiration. N.B. Harmancioglu, S.D. Ozkul, O. Fistikoglu and P. Geerders (eds), *Integrated Technologies for Environmental Monitoring and Information Production*. Dordrecht, Kluwer Academic Publishers. pp. 55–62. (NATO Science Series: IV Earth and Environmental Sciences Volume 23).

Roy, B. 1990. Decision aid and decision-making. C. Bana e Costa (ed), *Readings in Multiple Criteria Decision Aid*. Berlin, Springer, pp. 17–35.

Vogt, K., and Petersen, V. 2003. Surface water quality monitoring - requirements of the EU Water Framework Directive. Paper presented at the International Workshop on Hydrological Networks for Integrated and Sustainable Water Resources Management, 22–23 October 2003, Koblenz, Germany, pp. 103–13. German IHP/OPHP National Committee and UNESCO IHP.

Chapter 4

Selecting variables to monitor

A. Deletić and T.D. Fletcher

Department of Civil Engineering (Institute for Sustainable Water Resources), Building 60, Monash University, Melbourne 3800, Australia

4.1 INTRODUCTION

Having defined the objectives, the next step in designing a monitoring programme is to specify the variables to be monitored. As outlined in Chapter 3, this selection is dependent entirely on the objectives of monitoring. Without clearly thought-out objectives based on the best available conceptual model of the system to be monitored, there are likely to be major omissions or flaws in any specification of monitoring variables.

Variable selection will be governed by the specifics of the system component that is to be monitored, *as well as how the data collected on one component will be used for analyses of other components of the IUWM system.* There are several categories of variables to be considered within the urban water system:

- site characteristics (natural environment of the system)
- infrastructure characteristics
- urban meteorology
- water quantity
- water quality
- water body and aquatic ecosystem characteristics
- socio-economic indicators.

4.2 SITE CHARACTERISTICS

Even before being able to define the objectives, it is necessary to have some understanding of how the urban water system component(s) integrate within their environment. Site variables may relate to:

- Water distribution area, for example, topography of the distribution area, includes the size, elevation, major natural barriers
- Catchment area, such as the characteristics of both wastewater and stormwater collection areas, including layout, size, slope, geological and soil characteristics, vegetation type and cover, land-use.

The majority of these variables usually involve static measurements, for example, the size or slope of catchment typically constant over the duration of the monitoring program and are measured at the start of the monitoring programme. However, some of the listed variables may change over time and this should be taken into account when designing the monitoring programme. For example stream profile may change as a result of increased discharges, which are direct results of an increase in the fraction of impervious areas. *The dynamics of urban development (over the duration of the monitoring programme) usually dictate the degree and rate of change in site variables.*

In some cases it is also important to collect historic data, in order to interpret current system condition and/or behaviour. In developed areas some surface water bodies may have been drained or their course changed. If the aim is to rehabilitate the area, historic data are the key requirement. For example, elevated pollutant levels in streams in urban areas may often be at least partly a legacy from previous land-use, such as intensive agriculture (e.g. Taylor et al., 2005).

Existing maps (including historic and up-to-date maps of the area, geological and geomorphological maps etc) of different resolution, aerial photographs, and satellite images are the most common data source. Generally, such maps and imagery are readily available at large scales, but small-scale detailed mapping or images will generally be obtained from relevant local authorities and geographical survey agencies. Digital Terrain Models (DTM) and Geographic Information Systems (GIS) can be used to analyse the survey data collected and extract information about the hydrological consequences of the observed land-form.

It is also common practice to use models to estimate some of the physical characteristics that are hard to measure. For example, catchment areas for urban stormwater collection may be adjusted in the process of calibration of a stormwater model (e.g. SWMM, MOUSE, etc). This approach is not ideal but is common if the availability of real data is limited.

4.3 INFRASTRUCTURE CHARACTERISTICS

IUWM systems, often very complex systems, can include different pipe and/or channel networks and small and large water system structures (such as treatment plants, pumping stations, storage basins, etc). Important physical characteristics may include:

- layout, such as size, slope, elevation, etc
- designed capacity
- materials, such as type, roughness, durability, strength
- age and condition
- energy requirements
- maintenance practices.

Many of these variables usually do not change over the duration of the monitoring period. However, this is not always the case, and there may be changes due to infrastructure modifications, or simply due to degradation of infrastructure.

A critical requirement is to ensure that the geometry of infrastructure is accurately measured, and not assumed based on previous experience. Calibrating a flow meter, for example, for an oval pipe based on an assumed round shape will lead to substantial errors.

Infrastructure data is usually collected from the authorities who maintain information on the urban infrastructure, such as local or regional authorities and water agencies or private companies, but also developers and construction companies. It is becoming common that the data on infrastructure are organized in large databases (usually GIS based). However, it is common in many places for data on infrastructure to be lost, well before the end of the lifespan of the infrastructure. While there are no ready solutions to this problem, it is imperative that any new monitoring programme which collects data about infrastructure, has a clear plan (with accountability to relevant personnel or agencies) for the storage and long-term protection of such data (refer to Chapter 9).

Another common problem is the reliance on 'as-designed' rather than 'as-constructed' measurements. *If a thorough understanding of the infrastructure characteristics and dimensions is important to the objectives of the monitoring programme, a specific survey should be undertaken.* In the case of buried infrastructure, indirect measurements may be required.

4.4 URBAN METEOROLOGY

Climate is often a driving factor in understanding performance of an IUWM system and there are many climate variables which may be measured (refer to Chapter 14 for further details). Precipitation (e.g. rainfall intensity, snowfall), being the main link in the natural water cycle, is the most commonly measured variable. Others which drive hydrological processes include ambient temperature and evaporation or evapotranspiration.

For climatic variables, questions of spatial and temporal scale become dominant (see also Chapter 5). In the case of an integrated approach it is important that selection of the temporal resolution considers *all components* for which the data may be used. For example, for water storages (reservoirs) it may only be necessary to measure monthly precipitation depths, while for modelling of stormwater runoff, rainfall intensity measured at one minute interval may be required, for small catchments.

4.5 WATER QUANTITY

Practically all urban water monitoring programmes will involve the measurement of water flows. In the case of integrated systems, water discharged from one component usually becomes inflow into another. *Therefore, closure of the mass balance of water movement within IUWM systems is highly important*, and should be an integral question in the data validation process (Chapter 8).

The following variables are commonly monitored within different points of the system:

- flow rate (including peaks and minimums)
- volume
- velocity
- pressure
- level
- depth
- head losses.

These variables can be measured directly or indirectly (by recording some other variables). For example, monitoring of the level in front of a weir will be a common way of monitoring head over the weir and consequently flow rate. Monitoring of volumes is usually done by continuous monitoring of flow rates, which are then integrated over time.

In an integrated system, decisions will need to be made in relation to the monitoring of similar variables in all locations across the system. This decision will depend on the original objectives. For example, if the objective of a monitoring programme is to monitor losses within a small stormwater re-use system, designed to harvest, store and distribute water within a small urban development, the losses within the distribution system would usually be negligible compared to the losses occurring within the stormwater collection system (i.e. the percentage of runoff captured). Therefore monitoring losses within the distribution system may be able to be ignored, thus requiring only the measure of outflow rate from the storage area. If the storage area is covered and does not show signs of leaking, we may need to monitor only inflows into the storage area, and its volume (by simple measurement of level). These questions can only be appropriately answered by having a clear understanding of the objectives of the monitoring programme and of the behaviour and interactions within the system. Equipment suitable for monitoring some of the variables noted is discussed in Chapter 7.

4.6 WATER QUALITY

The definition of objectives will determine the water quality variables of importance for a given monitoring campaign. Therefore no attempt will be made here to list all physical, chemical or biological parameters to be measured to assess water quality within an IUWM system.

As discharge from one component of an IUWM system may become the input to another (e.g. treated wastewater may be used in dual-reticulation systems), it is important to understand levels of the same parameters that may occur in different water streams. Local guidance is necessary here, since values will be dependent on local context. In many situations, some pilot investigations are usually required, to determine the likely range. The potentially vastly different concentrations in each component of the urban water system will have implications for sensor selection (Chapter 7). Even within a single system (such as a wastewater treatment plant), the concentration of sediment or dissolved nitrogen, for example, will vary dramatically between the inlet and the outlet.

The selection of quality parameters for monitoring will depend highly on our understanding of the importance of that parameter in achieving the monitoring objectives. *Knowing whether the selected variable is the correct one is critical.* For example, it is often assumed that coliform bacteria may be an indicator of presence of pathogenic organisms in water. This will not always be the case (O'Shea and Field, 1992).

4.7 WATER BODIES AND AQUATIC ECOSYSTEM HEALTH

Integrated urban water management means that understanding (in order to minimize) impacts on waterways and their ecosystems becomes an important objective in the overall management of water resources. Important aspects of water bodies and their ecosystems to be monitored may include:

- Physical characteristics of:
  - Surface water bodies, for example, the geomorphology of reservoirs, streams, bays and lakes, that is, the layout, size, slope, water depth and distribution of

velocity, shear stress, as well as bed and bank conditions (e.g. particle size distribution of benthos, presence of erosion, etc.)
- ○ Habitat diversity (e.g. proportions of pools and riffles, benthos distribution, etc) (see for example Gorman and Karr, 1978; Karr and Chu, 1999).
- ○ Quantity, distribution and condition of vegetation (aquatic and riparian)
- ○ Local groundwater aquifer, such as aquifer size, aquifer material characteristics
- Water quality, that is the concentrations and loads of physical, chemical and biological parameters
- Ecosystem health and functioning
  - ○ Composition and diversity of organisms such as macroinvertebrates (using indices of biological integrity, such as the *Indice Biologique Diatomées* (IBD) or a benthic macroinvertebrate index, e.g. SIGNAL (see for example Chessman, 1995; Karr, 1999; Karr and Chu, 1999; Metcalfe, 1989) and diatoms (Chessman et al. , 1999; Reid et al. , 1995; Round, 1991).
  - ○ Indicators of process rates and equilibriums, such as photosynthesis-to-respiration ratios and chlorophyll-*a* density (Karr and Chu, 1999).

The selection of variables relating to water bodies and aquatic ecosystem health will depend on the *values* that are to be protected and the identified *threats* to those values. Again, this determination is to be made at the *definition of objectives* stage. Typically, the water body health variables will need to be measured simultaneously with variables measuring potential causal factors (relating to water quality, water flow, catchment characteristics). For example, substantial recent progress has been made in identifying and measuring the key catchment and urban land-use characteristics which can explain degradation in aquatic ecosystem health in receiving waters (e.g. Hatt et al., 2004; Walsh et al., 2005). Chapter 20 covers in depth the data needs in this area, as well as methods used in the collection of the listed variables.

4.8 SOCIO-ECONOMIC INDICATORS

Processes for collecting and assessing economic indicators are generally well established, and include indicators such as capital cost (typically made up of design, land acquisition and construction costs), plus operating costs (such as interest, depreciation, consumables, maintenance, modification and renewal, etc). For infrastructure with commercial imperatives (there will typically also be indicators on the 'return on investment', such as earnings and/or profits). A *life-cycle cost* analysis, or one of its variants (see Gluch and Baumann, 2004) will generally be undertaken to calculate the sum of all expenses (and potentially revenues). The advantage of such an approach is that it forces the consideration of long-term costs, so that ongoing commitments such as maintenance are taken into account when the original purchasing decisions are made. However, for effective use of such tools, it is critical that reliable data on maintenance costs be collected (Taylor and Fletcher, 2004).

With the advent of integrated urban water management, there is an increasing interest in tools that consider more than just financial aspects (Gluch and Baumann, 2004). As a result, techniques such as Triple-Bottom-Line accounting (TBL), Multi-Criteria Analysis (MCA) and Cost-Benefit Analysis (CBA) have been employed in an attempt to satisfy *multiple objectives* and to quantify (or at least qualitatively assess) *externalities* (Taylor

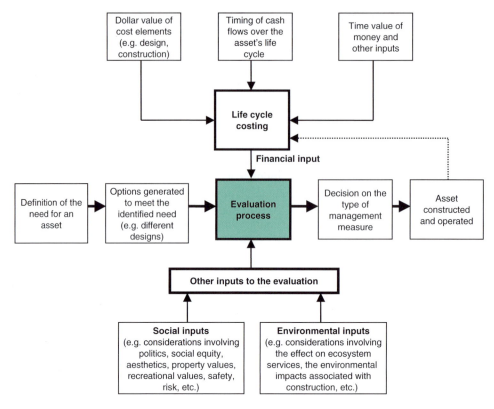

Figure 4.1 Conceptual framework of triple-bottom-line analysis using a multiple criteria assessment approach

Source: Adapted from: Taylor, 2003.

et al., 2006). A typical conceptual framework for MCA is shown in Figure 4.1. By definition, such techniques require data on environmental, social and economic indicators, such as:

* ecosystem health and function
* social impacts and responses (positive and negative) (for example, the community may be surveyed to determine their preference for utilization of alternative water sources such as recycled wastewater or stormwater collected from impervious surfaces.)
* economic aspects (such as the life-cycle cost).

The collection of such data provides a focus for systems that can provide multiple uses and benefits and follow principles of sustainability (Moura et al., in press; Taylor et al., 2006). However, there is also a challenge in collecting this sort of data (e.g. Table 4.1), because it involves social preference surveys which require specific expertise. Useful guidance on the collection of data and the use of multi-criteria analysis for decision support is provided by Ellis et al. (2006).

Table 4.1 Example of data requirements for multiple-criteria analysis

Criteria	Cost or benefit
Financial	1. Life-cycle cost of the water management measures. 2. Costs associated with occupying developable land. 3. Effect on local property values. 4. Maintenance burden/costs. 5. Up-front capital cost to developers (passed on to home buyers).
Ecological	6. Value of the protection and enhancement of *local* environmental conditions (i.e. the health of nearby creeks, wetlands and/or rivers). 7. Value of the protection and enhancement of *regional* environmental. 8. Environmental impacts associated with obtaining the raw materials, construction and maintenance. 9. Impact on the mains water supply system.
Social	10. The health and safety of residents and maintenance staff. 11. Effects on the neighbourhood's liveability/amenity values. 12. The ease of maintenance by home-owners. 13. Social equity (e.g. between and within generations). 14. Education and awareness opportunities/benefits.

Source: Taylor et al., 2006.

Other important variables may include those related to industry attitudes and uptake, and institutional impediments to the adoption of IUWM (Brown, 2005; Chocat et al., 2001). Local guidance on the collection of such data will be required.

REFERENCES

Brown, R.R. 2005. Impediments to integrated urban stormwater management: the need for institutional reform. *Environmental Management*, Vol. 36, No. 3, pp. 455–68.

Chessman, B.C. 1995. Rapid assessment of rivers using macroinvertebrates: a procedure based on habitat-specific sampling, family level identification and a biotic index. *Australian Journal of Ecology*, Vol. 20, No. 1, pp. 122–29.

Chessman, B., Growns, I., Curreys, J. and Plunkett-Cole, N. 1999. Predicting diatom communities at the genus level for rapid bioassessment of rivers. *Freshwater Biology*, Vol. 41, pp. 317–31.

Chocat, B., Krebs, P., Marsalek, J., Rauch, W. and Schilling, W.M. 2001. Urban drainage redefined: From stormwater removal to integrated management. *Water Science and Technology*, Vol. 43, No. 5, pp. 61–68.

Ellis, J.B., Deutsch, J.C., Legret, M., Martin, C., Revitt, D.M., Scholes, L., Seiker, H. and Zimmerman, U. 2006. The DayWater decision support approach to the selection of sustainable drainage systems: a multi-criteria methodology for BMP decision makers. *Water Practice and Technology*, Vol. 1, No. 1. Available at: http://www.iwaponline.com/wpt/default.htm (Accessed 02 July 2007.)

Gluch, P. and Baumann, H. 2004. The lifecycle costing (LCC) approach; a conceptual discussion of its usefulness for environmental decision-making. *Building and Environment*, Vol. 39, No. 5, pp. 571–80.

Gorman, O.T. and Karr, J.R. 1978. Habitat structure and stream fish communities. *Ecology*, Vol. 59, No. 3, pp. 507–15.

Hatt, B.E., Fletcher, T.D., Walsh, C.J. and Taylor, S.L. 2004. The influence of urban density and drainage infrastructure on the concentrations and loads of pollutants in small streams. *Environmental Management*, Vol. 34, No. 1, pp. 112–24.

Karr, J.R. 1999. Defining and measuring river health. *Freshwater Biology*, Vol. 41, No. 2, pp. 221–34.

Karr, J.R. and Chu, E.W. 1999. *Restoring Life in Running Waters: Better Biological Monitoring.* Washington DC, Island Press.

Mallory, C.W. 1973. *The Beneficial Use of Stormwater*. Washington DC, US Environmental Protection Agency. (EPA-R2-73-193)

Metcalfe, J. L. 1989. Biological water quality assessment of running waters based on macroin-vertebrate communities, history and present status in Europe. *Environmental Pollution Bulletin*, Vol. 60, No. 1–2, pp. 101–39.

Moura, P., Barraud, S. and Baptista, M. (*in press*). Multicriteria procedure for the design and the management of infiltration systems. *Water Science and Technology*.

NHMRC and NRMMC. 2004. *Australian Drinking Water Guidelines*. Canberra, National Health and Medical Research Council and Natural Resource Management Ministerial Council, Government of Australia.

O'Shea, M. L. and Field, R. 1992. Detection and disinfection of pathogens in storm-generated flows. *Canadian Journal of Microbiology*, Vol. 38, pp. 267–76.

Queensland EPA. 2003. *Queensland Guidelines for the Safe Use of Recycled Water, Public Consultation Draft.* Brisbane, EPA.

Reid, M.A., Tibby, J.C., Penny, D. and Gell, P.A. 1995. The use of diatoms to assess past and present water quality. *Australian Journal of Ecology*, Vol. 20, No. 1, pp. 57–64.

Round, F.E. 1991. Diatoms in river water-monitoring studies. *Journal of Applied Phycology*, Vol. 3, No. 2, pp. 129–45.

Taylor, A.C 2003. *An Introduction to Life Cycle Costing Involving Structural Stormwater Quality Management Practices*. Melbourne, Cooperative Research Centre for Catchment Hydrology. http://www.toolkit.net.au/cgi-bin/WebObjects/toolkit.net.au (Accessed 03 July 2007.)

Taylor, A.C. and Fletcher, T.D. 2004. Estimating life-cycle costs of structural measures that improve urban stormwater quality. Paper presented at the *International Conference on Water Sensitive Urban Design*, 21–24 November 2004 Adelaide, Australia.

Taylor, A.C., Fletcher, T.D. and Peljo, L. 2006. Triple-bottom-line assessment of stormwater quality projects. *Urban Water*, Vol. 3, No. 2, pp. 79–90.

Taylor, G.D., Fletcher, T.D., Wong, T.H.F. and Breen, P.F. 2005. Nitrogen composition in urban runoff – implications for stormwater management. *Water Research*, Vol. 39, No. 10, pp. 1982–89.

US EPA. 2004. *Guidelines for Water Reuse*. Washington DC, US Environmental Protection Agency (US EPA, EPA/625/R-04/108, September 2004).

Walsh, C.J., Roy, A.H., Feminella, J. W., Cottingham, P.D., Groffman, P.M. and Morgan, R.P. 2005. The urban stream syndrome: current knowledge and the search for a cure. *Journal of the North American Benthological Society*, Vol. 24, No 3, pp. 706–23.

Chapter 5

Spatial and temporal scale considerations

J-L. Bertrand-Krajewski[1], T.D. Fletcher[2] and V.G. Mitchell[2]

[1]Laboratoire LGCIE, INSA-Lyon, 34 avenue des Arts, Villeurbanne CEDEX, France
[2]Department of Civil Engineering (Institute for Sustainable Water Resources), Building 60, Monash, Monash University, Melbourne 3800, Australia

5.1 INTRODUCTION

Once the objectives of the monitoring programme have been defined, and the variables that are to be monitored have been decided upon, the next step is to consider the spatial and temporal scales at which these variables should be monitored. The *spatial scale* is comprised of two dimensions, the *size of the area* to be monitored (i.e. 1 m² or 1000 km²) and the *resolution* at which this area is monitored (fine or course density of monitoring sites). The *temporal scale* also has two dimensions, the *duration of monitoring* (e.g. 6 hours or 20 years) and the *frequency* of sampling (the time interval between sampling, from seconds to months). The different combinations of these dimensions result in quite different data monitoring requirements (See Figures 5.1 and 5.2).

For example, assessing the average annual residential water usage in a town may require the sampling to cover a large spatial area, with a selection of buildings being monitored (large area, course spatial resolution). Water meters (assuming these are already installed) could be read once every 3 months to 6 months for 18 months or more to provide an estimate of their average annual water usage; a longer period of sampling would be required if there is substantial outdoor water usage which varies with climate with infrequent samples over a longer period. Contrast this with the monitoring approach which would be required to determine the variation in the amount of

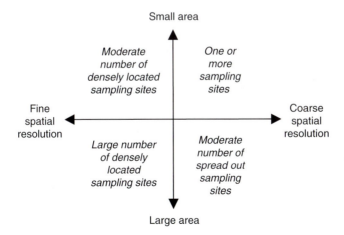

Figure 5.1 Spatial dimensions of a monitoring programme

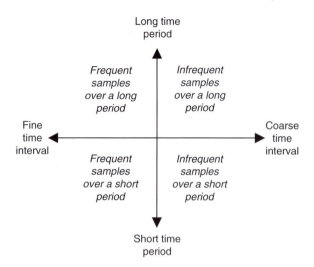

Long time
period

Frequent
samples
over a long
period

Infrequent
samples
over a long
period

Fine
time
interval

Coarse
time
interval

Frequent
samples
over a short
period

Infrequent
samples
over a short
period

Short time
period

Figure 5.2 Temporal dimensions of a monitoring programme

water used each time the kitchen tap is used within all the houses in that same town. A large number of people in the whole town (i.e. large number of densely located sampling sites) could be surveyed within a single week (frequent samples over a short period).

Within an Integrated Urban Water Management (IUWM) framework, there may be a number of variables within an urban water system that need to be monitored simultaneously. However, it is unlikely that they all require the same spatial and temporal monitoring, although the manner in which each variable is monitored must match with that of the other variables. This may mean that the frequency of each variable's monitoring matches the maximal frequency required across the number of system variables. It may also mean that the items monitored at a finer scale can be lumped or aggregated into a form compatible with the other variables being monitored. The key consideration is that the representativeness of the monitored data across the selected time and space scale should be ensured for each variable.

This chapter outlines issues relating to temporal and spatial scales, and outlines some methods which may be used to determine appropriate monitoring scales. Since a detailed description of the wide range of statistical and analytical methods available for scale analysis is beyond the scope of this book, references to key publications are also provided. It is important to recognize that there is no simple prescription for determining appropriate temporal and spatial scales for monitoring. The appropriate scale will be entirely case-specific and will require some *prior knowledge* of the systems' behaviours and functioning.

5.2 TEMPORAL AND SPATIAL REPRESENTATIVENESS OF SAMPLES AND DATA

In urban water systems, water quantity, water quality and pollutant loads are usually estimated by means of point measurements in both time and space. Thus the key question is *how to ensure that these point measurements are effectively representative of the phenomena and processes of interest to the user.* Representativeness is usually *assumed*, but very rarely (if ever) either seriously analysed or proven. Experience shows that most investments and efforts are made in sensors and in laboratory analyses of collected samples, and not in assessing the representativeness of collected samples. Such an approach results in a high risk that one knows very well the exact concentration of a given sample, but that the probability of this sample representing the true *in situ* concentration is either low or even unknown. To avoid such difficulties, rigorous statistical methods are available, which, unfortunately, are not yet sufficiently applied. The following sections explore these questions, with references also to other publications which explain in detail how to account for representativeness.

5.2.1 Temporal variability

The phenomena under investigation in monitoring urban water systems are mostly not permanent, but rather highly unsteady and variable. Some temporal fluctuations may be periodic and thus relatively predictable (e.g. daily discharge patterns in sewer systems, drinking water needs between week days and weekends, seasonal fluctuations of groundwater levels, etc.). Other fluctuations are by nature random (e.g. occurrence and natural variability of storm events, incidents and unexpected phenomena, etc.), which renders monitoring and prediction thus much more difficult.

The choice of appropriate monitoring time scale and time step depends on both the objectives of the monitoring programme and the nature of the phenomena to be monitored. The *sampling theorem* (also called the Shannon or Nyquist theorem) indicates that, in order to observe phenomena and effects occurring with a frequency F, one should make measurements with a frequency at least equal to $2 \times F$. If, for example, the objective consists to evaluate daily variations, measurements have to be made at least twice a day. It is also important to note that the choice of the most appropriate time steps for a given objective is an iterative process, based on some prior knowledge. This knowledge may pre-exist from previous experience, antecedent monitoring or specific preliminary measurement campaigns. There is no universally applicable prescription or simplistic formula to define time steps and time scales for monitoring; each case is specific and should be investigated as such.

5.2.1.1 Examples of temporal variability

The following examples will help to illustrate these concepts.

Example 1: Monitoring COD concentrations during dry weather

Example 1 deals with dry weather monitoring of COD concentrations and pollutant loads in a combined sewer system. Discharge is calculated with data provided by an ultrasonic sensor for water depth and a Doppler sensor for flow velocity. COD is estimated by means of an online calibrated UV-visible spectrometer. Both data are stored in a data logger with a 2 minute time step. Monitoring results during four days (01 July to 04 July 2004) are shown in Figure 5.3 (for legibility, the figure shows the *moving average* calculated from seven adjacent 2 minute time steps).

Assuming that these time series represent the truth, the total COD load during the four day period is:

$$L_{COD} = \sum_{i=1}^{2880} C_i Q_i \Delta t = 5141 \text{ kg}$$

5.1

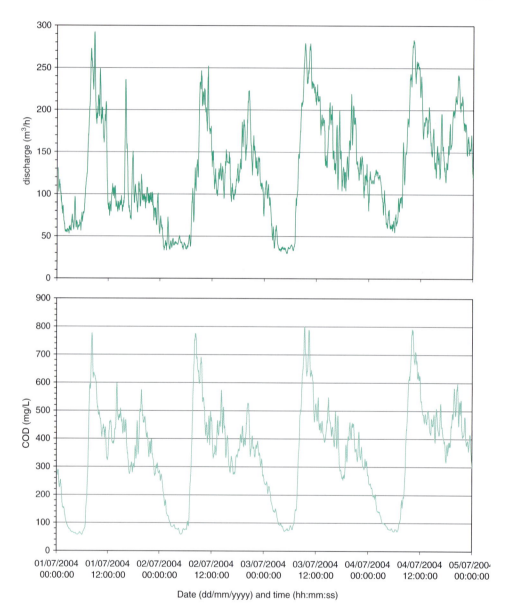

Figure 5.3 Hydrograph of water discharge and chemical oxygen demand (COD) concentrations, 01 July 2004 to 04 July 2004 (a 14-minute moving average derived from 2-minute time step monitoring)

Source: Adapted from de Bénédittis and Bertrand-Krajewski, 2005.

Where C_i is the concentration, Q_i the discharge, $\Delta t = 2$ minutes and i the index ranging from 1 to 2880.

Five sampling scenarios have been compared, where the total COD load was calculated with one COD measurement made

- every hour at 0.30, 01.30, 02.30, etc
- every 2 hours at 01.00, 03.00, 05.00, etc
- every 6 hours at 03.00, 09.00, 15.00, etc
- every 12 hours at 06.00, 18.00, 06.00, etc
- every 24 hours at 12.00 on 01 July, 02 July, etc.

All loads have been calculated by using the 2 min discharge values. Results are given in Figure 5.4. COD loads range from 3356 kg to 5685 kg (i.e. from −35% to +10% compared to the true value). Many other sampling scenarios could have been compared, leading to more spread results.

Example 2: Monitoring Total Suspended Solids (TSS) concentrations and loads during storm events

Example 1 presented an approach suitable for rather periodic or predictable phenomena. Such an approach will not be appropriate for sampling during storm events, where the behaviour will be stochastic and highly variable. Such phenomena have to be analysed with a different approach, based on some hypotheses about their occurrence frequency, duration, amplitude and internal dynamics. If random and transient phenomena (e.g. storm events, floods, intermittent and brief discharges of some specific pollutants, etc.) are to be monitored effectively, a very short observation time step

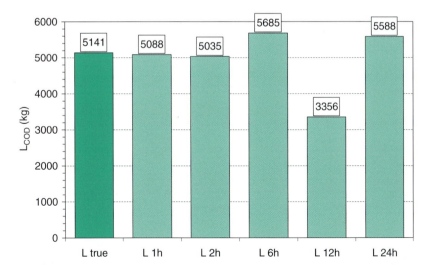

Figure 5.4 Total chemical oxygen demand load (L$_{COD}$) from 01 July to 4 July 2004 for various sampling scenarios (true value is shown at far left)

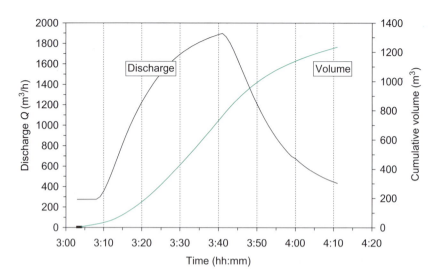

Figure 5.5 Case study hydrograph illustrating discharge (*Q*) and cumulative volume (*V*)

Source: Adapted from Bertrand-Krajewski et al., 2000.

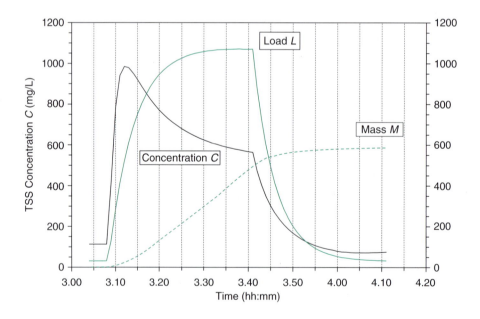

Figure 5.6 Case study *reference* total suspended solids (TSS) concentration (*C*), load (*L*) and cumulative mass (*M*)

Source: Adapted from Bertrand-Krajewski et al., 2000.

is necessary, which usually could be hardly compatible with traditional sampling and would preferably require the use of online sensors or detectors. An alternative approach may consist of using integrated sampling devices.

Consider the estimation of both the *TSS pollutograph* and the *TSS pollutant load* at the outlet of a sewer system during a storm event. The determination of the pollutograph requires a series of individual samples, either simple or composite, which are used for TSS analyses in the laboratory. Many sampling strategies are possible. The following four strategies are among the most frequently used, with Q being the discharge in the sewer and V the corresponding cumulative volume:

- Strategy A: sampling of a *constant volume* at a *constant time step*
- Strategy B: sampling of a *volume proportional to cumulative flow volume, V,* at a *constant time step*
- Strategy C: sampling of a *volume proportional to instantaneous flow rate, Q,* at *constant time step*
- Strategy D: sampling of a *constant volume* at *variable time step proportional to V.*

These strategies are illustrated with a case study derived from a real storm event. The hydrograph (discharge Q and cumulative volume V) and pollutograph (total suspended solids TSS concentration C, load L and cumulative mass M) given in Figure 5.5 and Figure 5.6 are considered *as reference (or true) values.* We will calculate the estimated TSS event mean concentration (EMC) from each of the above four sampling strategies, and then compare them to the *reference (or true)* TSS EMC. This example aims to show how sampling strategies may possibly lead to results differing from the true values. At issue is the definition of sampling strategies ensuring representative and unbiased results.

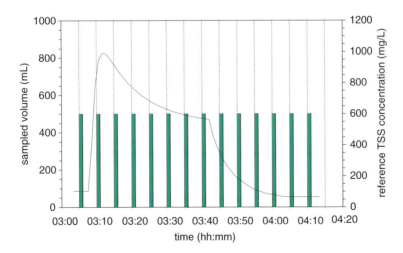

Figure 5.7 Profile of sampling volumes of 500 mL (constant volume) taken every 5 minutes against the reference TSS pollutograph. Sample times and volumes (in mL) are represented by bars, e.g. at 03:20, the sample volume is 500 mL

Source: Adapted from Bertrand-Krajewski et al., 2000.

Reference values: the data illustrated in Figures 5.5 and 5.6 correspond to the following values. The event started at 03:03 and ended at 04:11, corresponding to 68 minutes. All data are recorded with a 1 minute time step. The total cumulative volume $V = 1234.2\,\mathrm{m}^3$ and the total cumulative mass of TSS $M = 591.7\,\mathrm{kg}$. Consequently, the reference TSS EMC is equal to $M/V = 479\,\mathrm{mg/L}$.

Strategy A: sampling of a constant volume at constant time step (e.g. sampling of $V_p = 500\,\mathrm{mL}$ every 5 minutes). This strategy, illustrated in Figure 5.7, gives an accurate description of the pollutograph, *provided that the time step is short enough*, typically from 2 minutes to 10 or 15 minutes. The minimum allowable time step will increase with catchment size (and thus time of concentration).

Advantages: (i) it is easy to identify the exact time of each sampling, either with the sampler itself or with the data logger, (ii) it is easy to select the appropriate volume to make all required analyses (i.e. to ensure that there is enough water provided to the laboratory for all tests to be conducted).

Disadvantages: the determination of the pollutant load in kg/h or kg is not direct and needs some calculations, assuming that the sample is representative of the complete period since the previous sample (this assumption is always made for all discrete samples).

Strategy B: sampling of a variable proportional volume V at constant time step (e.g. sampling of $V_p = 25\,\mathrm{mL}$ for each $5\,\mathrm{m}^3$ volume increment, every 5 minutes). At each time step, the sample taken is proportional to the cumulative volume since the previous sample, as shown in Figure 5.8. An alternative strategy, to improve the representativeness, is to have a sample based on sub-samples taken at each increment of the cumulative volume.

Advantages: (i) the determination of the TSS load in kg/h is direct, (ii) it is easy to identify the exact time of each sampling, either with the sampler itself or with the data logger.

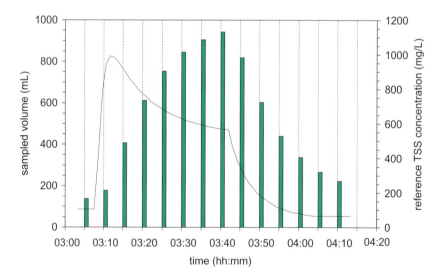

Figure 5.8 Profile of sampling volumes of 25 mL for each $5\,\mathrm{m}^3$ cumulative volume increment taken every 5 minutes against the reference TSS pollutograph. The sample times and volumes (in mL) are represented by bars: e.g. at 03:20, the sample volume is 450 mL ($18 \times 25\,\mathrm{mL}$) because a cumulative volume of $90\,\mathrm{m}^3$ ($18 \times 5\,\mathrm{m}^3$) was discharged between 03:15 and 03:20 as shown in Figure 5.5

Source: Adapted from Bertrand-Krajewski et al., 2000.

Disadvantages: sample volumes are not constant, and sometimes some samples are not large enough to make all required laboratory analyses.

Strategy C: sampling of a variable volume proportional to Q at constant time step, for example, sampling of $V_p = 0.5$ mL per m³/h, every 5 minutes (Figure 5.9). As discussed for strategy B, each sample could also be composed of sub-samples taken within each sampling interval, to increase the temporal resolution of sampling, and increase the representativeness of the overall sample.

Advantages: (i) it is easy to identify the exact time of each sampling, either with the sampler itself or with the data logger, (ii) the strategy is easily implemented, as only the instantaneous discharge values are used, and not cumulative volume.

Disadvantages: as sample volumes are not constant, it can sometimes occur that some samples are not large enough to make all required analyses in the laboratory.

Strategy D: sampling of a constant volume at variable time step proportional to V (e.g. sampling of $V_p = 500$ mL per 100 m³ volume increment as in Figure 5.10). As for strategies B and C, each sample could also be composed of sub-samples (e.g. four sub-samples of 125 mL taken for each 25 m³ volume increment).

Advantages: (i) the determination of the TSS load in kg/h is direct and thus straightforward (ii) an accurate description of the pollutograph is obtained, especially during peaks, provided that the volume increment is appropriate, (iii) the constant sample volume ensures that there is always adequate water to make all required laboratory analyses.

Disadvantages: (i) it is more difficult to identify the exact time of sampling, (ii) if the volume increment is too large, very few samples are taken, which may be particularly critical at the beginning and at the end of storm events, (iii) as sample volumes are not

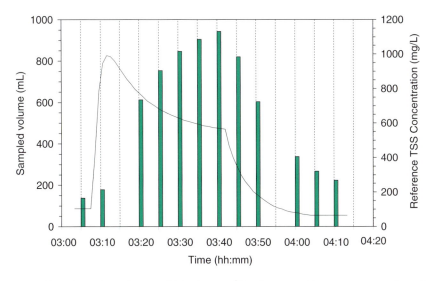

Figure 5.9 Profile of sampling volumes of 0.5 mL per m³/h taken every 5 minutes against the reference TSS pollutograph. The sample times and volumes (in mL) are represented by bars: e.g. at 03:20, the sample volume is approximately 610 mL ($\approx$ 1225 × 0.5 mL) because the discharge at 03:20 was 1225 m³/h as shown in Figure 5.5

Source: Adapted from Bertrand-Krajewski et al., 2000.

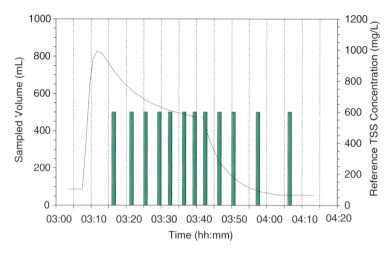

Figure 5.10 Profile of sampling volumes of 500 mL (constant volume) taken every 100 m³ of cumulative volume increment against the reference TSS pollutograph. The sample times and volumes (in mL) are represented by bars: e.g. one 500 mL sample is taken at 03.21 because 100 m³ of cumulative volume have been discharged since 03.16 as shown in Figure 5.5

Source: Adapted from Bertrand-Krajewski et al., 2000.

constant, it can sometimes occur that some samples are not large enough to make all required analyses in the laboratory.

Calculation of the event mean concentration (EMC): the EMC has been calculated for the reference situation and for 20 sampling strategies with varying time steps defined as follows (more details about EMC calculations are given in Bertrand-Krajewski et al., 2000):

- four strategies **An**: the strategy An is the above strategy A with a time step of n minutes, with all samples taken proportionally to the volume *V* accumulated between samples
- four strategies **Ans**: as above, but simplified with direct mixing of all samples without accounting for the volume *V* accumulated between samples (this strategy is of course theoretically incorrect, but is sometimes applied simply for convenience)
- four strategies **Bn**: the strategy Bn is the above strategy B with a volume increment of n m³ and a time step of n minutes
- four strategies **Cn**: the strategy Cn is the above strategy C with a time step of n minutes
- four strategies **Dn**: the strategy Dn is the above strategy D with a volume increment of n m³.

Strategies A5, B5, C5 and D100 correspond respectively to the examples shown in Figure 5.7 through Figure 5.10. All resultant EMCs are given in Figure 5.11. Compared to the reference value of 479 mg/L, the simulated strategies give EMCs ranging from 395 to 546 mg/L (i.e. from −7.5% to +14%). Strategies An and Bn are equivalent, with a

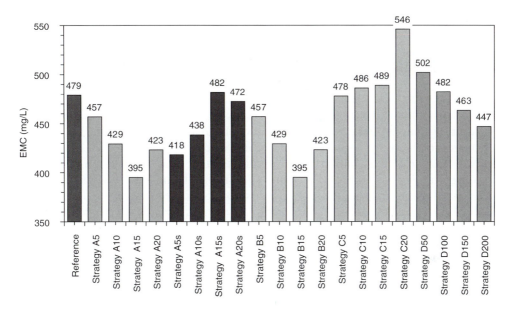

Figure 5.11 Comparison of 21 event mean concentration (EMC) values calculated according to outlined sampling strategies (reference and 20 sampling strategies) (see also colour Plate 2)

significant difference between strategies An and the incorrect simplified strategies Ans. Strategies An, Bn and Dn give decreasing EMCs when the time step increases while strategies Cn give increasing EMCs. It is clear that the choice of sampling strategy can have significant impacts on the estimated EMC of TSS. More generally, the strategy itself and the number of samples and sub-samples are important parameters that should be determined carefully according to the specified objectives.

Example 3: Monitoring long-term TSS loads

Example 2 used the sampling of TSS concentrations in a combined-sewer outlet, to determine *pollutographs* and *pollutant loads*. Consider the case now where a monitoring programme is being established *to determine the long-term mean annual load* of TSS coming from a particular catchment. A number of approaches could be taken:

- Routine sampling every n days
- Routine sampling every n days plus grab-sampling during storms
- Routine sampling every n days plus auto-sampling during storms
- Continuous monitoring of a surrogate variable, such as turbidity.

Given these possible approaches, a number of questions arise:

- Can TSS loads be estimated using continuously-monitored turbidity as a surrogate?
- If only routine sampling is used, what frequency of sampling is required?

- Are auto-samplers necessary to effectively capture long-term Site Mean Concentrations (SMCs) for storm events?

The questions were examined by Fletcher and Deletic (2006, 2007), and the key implications of their findings, for the understanding of temporal variability, are presented here.

Use of a continuously monitored surrogate: Using data from ten catchments, Fletcher and Deletic (2006, 2007) found that correlations between simultaneously sampled TSS (laboratory analysis) and turbidity (*in situ* probe) had an average R^2 of 0.63. Using linear regressions between TSS and turbidity, the TSS load was calculated for three catchments. Assuming that only random errors existed, that is, that no systematic measurement errors occurred (see Chapter 6 for further discussion of this concept), they determined uncertainties in the *long-term* load of TSS to be less than 10% in all cases (Table 5.1). The results show that the *use of a surrogate variable which can be continuously measured (such as turbidity) has great advantages for measuring long-term loads*, because it ensures that samples are taken *frequently enough to represent the natural variability of the phenomenon*. This may be one approach to overcoming the difficulty in achieving an adequate sampling frequency for highly variable phenomena.

However, continuously measured surrogates are not appropriate for many variables. For example, the same study showed that correlations between turbidity and many other variables were very poor. The results also depend strongly on the adequate calibration of the turbidity sensors.

Required frequency of routine grab-sampling: The same continuously measured surrogate estimates of TSS loads were used to assess the implications of varying sampling frequency, assuming that grab samples were taken on a uniform basis (i.e. constant time-interval). It is clear (Figure 5.12) that a sampling interval of greater than three days results in very large errors in the calculated load of TSS. *It is recommended that an initial high-resolution sampling pilot study be conducted for a one-year period in order to determine the most appropriate ongoing sampling interval.*

Requirement for auto-samplers: Auto-samplers are used to ensure that variability in concentrations during a storm can be captured, by triggering the auto-sampler to take a sample according to some pre-determined strategy (see Example 1). However, installation

Table 5.1 Relative uncertainty in long-term estimates of total suspended solids (TSS) load using continuously measured turbidity as a surrogate

Catchment	Monitoring period		Number of years	Relative uncertainty (%)
	Start	End		
Bunyip	17 Aug 2000	13 Dec 2005	5.3	1
	17 Aug 2000	17 Aug 2001	1.0	2
Lang-Lang	17 Aug 2000	24 Nov 2005	5.3	4
	17 Aug 2000	17 Aug 2001	1.0	7
Cardinia	17 Aug 2000	24 Nov 2005	5.3	3
	17 Aug 2000	17 Aug 2001	1.0	6

Source: Adapted from Fletcher and Deletic, 2006.

of such equipment can often be very expensive, and they are also prone to vandalism. Fletcher and Deletić (2006, 2007) also investigated whether long-term loads could be adequately represented without the use of auto-samplers. Data from nine auto-sampled events (average of 20 discrete samples taken per event) were sub-sampled, using two strategies:

(1) One sample per storm taken at a random time within the storm event
(2) One sample taken at the same time, one hour after the start of every event.

In either case, the errors in the *annual load estimate* were relatively minor, typically around 10% (Table 5.2). However, it is critical to remember that for characterization of the pollutograph (see Example 2) auto-sampling is essential to capture the variability during a storm.

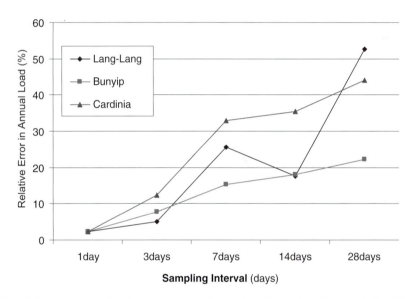

Figure 5.12 Influence of sampling interval on error in calculated load of total suspended solids (TSS)

Source: Adapted from Fletcher and Deletić, 2007

Table 5.2 Difference between 'true' load and load estimated from grab samples for selected parameters (percentages)

Timing of grab sample	Parameter	Total suspended solids	Total phosphorous	Total Nitrogen	Lead	Zinc
Random, within storm (20 replicates)	Mean error	8	12	11	12	10
	95 percentile error	10	16	12	14	17
1 hour after storm commencement	Error	9	8	5	13	7

Source: Fletcher and Deletić, 2006.

Example 4: Determining the appropriate sampling duration

Examples 1 to 3 have illustrated the importance of considering the implications of *temporal variability* on the representativeness of samples. However, capturing the *appropriate duration of a phenomenon* is also critical in ensuring representative sampling.

Consider for example the monitoring of a stormwater wetland to measure its treatment performance. Typically, we assume that the critical requirement is to capture the variability *during the storm event*, assuming that pollutant concentrations will be somewhat proportional to the flow hydrograph (i.e. pollutant concentrations will become elevated during a storm, and then recede in the post-storm process). However, to make such an assumption *without prior knowledge* to confirm its validity is dangerous. Taylor et al. (2006) showed that highly elevated concentrations of dissolved pollutants may enter a wetland in the post-storm period as a result of processes such as interflow and throughflow (Figure 5.13).

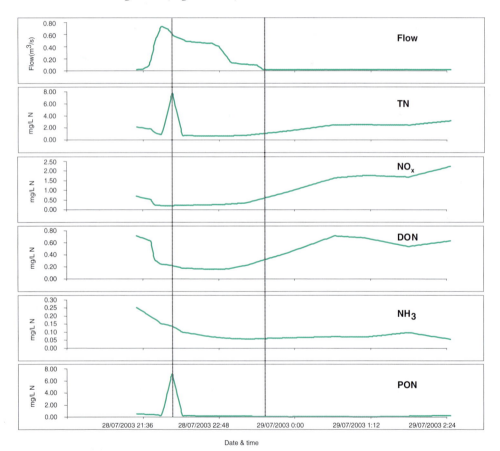

Figure 5.13 Example of post-storm elevated concentrations of various dissolved forms of nitrogen TN = Total Nitrogen, NO_x = oxides of nitrogen (nitrate + nitrite, DON = dissoled Organic Nitrogen, NH_3 = ammonia, PON = Particulate Organic Nitrogen)

Source: Adapted from Taylor et al., 2006.

As a consequence of this observed behaviour, Taylor (2006) recommended that a *variable resolution continuous monitoring* approach be used to conduct a pilot study in order to understand the various dynamics of a system. They proposed, for example, that sampling be conducted during several storm events *and* for the intervening dry weather period and that a *subset* (e.g. every fourth sample during storm events, and every eighth sample during dry weather) be sent to the laboratory for analysis. If results were highly consistent (static), remaining samples could be discarded. If not, intervening samples would be submitted (e.g. every other second sample during storms, and every other fourth sample during dry weather), and so on, until a thorough understanding of the system had been obtained. This could then be used to develop an appropriate sampling protocol for the monitoring programme.

5.2.1.2 Temporal variation: useful sources of information

For more specific needs, *ad hoc* sampling strategies have to be defined to avoid bias and to evaluate correctly measurement uncertainties. For example, the sampling strategy for benzotriazole, an ingredient in dishwasher detergents, should be specifically adapted to both time and space scales, due to its short and very intermittent emissions in households, by using modelling to define an appropriate sampling strategy (Ort et al., 2005; Ort and Gujer, 2005). It also appears in this case that temporal and spatial scales cannot be considered separately. The transient pollutant emission into a sewer with a measurable concentration at the house connection scale becomes a diffuse background emission (but sometimes below detection or quantification limits) at the catchment scale. Similar effects may also occur in rivers or in groundwater. If the objective is to detect point emissions, the sampling strategy and the observation time step will strongly differ compared to those applied if the objective was to evaluate global flux at catchment scale.

More detailed information about techniques and methods are beyond the scope of this document. However, some key references should be cited here for readers interested in further pursuing these essential questions. The most important overall guidance comes from the sampling theory elaborated by Gy (1988, 1992, 1996). The work provides the best theoretical and statistical approach for sampling of moving material (of which water is a prime example), defining in a rigorous way representativeness, heterogeneity and bias and using chronostatistics in the same way that geostatistics are used for spatial sampling. Some authors have already attempted to apply this theory in urban water systems (e.g. Rossi, 1998) for sampling of storm events in sewer systems. Further applications and adaptations of this theory should be made to improve sampling strategies in urban water systems.

Continuous *in situ* monitoring with sensors has rapidly expanded in recent times, and new techniques and methods, derived from signal analysis, should be applied to evaluate appropriate time scales of observation and to analyse trends which were impossible to detect with limited traditional sampling. Up to now, there are only very few applications of these methods in the field of urban water systems, and examples should be taken from other research fields (e.g. Labat et al., 2001, 2002; Mangin, 1984). These techniques are also potentially very useful for modelling applications. Another useful reference on scale issues is the book on *Scales in hydrology and water management* (Tchiguirinskaia et al., 2004).

5.2.2 Spatial variability

Where to measure is dictated by the monitoring objectives. In some cases, the locations can be easily defined: at the inlet and outlet of drinking water or wastewater treatment plants, at the interface between urban water components (like CSOs), at points of transfer between sub-systems. In other cases, for spatially distributed components, such as groundwater, lakes and rivers, the question may be more complex and specific analysis, including preliminary measurement may be necessary. In principle, space can be analysed in the same manner as time. That is, one can use a preliminary analysis based on a priori knowledge and an iterative process to better define the preferable sampling strategy for a given objective. We emphasize that *coupling of time and space in the analysis is fundamental.*

It is also important to recognize that spatial influences on the representativeness of samples occur not only at large scales. Even at very small scales, spatial heterogeneity may distort sampling results; a common example would be the sampling of TSS within a pipe cross section. The mass and particle size distribution of TSS may vary across the pipe cross section, and ideally a pilot study should be used to identify the degree of variation, so that it can be accounted for, in the eventual monitoring programme.

Regarding spatial sampling, there are many useful references which are applicable to urban drainage systems (e.g. Gilbert, 1987 or Keith, 1996). More broadly, the application of geostatistics to environmental problems, methods and models are described in many textbooks, such as Myers (1997).

Spatial variability is illustrated here with an example dealing with the heterogeneity of the characteristics of solids settled in a stormwater detention and settling basin (this basin is fully described in Chapter 23, which outlines the OTHU case study). The basin has a bottom surface of 10,000 m^2, and twelve sediment traps have been installed to collect sediments settling during storm events. The traps are designated with numbers increasing with their altitude, located as shown in Figure 5.14.

Distributions of settling velocities of sediments have been measured in the twelve traps according to the VICAS (*VItesse de Chute en ASsainissement*) protocol for laboratory analysis of settling velocity distributions (Gromaire and Chebbo, 2003) for various storm events. Figure 5.15 shows the result for a storm event on 14 June 2005: the twelve curves indicate median settling velocities (v_{50}) ranging from 6 m/h to 30 m/h. This high spatial variability is also coupled to very significant temporal variability; for another storm event on 24 March 2006 (Figure 5.16), v_{50} ranges from 0.7 m/h to 7 m/h. Several measurements in space and time are necessary in order to obtain information about the heterogeneity of the variables, their distribution and their uncertainties. If only one trap had been used, its representativeness would have been very limited and could have led to a strongly biased image of the real phenomena.

5.3 CONCLUSIONS

This chapter has highlighted and illustrated the importance of considering the appropriate temporal and spatial scales at which monitoring programmes should be conducted. The most important underlying principle is that generic guidelines for accounting for scale are not useful. Instead, sufficient understanding of the spatial and temporal

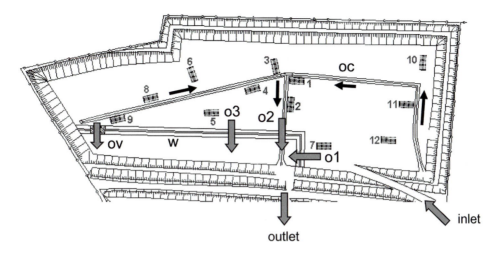

Figure 5.14 Sediment trap locations in a stormwater detention and settling basin. Small rectangles represent the 12 traps numbered in order of increasing altitude with arrows indicating flow pattern (see also colour Plate 3)

Source: Adapted from Torres A., Hasler M. and Bertrand-Krajewski J.-L. 2006. Hétérogénéité spatiale et événementielle des vitesses de chute des sédiments décantés dans un bassin de retenue d'eau pluviale. *Actes des 2° Journées Doctorales en Hydrologie Urbaine 'JDHU 2006'*, Nantes, France, 17–18 October, pp. 59–67.

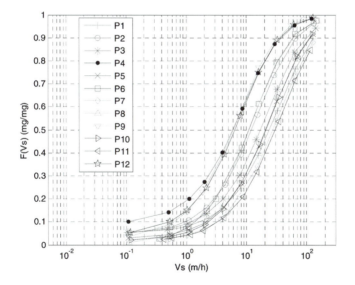

Figure 5.15 Spatial heterogeneity of settling velocities of sediments in a stormwater detention and settling basin, 14 June 2005. V_s = settling velocity, $F(V_s)$ = cumulative frequency of measured settling velocity. P1-P12 are sediment traps as detailed in Figure 5.14 (see also colour Plate 4)

Source: Adapted from Torres A., Hasler M. and Bertrand-Krajewski J.-L. 2006. Hétérogénéité spatiale et événementielle des vitesses de chute des sédiments décantés dans un bassin de retenue d'eau pluviale. *Actes des 2° Journées Doctorales en Hydrologie Urbaine 'JDHU 2006'*, Nantes, France, 17–18 October, pp. 59–67.

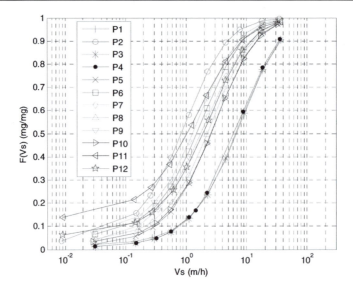

Figure 5.16 Spatial heterogeneity of settling velocities of sediments in a stormwater detention and settling basin, 24 March 2006. V_s = settling velocity, $F(V_s)$ = cumulative frequency of measured settling velocity. P1-P12 are sediment traps as detailed in Figure 5.14 (see also colour Plate 5)

Source: Adapted from Torres A., Hasler M. and Bertrand-Krajewski J.-L. 2006. Hétérogénéité spatiale et événementielle des vitesses de chute des sédiments décantés dans un bassin de retenue d'eau pluviale. *Actes des 2° Journées Doctorales en Hydrologie Urbaine 'JDHU 2006',* Nantes, France, 17–18 October, pp. 59–67.

dynamics of the system(s) of interest is needed. Such knowledge may come from prior information obtained from other systems or more commonly from pilot investigations. Consideration of temporal and spatial scale aspects needs to be coupled, since the two factors inherently interact.

REFERENCES

Bertrand-Krajewski, J.-L., Laplace, D., Joannis, C. and Chebbo, G. 2000. *Mesures en hydrologie urbaine et assainissement.* Paris, Editions Tec et Doc.

De Benedittis, J. and Bertrand-Krajewski, J. 2005. Infiltration in sewer systems: comparison of measurement methods. *Water Science and. Technology* 52, pp. 219–27.

Fletcher, T.D., and Deletić, A. 2006. *A review of Melbourne Water's Pollutant Loads Monitoring Programme for Port Phillip and Western Port.* Melbourne, Melbourne Water Corporation.

Fletcher, T.D. and Deletić, A. 2007. Observations statistiques d'un programme de surveillance des eaux de ruissellement; leçons pour l'estimation de la masse de polluants [Statistical observations of a stormwater monitoring programme; lessons for the estimation of pollutant loads]. Paper presented at the NOVATECH 2007: Sixth Conference on Sustainable Techniques and Strategies in Urban Water Management, 25–28 June, 2007, Lyon, France.

Gilbert, R.O. 1987. *Statistical methods for environmental pollution monitoring.* New York, John Wiley.

Gromaire, M.C., and Chebbo, G. 2003. *Mesure de la vitesse de chute des particules en suspension dans les effluents urbains, protocole VICAS, manuel de l'utilisateur.* Marne-la-Vallée, France, CEREVE.

Gy, P. 1988. *Hétérogénéité, échantillonnage, homogénéisation – Ensemble cohérent de théories.* Paris, Masson.

——. 1992. *Sampling of heterogeneous and dynamic material systems – Theories of heterogeneity, sampling and homogenizing.* Amsterdam, Elsevier.

——. 1996. *L'échantillonnage des lots de matière en vue de leur analyse.* Paris, Masson.

Keith, L.H. 1996. *Principles of environmental sampling*, 2nd edn. Washington DC, American Chemical Society.

Labat, D., Ababou, R. and Mangin, A. 2001. Introduction of wavelet analyses to rainfall/runoffs relationship for karstic basin: the case of Licq-Atherey karstic system (France). *Ground Water*, Vol. 39, No. 4, pp. 605–14.

Labat, D., Mangin, A. and Ababou, R. 2002. Rainfall-runoff relations for karstic springs: multifractal analyses. *Journal of Hydrology*, Vol. 256, No. 3, pp. 176–95.

Mangin A. 1984. Pour une meilleure connaissance des systèmes hydrologiques à partir des analyses corrélatoire et spectrale. *Journal of Hydrology*, Vol. 67, No. 1, pp. 25–43.

Myers, J.C. 1997. *Error Management: Quantifying Uncertainty for Environmental Sampling and Mapping.* New York, John Wiley.

Ort, C. and Gujer, W. 2005. Sampling for representative micropollutant loads in sewer systems. Paper presented at Tenth ICUD Conference, Copenhagen, Denmark, 21–26 August 2005.

Ort C., Schaffner, C., Giger, W. and Gujer, W. 2005. Modeling stochastic load variations in sewer systems. *Water Science and Technology*, Vol. 52, No. 5, pp. 113–22.

Rossi, L. 1998. Qualité des eaux de ruissellement urbain. Ph.D. thesis, Ecole Polytechnique Fédérale de Lausanne, Switzerland.

Taylor, G.D. 2006. Improved effectiveness of nitrogen removal in constructed stormwater wetlands. Ph.D. thesis, Department of Civil Engineering, Monash University, Melbourne, Australia.

Taylor, G.D., Fletcher, T.D., Wong, T.H.F. and Duncan, H.P. 2006. Baseflow water quality behaviour: implications for wetland performance monitoring. *Australian Journal of Water Resources*, Vol. 10, No. 3, pp. 293–302.

Tchiguirinskaia, I., Bonell, M. and Hubert, P. 2004. *Scales in hydrology and water management.* Paris, IAHS-AISH/UNESCO (IAHS Publication 287).

Torres A., Hasler M. and Bertrand-Krajewski J.-L. 2006. Hétérogénéité spatiale et événementielle des vitesses de chute des sédiments décantés dans un bassin de retenue d'eau pluviale. *Actes des 2° Journées Doctorales en Hydrologie Urbaine 'JDHU 2006'*, Nantes, France, 17–18 October, pp. 59–67.

Chapter 6

Understanding and managing uncertainty

J.-L. Bertrand-Krajewski[1] and M. Muste[2]

[1]Laboratoire LGCIE, INSA-Lyon, 34 avenue des Arts, F-69621 Villeurbanne CEDEX, France
[2]IIHR-Hydroscience and Engineering, University of Iowa, Iowa City, IA 52245, USA

6.1 INTRODUCTION

Data uncertainty, used here interchangeably with data quality, is crucial for safety, reliability, and risk assessment in urban water management. Estimating data uncertainty, however, is difficult because of the unpredictable, complex, non-linear, dynamic systems involved in environmental systems. It is pragmatic to state that despite all the progress in science, uncertainty will never be fully eliminated, and any decision based on the best available predictions will always contain a certain risk of error (Singh et al., 2001). Both decision makers and technicians alike, however, have difficulties accepting uncertainty as one of the fundamental properties of natural processes.

Despite that environmental systems in the urban water cycle are subject to uncertainty, the planning, design, operation and management of these systems are often undertaken without taking it into account. Uncertainty in environmental analysis arises due to stochasticity of physical phenomena, as well as errors in observations and modelling. Repeated measurements in field or laboratory made on the same process using the same instruments and procedures do not match identically due to random fluctuations in physical processes. Besides the uncertainty produced by the randomness in nature, additional uncertainty is caused by the instruments, by data acquisition and reduction procedures, by the operator and by the context of measurements.

Environmental systems can be generically described by a set of variables $X_1, \dots, X_k$ related in p functional forms (Singh et al., 2001):

$$f_j(X_1, \dots, X_k; \alpha_1, \dots, \alpha_l) = 0, \quad j = 1, 2, \dots, p \qquad 6.1$$

depending on α_q parameters ($q = 1, 2, \dots, l$). If the variables are random, the functional relation reduces to a structural relation, which in environmental systems involves time-space distributed variables.

The problem posed in environmental sciences is to estimate α_q from a set of observations. If values of X could be observed without error, there would be no statistical problem, and the problem would merely be a mathematical one. However, the variables are unobservable, and what is observed (measured) are indices of mathematical variables, i.e. ($\xi_1, \dots, \xi_l$), being often lumped in space or time. One can ask about the practical need for Equation 6.1 if observable variables differ so much from their respective counterparts ($X_1, \dots, X_k$). The interest in environmental modelling is in the relation between observable variables, which is statistical in character. An application of statistical models is constrained to the same kind of data as used for their calibration. Therefore, in the case of substantial change in an observation network or in the measurement technique of the independent variables, the model should be recalibrated. Moreover, the model should be calibrated with the appropriate data range for which its application is intended (e.g. flood protection). The vexing problem of uncertainty in the decision-making process is that it is impossible to deal with every kind of uncertainty. In the present chapter, *only the kind of uncertainty that can be measured quantitatively*, at least in principle, will be addressed, that is, *the measurement uncertainty*.

Measurements are essential and integral tools for engineering and science in general. Of particular interest here are measurements conducted for:

- maintaining quality control and quality assurance
- complying with and enforcing laws and regulations
- tracing measurement accuracy to national standards
- establishing benchmark data for evaluation of instrument biases and validation of numerical simulations.

This chapter assumes that the processes involved in the urban water environment are known (see also Chapter 3), and thus focuses on the assessment of the uncertainties associated with the measurement process (i.e. data collection, reduction and processing).

The *measurement process* is the act of assigning a value to some physical variable, by operating sensors and instruments in conjunction with data acquisition and reduction procedures. Ideally the value assigned by the measurement would be the actual value of the physical variable measured. However, measurement processes and environmental errors introduce uncertainty regarding the correctness of the value. To give some degree of confidence to the measured value, measurement errors must be identified, and their probable effect on the result estimated. *Uncertainty is simply an estimate of a possible value for the error in the reported results of a measurement.* The process of systematically quantifying error estimates is known as 'uncertainty analysis'.

Monitoring of urban water processes should be governed *by the ability of the expected measurements to achieve the specific objectives within the allowable uncertainties.* Thus, the assessment of measurement uncertainty should be a key element throughout the entire monitoring program: description of the measurements, determination of error sources, estimation of uncertainties, and documentation of the results. Uncertainty considerations need to be integrated in all phases of the monitoring process, including planning, design, the decision whether to use specific instruments, and the actual measurement process. In essence, this means that uncertainty must be considered even at the *definition-of-objectives* stage; the objectives should include a specification of the allowable uncertainty.

Rigorous application and integration of uncertainty assessment methodology is an integral part of all monitoring phases. The most important benefits of uniform uncertainty analysis are:

- identification of the dominant sources of error, their effects on the result, and estimation of the associated uncertainties
- facilitation of meaningful and efficient communication of data quality
- facilitation of the selection of the most appropriate and cost effective measurement devices and procedures for a given measurement task

- consideration and reduction of the risks in decision making
- demonstration of compliance with regulations.

6.2 UNCERTAINTY ANALYSIS CONCEPTS, APPROACHES AND TERMINOLOGY

6.2.1 Uncertainty analysis concepts

Uncertainty analysis is a rigorous methodology for estimating uncertainties in measurements and in the results calculated from them. It combines statistical and engineering concepts. The analysis must be done in a manner that can be systematically applied to each step in the data uncertainty assessment determination.

The error in a measurement is the difference between the true and the measured value of the result. In most situations the true value is not known, hence the limits that bound the possible error need to be estimated. These estimated limits are called uncertainty or uncertainty intervals. Uncertainty defines the interval around the measured value within which the true value is believed to lie *with a pre-established level of confidence*. Frequently in engineering practice, since a true value cannot be determined, a number of results of measurements of a quantity are used to establish a 'conventional true value', that is, a value attributed to a particular quantity and accepted as having an uncertainty appropriate for a given purpose. In this context, the error is defined as the result of a measurement minus the conventional true value, and the relative error is defined as the error of the measurement divided by the conventional true value.

The conventional approach to categorizing error types in engineering is based on their effect on the measured value (AIAA, 1995; ASME, 1998). Accordingly, the total measurement error consists of two components: bias (fixed) and precision (random).

Bias errors are systematic, hence, difficult to detect and remove. Sensor calibration with links to primary or secondary standards is a way to evaluate and remove (by correction) bias errors. However, sensor calibration qualifies the sensor itself, and not necessarily its use in a given location under given conditions which may themselves be the source of additional bias. This aspect should be accounted for as much as possible, *as even relatively small bias or systematic errors may have dramatic effects on the final results from monitoring programmes* (Fletcher and Deletic, 2007). If bias errors can be detected and evaluated, they can be taken into account in uncertainty assessment (e.g. as described in AIAA, 1995). In other cases, correct information on bias is non-existent or very weak, and estimations are not possible. An alternative method in this case may be to simulate scenarios, i.e. to simulate the effects of possible bias on the final results, in order to answer questions like 'what if ...' (e.g. how would the discharge and its uncertainty change if the water level sensor had a bias of +2 cm?). *In all cases, investigation to identify and remove possible bias, even if it is difficult, is a very important task to be carried out with the highest degree of rigour and intellectual honesty.*

Precision errors are random and can be estimated by inspection of the measurement scatter. Bias and precision errors are generated in all phases of the experimental investigation. Bias and precision estimates are combined through the data reduction equation (i.e. the relationship defining the result where the individual measurements are used) to provide the total error in the result. Scientific standards, such as ISO (1993), use a different terminology for classifying uncertainties. Specifically, uncertainties are grouped into two categories according to the method used to estimate their numerical values (i.e. Type A evaluated by statistical methods, Type B by other means).

An alternative criterion to classify sources of uncertainties is related to where in the experimental process the uncertainties are generated. Accordingly, ASME (1998) classifies sources of uncertainties as: measurement system (calibration, data acquisition, data reduction), conceptual (methodological), and other sources. Conceptual errors are of the bias type, while the measurement system errors (calibration, data acquisition and reduction) can be of both bias and precision types. Measurement system errors are generated during the measurement process. In general, measurement system errors comprise bias and precision errors associated with the instrumentation, the simulation technique, the procedures for data acquisition and reduction and the operational environment.

Frequently instrumentation errors are the only ones addressed in estimating uncertainties. This is unfortunate, because in many situations errors such as those induced by flow-sensor interaction, flow characteristics, and measurement operation are frequently larger than the instrument errors. This is why, as much as possible, the location and conditions of use of sensors should be accounted for to evaluate the total uncertainty. For example, a water level sensor may have an instrument uncertainty (evaluated by means of an adequate calibration with certified standards) of ± 1 mm. If this sensor is used in a sewer system where the water is not still and perfectly horizontal, but moves downstream and generates small waves at the surface with possible secondary currents, leading to a non-horizontal, free surface, the final uncertainty may reach ± 1 cm or more. Conceptual bias errors (i.e. errors that might stand between concept and measurement) are generated during the test design and data analysis through idealizations (assumptions) in the data interpretation equations, through the use of equations which are incomplete and do not acknowledge all significant factors, or by not measuring the correct variable (Moffat, 1988). Despite the potential importance of conceptual bias errors and the challenge of assigning significance to what has been measured, this category of uncertainty is beyond the scope of this chapter and will not be further discussed.

6.2.2 Overview of relevant uncertainty analysis approaches

In the late 1970s, the world's highest authority in metrology, the *Comité International des Poids et Mesures* (CIPM), initiated the development of a detailed guide for establishing general rules for evaluating and expressing uncertainties in measurements. The fundamental requirements for the guide were: to be universal (applicable to all kinds of measurements), internally consistent (directly derivable from the components that contribute to the actual quantity), and transferable (able to be used for both primary and derived quantities). The 'Working Group on the Statement of Uncertainties', convened by CIPM, formulated the approach for the classification, estimation, and expression of experimental uncertainties in the document Recommendation INC-1 (BIPM, 1980).

Based on Recommendation INC-1, a joint working group of international experts prepared the 'Guide to Expression of Uncertainty in Measurement' (ISO, 1993) that is the first set of internationally recognized guidelines for the conduct of uncertainty analysis. ISO (1993) differs considerably from previous uncertainty assessment methodologies in terminology, classification, and presentation (Herschy, 2002). The Standard's procedure is based on sound principles of mathematical statistics and is simple to apply. Since its inception, ISO (1993) has been increasingly adopted worldwide for assessing the quality of the measurements in various scientific areas (Taylor and Kuyatt, 1994; ENV 13005, 1999).

The ISO guide provides general rules for evaluating and expressing uncertainty in measurement, rather than detailed, scientific- or engineering-specific instructions. The standard does not discuss how the uncertainty of a particular measurement result, once evaluated, may be used for different purposes, for example, to draw conclusions about the compatibility of that result with other similar results, to establish tolerance limits in a manufacturing process, or to decide if a certain course of action may be safely undertaken. Given its general formulation, the standard acknowledges that it might be necessary to develop particular standards based on the general guide framework to deal with the problems peculiar to specific fields of measurement. These standards may be simplified versions of the guide but should include detail that is appropriate to the level of accuracy and complexity of the measurements and uses addressed. Such relevant outgrowths of the ISO guide are the subsequently developed American Institute of Aeronautics and Astronautics (AIAA, 1995), American Society of Mechanical Engineers (ASME, 1998), and the European ENV 13005 (1999) standards that have successfully satisfied the needs of most engineering application users. While in harmony with the ISO (1993) guidelines, the standards differ slightly from the ISO guide in terminology and in the way they are applied in practical situations. The European standard ENV 13005 (1999) is, however, closer to the initial ISO (1993), being characterized by a more general framework.

The alignment of the AIAA and ASME standards with ISO (1993) has been achieved by combining the technology of the ISO guide with historical groupings of elemental uncertainties used in earlier versions of engineering uncertainty standards (ANSI/ASME, 1985), being systematic (bias) and random (precision) uncertainties. The AIAA (1995) standard is built on previous documents elaborated by the NATO Advisory Group on Aerospace Research and Development. The ASME (1998) builds on the ANSI/ASME 19.1-1985 'Measurement Uncertainty' standard. For the practicing engineer the most commonly used classification criterion is related to their effect rather than their evaluation type. That is, if an uncertainty source causes scatter in a test result, it is a precision (random) uncertainty being produced by random errors. If not, it is systematic uncertainty caused by systematic errors. The ISO (1993) classification of uncertainty as Type A (random) or Type B (bias), depending on the manner in which they are estimated, is made for convenience only, while the propagation and final uncertainty are evaluated in the same way as in the engineering uncertainty standards.

The integration of the ISO (1993) provisions into the AIAA and ASME standards has been made under the following common assumptions:

- Uncertainty sources are assumed to be normally distributed and are thus estimated by the standard deviation of the measurement population. Although this assumption is not always statistically rigorous, it has been found to work well in many engineering applications
- Elemental uncertainty estimates for individual variables are combined using the root-sum-of-squares (RSS) model

- Propagation of the elemental errors to the final measurement results are made using first-order Taylor series approximation
- Systematic and random uncertainties are RSS combined and multiplied by a coverage factor to obtain confidence of the result. Special procedures are provided for handling small statistical samples.

Additionally, the following assumptions are implicit:

- The measurement process is understood, critically analysed, and well defined
- The measurement system and process are controlled
- All appropriate calibration corrections have been applied
- The measurement objectives are specified
- The instrument package and data reduction procedures are defined
- The uncertainties quoted in the analysis of a measurement are obtained under conditions of full intellectual honesty and professional skill.

6.2.3 Terminology

The *accuracy* of a measurement indicates the closeness of agreement between a measured value of a quantity and its true value. *Error* is the difference between the measured and true values. Accuracy increases as error approaches zero. In practice, the true values of measured quantities are rarely known. Thus, one must estimate error and that estimate is known as *uncertainty*. In accordance with their effect (AIAA, 1995; ASME, 1998), the total error in a measurement, δ, is composed of two components: bias error, β, and precision error, ε (see Figure 6.1a). An error is classified as precision (random) error if it contributes to the scatter of the data, otherwise, it is considered to be bias (systematic) error. The effects of such errors on multiple readings of a variable are illustrated in Figure 6.1.b. The qualitative influence of various combinations of large and small precision and bias errors is shown in Figure 6.2. Accurate measurement requires minimizing both types of errors.

If we make N measurements of some variable, the bias error is the difference between the average (mean) value of the readings, μ, and the true value of that variable. For a single instrument measuring some variable, the bias errors, β, are fixed, systematic, or constant errors (e.g., scale resolution). Being of fixed value, bias errors cannot be determined statistically. The estimate for β is called the *bias limit, B*. A useful approach to estimating the bias limit is to assume that the bias error for a given case *is a single realization drawn from some statistical parent distribution of possible bias errors*. The interval defined by $\pm B$ includes 95% of the possible bias errors that could be realized from the parent distribution (see AIAA, 1995).

Precision errors, ε, are random errors and will have different values for each measurement. When repeated measurements are made for fixed test conditions, precision errors are observed as the scatter of the data. Precision errors are due to limitations on repeatability of the measurement system and to facility and environmental effects. The estimate for ε is called the *precision limit, P*. Precision limits are estimated using statistical analysis by inspection of the measurement scatter. The population standard deviation, σ, is a measure of the scatter about the true population mean, μ, caused by precision error. For a normal distribution the interval $\mu \pm 1.96\sigma$ ($\approx \mu \pm 2\sigma$) will

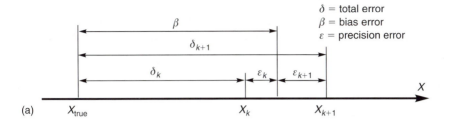

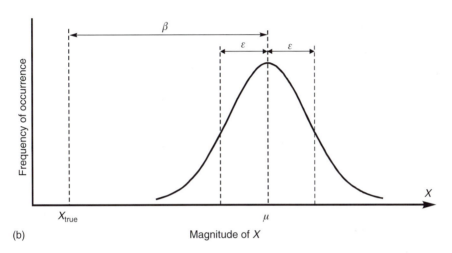

Figure 6.1 Errors in the measurement of a variable for (a) two readings or (b) an infinite number of readings with μ as the mean of the measurement population

Source: Coleman and Steele, 1995.

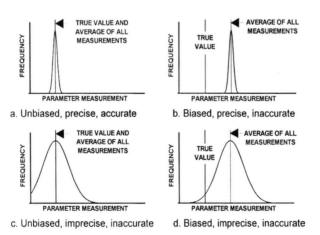

Figure 6.2 Illustration of measurement error effect

Source: AIAA, 1995.

include approximately 95% of the population. An estimate of the population standard deviation is the standard deviation of a data sample of N measurements.

The estimate of the total error, δ, is the total uncertainty, $\underline{U}$. The total uncertainty is defined as an interval about the average measured value, $\overline{X}$, that has a pre-assigned probability (percent-confidence level) of containing the true value.

While measurement uncertainties characterize the dispersion of values that could reasonably be attributed to the 'measurand' (the quantity that is measured), the underlying concept of the ISO (1993) standard is that there is no inherent difference between the uncertainty components arising from a random effect and one arising from a systematic effect. Uncertainties are also estimated using probability density functions of frequency distributions, but their classification is based on the method used to estimate their numerical values (i.e. Type A evaluated by statistical methods, Type B by other means). Uncertainties in category A are evaluated by statistical analysis of repeated *observations* to obtain statistical estimates. Uncertainties in category B are evaluated by other means. For example, *assumed* probability distributions based on scientific judgment and consideration of a pool of comparatively reliable (albeit often partly subjective) information may be used. Such information may come from previous measurements, calibrations, and experience or general knowledge of the behaviour and properties of relevant instruments. There is not always a simple correspondence between the commonly used classification into 'random' and 'systematic' uncertainty. By proper consideration of correlations, either Type A or Type B method of evaluation of uncertainty can be used for evaluation of either random or systematic uncertainty components.

The ISO (1993) guide quantifies uncertainty components of either type by the standard deviations of their probability distributions, which are termed *standard uncertainties*. The standard uncertainty of the result of a measurement, when that result is obtained from the values of a number of other quantities, is termed *combined standard uncertainty*. It is the standard deviation associated with the result and it is obtained using the RSS of all variance and covariance components, however evaluated. The combined uncertainty takes all elemental sources and components of uncertainties into account. To meet the needs of some area-specific applications, an *expanded uncertainty* is obtained by multiplying the combined standard uncertainty by a coverage factor. The intended purpose of the expanded uncertainty is to provide an interval about the result of a measurement that may be expected to encompass a large faction of the distribution of values that could reasonable be attributed to the measurand. The choice of the coverage factor, usually in the range 2 to 3, is based on the level of confidence required in the interval.

As mentioned previously, the two engineering standards (AIAA, 1995 and ASME, 1998) are offspring of the ISO (1993) Guide, while the European standard ENV 13005 (1999) closely resembles it. While providing practically the same output, the three standards differ slightly in their terminology as illustrated in Table 6.1.

6.3 UNCERTAINTY ANALYSIS IMPLEMENTATION

The uncertainty analysis methodology presented below closely follows the ISO (1993) guide. To ensure proper application, the standard is reproduced roughly verbatim, with minor modifications.

Table 6.1 Comparison of standard terminology for uncertainty assessment

International Standards Organization (ISO) (1)	*American Institute of Aeronautics and Astronautics (AIAA)* (2)	*American Society of Mechanical Engineers (ASME)* (3)
Uncertainty of a measurement • input quantity • type A standard uncertainty • type B standard uncertainty • combined standard uncertainty	• individual variable • bias limit • precision limit (differently estimated for: single and multiple test measurements) • total uncertainty	• independent parameter • systematic uncertainty • random uncertainty (differently estimated for single and multiple test measurements) • measurement uncertainty
Uncertainty of a result • functional relationship	• data reduction equation • bias limit (accounts for correlated errors) • precision limit (differently estimated for: single test with single readings, single test with averaged readings, multiple tests; accounts for correlated precision errors)	• derived result • systematic uncertainty (accounts for correlated errors) • random standard deviation (differently estimated for: single and multiple tests)
• sensitivity coefficients • combined standard uncertainty (accounts for correlated input quantities) • expanded uncertainty (accounts for the level of confidence) • coverage factor f(degrees of freedom and t-distribution)	• sensitivity coefficients • combined standard uncertainty • uncertainty at specified confidence level (coverage factor) • coverage factor f(degrees of freedom and t-distribution)	• sensitivity coefficients • uncertainty of the result (at a specified confidence level) • coverage factor f(degrees of freedom and t-distribution)

Source: Col 1: ISO, 1993; Col. 2: AIAA, 1995; Col. 3: ASME, 1998.

6.3.1 The measurement process

In most cases a measurand (physical quantity subject to measurement), Y, is not measured directly, but is determined from N other quantities $X_1, X_2, \ldots, X_N$ through a functional relationship f:

$$Y = f(X_1, X_2, \ldots, X_N) \qquad 6.2$$

For economy of notation, the same symbol is used for the physical quantity and for the random variable which represents the possible outcome of an observation of that quantity. In a series of observations, the kth observed value of X_i is denoted by X_{ik}. The estimate of X_i (strictly speaking, of its expectation) is denoted by x_i. The *input quantities* $X_1, X_2, \ldots, X_N$ upon which the *output quantity* Y depends may themselves be viewed as measurands and may themselves depend on other quantities, including corrections and correction factors for systematic effects, thereby leading to a complicated functional relationship f that may never be written down explicitly. Further, f may be determined experimentally or exist only as an algorithm that must be evaluated numerically.

The set of input quantities $X_1, X_2, \ldots, X_N$ may be categorized as quantities whose values and uncertainties are directly determined in the current measurement. These values and uncertainties may be obtained from, for example, a single observation, repeated observations (similar to precision errors) or judgment based on experience, and may involve the determination of corrections to instrument readings and corrections for influence quantities, such as ambient temperature, barometric pressure and humidity. Alternatively, values of the quantities of the uncertainties are brought into the measurement from external sources, such as quantities associated with calibrated measurement standards, certified reference materials, and reference data obtained from handbooks (similar to bias errors).

An estimate of the measurand Y, denoted by y, is obtained from Equation 6.2 using input estimates $x_1, x_2, \ldots, x_N$ for the values of the N quantities $X_1, X_2, \ldots, X_N$. Thus the output estimate y, which is the result of the measurement, is given by:

$$y = f(x_1, x_2, \ldots, x_N) \qquad\qquad 6.3$$

The estimated standard deviation associated with the output estimate or measurement result y, termed combined standard uncertainty and denoted by $u_c(y)$, is determined from the estimated standard deviation associated with each input estimate x_i, termed standard uncertainty and denoted by $u(x_i)$.

Each input estimate x_i and its associated standard uncertainty $u(x_i)$ are obtained from a distribution of possible values of the input quantity, X_i. This probability distribution may be frequency based, that is, based on a series of observations $X_{i,k}$ of X_i or it may be a priori, a distribution. Type A evaluations of standard uncertainty components are founded on frequency distributions while Type B evaluations are founded on a priori distributions. It must be recognized that in both cases the distributions are models that are used to represent the state of knowledge.

6.3.2 Type A evaluation of standard uncertainty

In most cases, the best available estimate of the expected value μ_q of a random variable q measured by n independent observations, q_k obtained under the same measurement conditions is the arithmetic mean or average $\bar{q}$ of the n observations:

$$\bar{q} = \frac{1}{n} \sum_{k=1}^{n} q_k \qquad\qquad 6.4$$

Thus for an input quantity X_i estimated from n independent repeated observations $X_{i,k}$, the arithmetic mean $\bar{X}_i$ obtained from Equation 6.4 is used as the input estimate x_i in Equation 6.3 to determine the measurement result y; that is, $x_i = \bar{X}_i$. Those input estimates not evaluated from repeated observations must be obtained by other methods (external sources).

The individual observations q_k differ in value because of random variations in the influence quantities, or random effects. The experimental variance of the observations, which estimate the variance σ^2 of the probability distribution of q, is given by:

$$s^2(q_k) = \frac{1}{n-1} \sum_{k=1}^{n} (q_k - \bar{q})^2 \qquad\qquad 6.5$$

This estimate of variance and its positive square root $s(q_k)$, termed the experimental standard deviation, characterize the variability of the observed values q_k, or more specifically, the dispersion about their mean $\bar{q}$.

The best estimate of $\sigma^2(\bar{q}) = \sigma^2/n$, the variance of the mean, is given by:

$$s^2(\bar{q}) = \frac{s^2(q_k)}{n} \qquad\qquad 6.6$$

The experimental variance of the mean $s^2(\bar{q})$ and the experimental standard deviation of the mean $s(\bar{q}) = [s^2(\bar{q})]^{1/2}$ quantify how well $\bar{q}$ estimates the expectation μ_q of q, and either of these may be used as a measure of the uncertainty of $\bar{q}$.

Thus, for an input quantity X_i determined from n independent repeated observations $X_{i,k}$, the standard uncertainty $u(x_i)$ of its estimate $x_i = \bar{X}_i$ is $u(x_i) = s(\bar{X}_i)$ with $s^2(\bar{X}_i)$ calculated according to Equation 6.6. For convenience, $u^2(x_i) = s^2(\bar{X}_i)$ and $u(x_i) = s(\bar{X}_i)$ are sometimes called a Type A variance and a Type A standard uncertainty.

6.3.3 Type B evaluation of standard uncertainty

For an estimate x of an input quantity X_i that has not been obtained from repeated observations, the associated estimated variance $u^2(x_i)$ or the standard uncertainty $u(x_i)$ is evaluated by scientific judgement based on all of the available information on the possible variability of X_i. The pool of information may include previous measurement data, experience with or general knowledge of the behaviour and properties of relevant materials and instruments, manufacturer's specifications, data provided in calibration and other certificates and uncertainties assigned to reference data taken from handbooks. For convenience $u^2(x_i)$ and $u(x_i)$ evaluated in this way are sometimes called Type B variance and Type B standard uncertainty, respectively.

6.3.4 Determining the combined standard uncertainty

6.3.4.1 Uncorrelated input quantities

This situation refers to the case where all input quantities and their uncertainties are independent. The case where two or more input quantities are related is referred to as correlated input quantities. The standard uncertainty of y, where y is the estimate of the measurand Y and thus the results of the measurement, is obtained by appropriately combining the standard uncertainties of the input estimates $x_1, x_2, \ldots, x_n$. This combined standard uncertainty of the estimate y is denoted by $u_c(y)$.

The combined standard uncertainty $u_c(y)$ is the positive square root of the combined variance $u_c^2(y)$, which is given by:

$$u_c^2(y) = \sum_{i=1}^{N} \left(\frac{\partial f}{\partial x_i} \right)^2 u^2(x_i) \qquad\qquad 6.7$$

where f is the function given in Equation 6.2. Each $u(x_i)$ is a Type A or B standard uncertainty evaluated as described above. The combined standard uncertainty $u_c(y)$ is an estimated standard deviation and characterizes the dispersion of the values that could reasonably be attributed to the measurand Y. Equation 6.7 is based on a first-order Taylor series approximation of $Y = f(X_1, X_2, \ldots, X_N)$.

These derivatives, called sensitivity coefficients, describe how the output estimate y varies with changes in the values of the input estimates $x_1, x_2, \ldots, x_N$. Strictly speaking, partial derivatives are $\partial f / f x_i = \partial f / \partial X_i$ evaluated at the expectations of the X_i. However, in practice, the partial derivatives are estimated by

$$\frac{\partial f}{\partial x_i} = \frac{\partial f}{\partial X_i}\bigg|_{x_1, x_2, \ldots, x_N} \tag{6.7a}$$

6.3.4.2 Correlated input quantities

Equation 6.7 is valid only if the input quantities X_i and their uncertainties are independent or uncorrelated. If some of the X_i are significantly correlated, the correlation must be taken into account. There may be significant correlation between two input quantities if the same measuring instrument, physical measurement standard or reference datum having a significant standard uncertainty is used in their determination.

The appropriate expression for the combined variance $u_c^2(y)$ associated with the results of a measurement when some of the input quantities are correlated is:

$$u_c^2(y) = \sum_{i=1}^{N} \left(\frac{\partial f}{\partial x_i}\right)^2 u^2(x_i) + 2\sum_{i=1}^{N-1}\sum_{j=i+1}^{N} \frac{\partial f}{\partial x_i}\frac{\partial f}{\partial x_j} u(x_i, x_j) \tag{6.8}$$

where x_i and x_j are estimates of X_i and X_j and $u(x_i, x_j) = u(x_j, x_i)$ is the estimated covariance associated with x_i and x_j.

Operators who are not familiar with partial differential equations (PDE) as used in Equations 6.7 and 6.8 can replace them by second order numerical approximations which are easily determined by calculation spreadsheets, with the following equation:

$$\frac{\partial f}{\partial x_i} \approx \frac{f(x_i + \delta x_i) - f(x_i - \delta x_i)}{2\delta x_i} \tag{6.8a}$$

where δx_i is a very small variation of the quantity x_i (typically 1000 times smaller than the value of x_i).

6.3.5 Determining expanded uncertainty

Although $u_c(y)$ can be universally used to express the uncertainty of a measurement result, in some commercial, industrial, and regulatory applications it is often necessary to give a measure of uncertainty that defines an interval about the measurement result that may be expected to encompass a large fraction of the distribution of values that could reasonably be attributed to the measurand. The additional measure of uncertainty that meets the requirement of providing an interval of the kind indicated above is termed 'expanded uncertainty', denoted by U. The expanded uncertainty U is obtained by multiplying the combined standard uncertainty $u_c(y)$ by a *coverage factor k*:

$$U = ku_c(y)$$ 6.9

The result of a measurement is then conveniently expressed as $Y = y \pm U$, which is interpreted to mean that the best estimate of the value attributable to the measurand Y is y, and that $y - U$ to $y + U$ is an interval that may be expected to encompass a large fraction of the distribution of values that could reasonable by attributed to Y. Such an interval is also expressed as $y - U \leqslant Y \leqslant y + U$. The value of the coverage factor k is chosen on the basis of the required level of confidence.

Ideally, one would like to be able to choose a specific value of the coverage factor k that would provide an interval $Y = y \pm U = y \pm ku_c(y)$ corresponding to a particular level of confidence p, such as 95% or 99%; equivalently, for a given value of k, one would like to be able to state unequivocally the level of confidence associated with that interval. However, this is not easy to do in practice because it requires extensive knowledge of the probability distribution characterized by the measurement result y and its combined standard uncertainty $u_c(y)$. Although these parameters are of critical importance, they are by themselves insufficient for the purpose of establishing intervals having exactly known levels of confidence. The simplest approach, often adequate in measurement situations, is to assume that the probability distribution characterized by y and $u_c(y)$ is approximately normal and the effective degrees of freedom of $u_c(y)$ is of significant size (at least 30 repeated observations). When this is the case, which frequently occurs in practice, one can assume that taking $k = 2$ produces an interval having a level of confidence of approximately 95%, and that taking $k = 3$ produces an interval having a level of confidence of approximately 99%. For smaller sample sizes, the distribution of y and $u_c(y)$ may be approximated by a t-distribution with an effective degrees of freedom, v_{eff}, obtained from the Welch-Satterthwaite formula:

$$v_{eff} = \frac{u_c^4(y)}{\displaystyle\sum_{i=1}^{N} \frac{u_i^4(y)}{v_i}}$$ 6.10

where, $v_i = n - 1$ with n being the number of repeated measurements for each X_i.

6.3.6 Calibration

Calibration is conducted to exchange the large bias error of the uncalibrated instrument with the hopefully smaller uncertainty of the working standard and the error associated with the calibration process (actually another experiment). Calibrations can be conducted to reference (trace) the instruments to primary or secondary standards. Primary standards are those documented through very carefully established procedures with specialized facilities and instrumentation, usually in specialized institutes (e.g. National Bureau of Standards). An instrument that has been satisfactorily calibrated against a primary standard can be used as a secondary or transfer-standard to calibrate other instruments.

In the following some essential aspects of the calibration of an instrument at a single evaluation set point are highlighted. Calibrations are conducted using the best available instruments (traceable to primary or secondary standards), very well documented procedures, and good facilities and environmental conditions. Calibration uncertainties are estimated using the end-to-end approach for multiple tests (AIAA, 1995). The instrument to be calibrated and the working standard (primary or secondary) are measured in parallel in the same facility using the same flow and experimental procedures. For such conditions we define:

the individual sample errors, E_i, as the differences between individual measurements, M_i, and the working standard WS_i , i.e.:

$$E_i = M_i - WS_i \qquad\qquad 6.11$$

the mean error of the sample

$$\bar{E} = 1/N \sum_{i=1}^{N} E_i \qquad\qquad 6.12$$

where N is the number of repeated measurements for the set point ($N \geqslant 10$). Before calculating $\bar{E}$, the sample outliers (spurious measurements due to a temporary or intermittent malfunction which would not occur during the testing process) need to be removed (AIAA, 1995).

the standard deviation of the individual sample errors

$$S_E = \pm \left(\sum_{i=1}^{N} \frac{(E_i - \bar{E})^2}{N - 1} \right)^{1/2} \qquad\qquad 6.13$$

Since the calibration of the instrument usually accounts for any large errors in the system (carefully designed, controlled and conducted experiment), $\bar{E}$ consists of small known biases too difficult or too costly to correct and small unknown biases assumed to possess an equal probability of being negative or positive.

6.4 IMPLEMENTATION EXAMPLES

6.4.1 Example 1: Calibration of a water level piezoresistive sensor

Consider a piezoresistive sensor with an aluminium oxide (Al_2O_3) ceramic membrane used to measure water level in the range 0 to 2 m. This sensor shall be checked and calibrated before its installation at the monitoring site. The two most frequently applied techniques to calibrate a piezoresistive sensor are: (i) the measurement in a column of pre-established water levels measured independently by means of a precise meter used as a secondary standard (Figure 6.3), (ii) the measurement of pre-established air pressures applied to the sensor by means of a high precision pressure generator used as a secondary standard after its own calibration has been carried out at least once a year by

Figure 6.3 Water column used to calibrate piezoresistive sensors (see also colour Plate 6)

Source: J.-L. Bertrand-Krajewski.

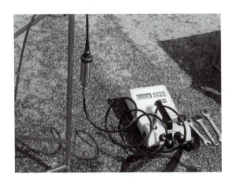 

Figure 6.4 High precision air pressure generator (left) used to calibrate piezoresistive sensors on site (right) (see also colour Plate 7)

Source: LCPC, France.

an independent certified organization using primary standards (Figure 6.4). In both cases, from the end user point of view, the complete instrumental chain has to be calibrated, from the transducer in contact with water or air to the final output signal (e.g. 4 mA to 20 mA output) delivered by the transmitter, and not only the transducer (i.e. the ceramic membrane alone and its direct electric output). This is particularly important because, for some sensors, the main source of global instrumental uncertainty is not necessarily the transducer itself, but other elements such as converters, amplifiers, data pre-processors, and also human mistakes in programming or setting the sensor and its transmitter.

In this example, only the first technique with the water column is presented. More detailed information on calibration methods and techniques are given in Bertrand-Krajewski et al. (2000). A 3.5 m high Plexiglas® column is equipped with a 4 m long certified class II meter, with a maximal uncertainty of 0.5 mm. This uncertainty of the secondary standard is acceptable as it is approximately ten times smaller than the expected in situ uncertainty which is evaluated, according to experience, at 5 mm. The meter is attached at a known position at the bottom of the column. The sensor is attached in the column in such a way that the zero of the meter and the zero level of the sensor are exactly in the same position.

The calibration starts with the column empty. It is then progressively filled with tap water, with $m = 5$ levels of x_i ranging from 0 to 2 m corresponding approximately to 20%, 40%, 60%, 80% and 100% of the measurement range (as piezoresistive sensors are affected by hysteresis, a second calibration with decreasing water levels is carried out, but it is not presented here). At each defined level x_i, n_i readings y_{ik} (here with $k = 1$ to $n_i = 12$) are made every 30 seconds, including both the water level displayed on the sensor transmitter and the 4 mA to 20 mA output. It is important to check that both values are identical. For this purpose, a high precision calibrated amp meter is used. If both values are not identical, the transmitter should be adjusted first and the calibration should restart from the beginning. In this example, the relationship between the 4 mA to 20 mA intensity I_c (mA) and the water level h (mm) is:

$$h = 125 \, I_c - 500 \qquad\qquad 6.14$$

Experimental data are given in Table 6.2. For each x_i, the table shows the $n_i = 12$ readings y_{ik}, the mean value $\bar{y}_i$ and the standard deviation s_i. The s_i values are shown on Figure 6.5. As a first approximation, one can reconize that they are almost constant over the measurement range (except for $x_i = 800$ mm where all readings were identical, leading to a standard deviation equal to zero).

The calibration curve is determined by linear regression from the $N = m \times n_i = (5 \times 12) = 60$ values given in Table 6.2. The regression results are given in Table 6.3. The calibration equation, as shown in Figure 6.6, is then:

$$y = a + bx = 0.508854 + 1.000395 \, x \qquad\qquad 6.15$$

where $\bar{x} = \dfrac{\sum_i n_i x_i}{N}$, $\bar{y} = \dfrac{\sum_i \sum_k y_{ik}}{N}$, $b = \dfrac{\sum_i \sum_k (x_i - \bar{x})(y_{ik} - \bar{y})}{\sum_i n_i (x_i - \bar{x})^2}$, $a = \bar{y} - b\bar{x}$,

$$6.15a,b,c,d$$

Table 6.2 Calibration experimental data (increasing water level)

Sensor XXX n° 839 NBP 1083
Transmitter XXX n° 839 OGH 03
Calibration date: 15 June 1999 by CR, AJR, MM, JLBK

x_i	y_{i1}	y_{i2}	y_{i3}	y_{i4}	y_{i5}	y_{i6}	y_{i7}	y_{i8}	y_{i9}	y_{i10}	y_{i11}	y_{i12}	Mean y_i	Standard deviation $s\,(y_{ik})$
399	399	400	400	400	400	400	400	399	399	399	399	399	399.50	0.5222
799	800	800	800	800	800	800	800	800	800	800	800	800	800.00	0.0000
1200	1201	1201	1202	1202	1202	1202	1201	1201	1201	1201	1201	1201	1201.33	0.4924
1600	1601	1601	1601	1601	1601	1602	1600	1600	1600	1600	1600	1600	1600.58	0.6686
2000	2002	2002	2002	2002	2002	2002	2001	2001	2001	2001	2001	2001	2001.50	0.5222

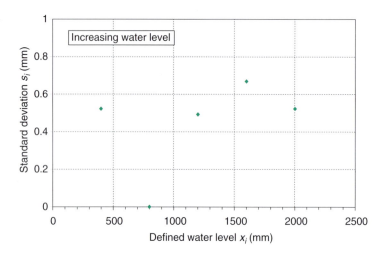

Figure 6.5 Standard deviation of the calibration readings (increasing water level)

Table 6.3 Numerical results of the linear regression analysis (increasing water level)

Variable	Value	95% confidence interval	Standard deviation, s
N	60		
$\overline{x}$	1199.600		
$\overline{y}$	1200.583		
A	0.508854	0.153557 – 0.864150	$s(a) = 0.177522$
B	1.000395	1.000128 – 1.000663	$s(b) = 0.000134$
s_i^2	0.344398		

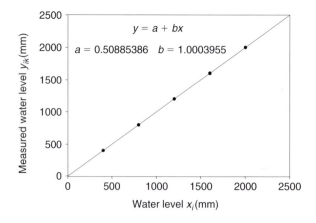

$$y = a + bx$$

$a = 0.50885386 \quad b = 1.0003955$

Figure 6.6 Calibration curve based on linear regression analysis

$$s_l^2 = \frac{\sum_i \sum_k (y_{ik} - \hat{y}_i)^2}{N - 2} = \frac{1}{N - 2}\left(\sum_i \sum_k (y_{ik} - \bar{y})^2 - b^2 \sum_i n_i (x_i - \bar{x})^2\right),$$

$$s_a^2 = s_l^2 \left(\frac{1}{N} + \frac{(\bar{x})^2}{\sum_i n_i (x_i - \bar{x})^2}\right) \quad \text{and} \quad s_b^2 = \frac{s_l^2}{\sum_i n_i (x_i - \bar{x})^2} \qquad \text{6.16a,b,c}$$

According to the confidence intervals given in Table 6.3, it appears that the offset value a is significantly different from zero and that the slope b is significantly greater than 1. A further analysis based on statistical tests, not presented here, indicates that a second order polynomial equation does not significantly improve the calibration equation. A first order linear equation as per Equation 6.15 is the most appropriate for this sensor.

Based on Equation 6.15, it is then possible to transform any measured value y_0 into the corresponding most likely true value of the water level x_0, and also to evaluate its standard uncertainty $u(x_0)$.

Consider one single measured value $y_0 = 600$ mm. The most likely true value x_0 is derived from Equation 6.17:

$$x_0 = \frac{y_0 - a}{b} = \frac{600 - 0.508854}{1.000395} \approx 599.2\,\text{mm} \qquad 6.17$$

The standard deviation $s(x_0)$ is due to two independent contributions: (i) the uncertainty in the measured value y_0, and (ii) the uncertainty in the calibration curve expressed by the uncertainties in both coefficients $s(a)$ and $s(b)$. The value of $s(x_0)$ is calculated by (see Bertrand-Krajewski et al. (2000) for details):

$$s(x_0)^2 = \frac{s_l^2}{b^2}\left[1 + \frac{1}{N} + \frac{(x_0 - \bar{x})^2}{\sum_i n_i (x_i - \bar{x})^2}\right] = \frac{0.344397}{(1.000395)^2}\left(\frac{61}{60} + \frac{360415.17}{19228814.4}\right) = 0.3564 \quad 6.18$$

Accordingly, adopting $u(x_0) = s(x_0)$ produces $u(x_0) = 0.6$ mm. The 95% confidence interval (coverage factor equal to 2) for x_0 is then given by $[x_0 - 2u(x_0)$, $x_0 + 2u(x_0)] \approx [598.0, 600.4]$. The final result is expressed $x_0 = 599.2 \pm 1.2$ mm.

The above result means that the sensor uncertainty, for $y = 600$ mm, is equal to 0.6 mm, under stable and laboratory conditions. In the field (e.g. in a real sewer) the *in situ* measurement uncertainty will increase because the water level is not flat and stable but uneven with at least some small waves, the free surface is not strictly horizontal due to turbulence and secondary flows, the position of the sensor is not known with a very high precision, etc. These sources of uncertainty, independent from the sensor uncertainty, are evaluated to approximately $u_f = 5$ mm. As a consequence, the in situ measurement standard uncertainty $u(h)$, with the above sensor, is assumed to be equal to

$$u(h) = \sqrt{u(x_0)^2 + u_f^2} = \sqrt{0.6^2 + 5^2} = \sqrt{0.36 + 25} \approx 5\,\text{mm} \qquad 6.19$$

In other words, in this case, the final uncertainty in water level measurement is mainly due to *in situ* conditions rather than to the sensor uncertainty. However, this is not always the case: the authors have observed sensors with $s(x_0)$ greater than 2.5 mm. In such a case, it is recommended to change the sensor if the resulting *in situ* uncertainty is too high for the user.

6.4.2 Example 2: Uncertainty in discharge measurement

Consider a circular pipe with a radius R(m) equipped with a piezoresistive sensor measuring the water level h(m) and with a Doppler sensor measuring the mean flow velocity U(m/s) through the wet section S(m²), as shown in Figure 6.7.

Assuming that the pipe is actually circular, the discharge Q(m³/s) is calculated by

$$Q = f(h, U) = S(h)U = \left[R^2 \, \text{Arc} \cos\left(1 - \frac{h}{R}\right) - (R - h)\sqrt{2hR - h^2} \right]U. \qquad 6.20$$

With $R = 0.5\,\text{m}$, $h = 0.7\,\text{m}$ and $U = 0.8\,\text{m/s}$, the result is that $Q = 0.4697\,\text{m}^3/\text{s}$.

The variables h, U and R, measured with different instruments, are independent and not correlated. Under these conditions, the law of propagation of uncertainty (i.e. Equation 6.7) can be used to estimate the combined standard uncertainty $u_c(Q)$:

$$u_c(Q)^2 = u(R)^2 \left(\frac{\partial Q}{\partial R}\right)^2 + u(h)^2 \left(\frac{\partial Q}{\partial h}\right)^2 + u(U)^2 \left(\frac{\partial Q}{\partial U}\right)^2 \qquad 6.21$$

where:

$$\left(\frac{\partial Q}{\partial R}\right) = 2UR \, \text{Arccos}\left(1 - \frac{h}{R}\right) - 2U\sqrt{2hR - h^2} \qquad 6.22$$

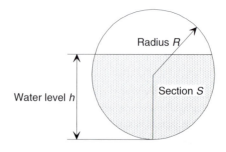

Figure 6.7 Schematic view of parameters used to calculate water flow using water level (from a level piezoresistive sensor) and flow velocity (from a Doppler sensor)

$$\left(\frac{\partial Q}{\partial h}\right) = 2U\sqrt{2hR - h^2} \qquad\qquad 6.23$$

$$\left(\frac{\partial Q}{\partial U}\right) = R^2 \operatorname{Arccos}\left(1 - \frac{h}{R}\right) - (R - h)\sqrt{2hR - h^2} \qquad\qquad 6.24$$

The true (but unknown) discharge Q_t has an approximately 95% probability of being within the interval $[Q - 2u(Q), Q + 2u(Q)]$. The standard uncertainties in the three variables h, U and R have to be evaluated separately, either by means of Type A or Type B methods, or by expertise when necessary.

$u(h)$ is estimated by combining the independent instrumental uncertainty $u_1(h) = 0.0006$ m for $h = 0.7$ m as calculated in the Example 1 and the uncertainty $u_2(h)$ due to the *in situ* operational conditions. As the sensor is used in a sewer system where the water is not still and perfectly horizontal but moves downstream and generates small waves at the surface with possible secondary currents leading to a non horizontal free surface, one estimates that $u_2(h) = 0.005$ m. As $u_1(h)$ and $u_2(h)$ are independent, the resulting global uncertainty $u(h)$ is given by:

$$u(h) = \sqrt{u_1(h)^2 + u_2(h)^2} = \sqrt{0.0006^2 + 0.005^2} \approx 0.005\,\mathrm{m} \qquad\qquad 6.25$$

Repeated measurements (Type A uncertainty) of the pipe radius with a Class 2 certified meter indicate that $u(R) = 0.002$ m.

As it is impossible to adequately and fully calibrate a Doppler sensor with certified standards in a way that is representative of actual conditions of use in a sewer, the value of $u(U)$ is estimated by means of both comparison with other known measurement devices and expertise. In this case, one assumes that $u(U) = 0.05$ m/s.

Substituting these values in Equations 6.12 through 6.24 gives:

$$\left(\frac{\partial Q}{\partial R}\right) = 0.8526\,\mathrm{m^2/s}$$

$$\left(\frac{\partial Q}{\partial h}\right) = 0.7332\,\mathrm{m^2/s}$$

$$\left(\frac{\partial Q}{\partial U}\right) = 0.5872\,\mathrm{m^2}$$

Equation 6.21 gives $u_c(Q)^2 = 1.344 \times 10^{-5} + 2.908 \times 10^{-6} + 8.621 \times 10^{-4} = 8.784 \times 10^{-4}\,\mathrm{m^6/s^2}$ and then $u_c(Q) = 0.0296\,\mathrm{m^3/s}$. With the coverage factor $k = 2$, the final result is expressed by an approximately 95% confidence interval: $Q = 0.4697 \pm 2 \times u_c(Q) = 0.4697 \pm 0.0593\,\mathrm{m^3/s}$, which corresponds to a relative uncertainty equal to $0.0593/0.4797 = 12.6\%$. In practice, only significant digits are used, which gives $Q = 0.47 \pm 0.03\,\mathrm{m^3/s}$.

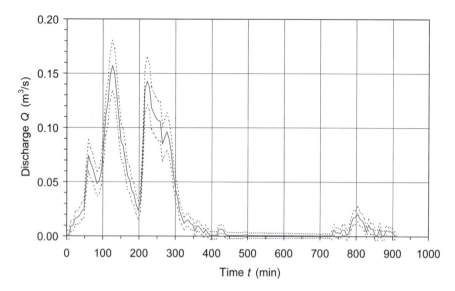

Figure 6.8 Hydrograph of water flows (solid line) in sewer system during a storm event and its 95% confidence interval (dotted lines) (see also colour Plate 8)

In this example, one observes that the major contribution in the combined uncertainty $u_c(Q)$ is the uncertainty in the mean flow velocity $u(U)$. Figure 6.8 shows an example of results obtained when the above calculations are applied to a hydrograph.

The continuous line represents the calculated discharge Q and the dashed lines represent the 95% confidence interval obtained from Equation 6.15 as explained above. More information about calculating uncertainty in volumes and pollutants loads can be found in Bertrand-Krajewski and Bardin (2001, 2002).

6.4.3 Example 3: Uncertainty in turbidity measurement

Consider a nephelometric turbidimeter to be used in a sewer system, with a measurement range of 0 to 2000 NTU (nephelometric turbidity units). Its measurement uncertainties have been evaluated according to the procedure described above in this chapter. Four certified reference formazine solutions of 20 NTU, 200 NTU, 1000 NTU and 1800 NTU have been used as standards for calibration. For each reference solution, twelve values measured by the turbidimeter have been recorded and processed to establish three equations. The first one gives the measured turbidity T_1 as a function of the reference turbidity T_0, directly based on calibration data, similarly to the above Example 1:

$$T_1 = 1.0144\, T_0 - 3.5245 \qquad\qquad 6.26$$

The true value T_0 for a given measured value T_1 is then estimated by the reciprocal of Equation 6.26):

$$T_0 = 0.9858\, T_1 + 3.4745 \qquad\qquad 6.27$$

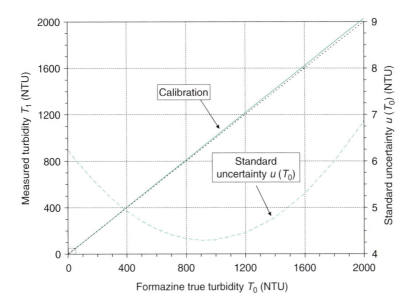

Figure 6.9 Calibration curves of the turbidimeter

its empirical standard uncertainty $u(T_0)$ (approximately equivalent to standard deviation), after evaluation carried out for the whole measurement range, is given by the following empirical equation:

$$u(T_0) = 2.25.10^{-6}\ T_0^2 - 0.0042\ T_0 + 6.251 \qquad\qquad 6.28$$

Consequently, there is a 95% probability that the true value of T_0 lies in the range $[T_0 - 2u(T_0),\ T_0 + 2u(T_0)]$. According to Equations 6.26 and 6.28 illustrated in Figure 6.9, the turbidimeter has been considered as valid for *in situ* measurements as its features comply with the user's requirements.

The above results allow the evaluation of sensor uncertainty expressed in NTU. If the sensor is used to estimate total suspended solids (TSS) concentrations in mg/L, for example, from turbidity measurements in NTU, a specific method is necessary to evaluate uncertainties in TSS concentration to account for uncertainties in both turbidity measurements and laboratory TSS measurements used to establish the correlation between NTU and TSS values (Bertrand-Krajewski, 2005).

6.5 RECOMMENDATIONS FOR UNCERTAINTY ANALYSIS IMPLEMENTATION

It is recommended that a preliminary uncertainty analysis be done before any main measurement campaign commences. This procedure allows corrective action to be taken to reduce uncertainties to within the limits determined by the objectives. This preliminary uncertainty analysis is based on data and information that exist before the test, such as calibration histories, previous tests with similar instrumentation, prior measurement uncertainty analysis, expert opinions, and, if necessary, special tests. Preliminary analysis determines if the measurement result can be measured with sufficient accuracy, to compare alternative instrumentation and experimental procedures, and to determine corrective actions. Corrective action resulting from preliminary analysis may include:

- improvements to instrument calibrations if the level of systematic uncertainties is unacceptable
- selection of a different measurement methods
- repeated testing and/or increased sample size if uncertainties are unacceptable.

Cost and time may dictate the choice of the corrective actions. If corrective actions cannot be taken, there may be a high risk that test objectives will not be met because of the large uncertainty level, and cancellation of the monitoring campaign should, therefore, be a genuine consideration.

Post-test analysis validates pre-test analysis, provides data for validity checks, and provides a statistical basis for comparing test results. For each measurement situation, the data reduction Equation 6.2 is determined first. Then a block diagram of the measurements is constructed to help organize the individual measurement systems and illustrate elemental error propagation. Data-stream diagrams (showing data flow from sensor to result) are constructed next, to identify and organize the assessment of the bias and precision limits of the individual variable. Once the sources of uncertainty have been identified, their relative significance should be established based on order of magnitude estimates. A 'rule of thumb' is that those uncertainty sources that are smaller that one fourth or one fifth of the largest sources are usually considered negligible.

Simplified analyses by using prior knowledge (e.g. data base), tempered with engineering judgment and with effort concentrated on dominant error sources and use of end-to-end calibrations (to avoid needlessly determining uncertainty contributions of every element in a process) and/or bias and precision limit estimation are highly recommended. The precision limit, bias limit, and total uncertainty for the measured result are then found. For each measured result, the bias limit, precision limit, and total uncertainty should be reported. Details of the uncertainty analysis should be documented either in an appendix to the primary test report or in a separate document.

6.6 UNCERTAINTY ANALYSIS AND THE DECISION-MAKING PROCESS

One of the axioms of science is the necessity of getting reproducible results in experiments, thus enabling the confirmation of scientific hypotheses and the possibility of predicting results. It is more appropriate perhaps to regard prediction as limited but not totally impossible. The thesis of limited predictability is very disheartening for those trying to make decisions. Decision makers should learn that uncertainty always exists and that it limits a prediction and cannot be eliminated completely. Limiting uncertainty improves the decision making, but it is difficult to be achieved when prediction is considered entirely as deterministic. Problems in operations research and control theory are usually deterministic, that is, the availability of full knowledge is assumed to achieve the goal of carrying out an activity. Optimization (i.e. searching for the extremity of the function) does not involve decision making. It can be shown that the necessity of decision making results totally from uncertainty (Singh et al., 2001). That is, if the uncertainty did not exist, there would be no need for decision making.

Decisions about environmental systems encompassed in urban water management, such as wastewater treatment plants, water purification, and so on, are often made under conditions of uncertainty. The immediate question, according to Singh et al. (2001), is what makes a decision rational. There can be more than one criterion of rationality. One criterion of rationality is that the decision must aim at a goal one wants to achieve. For example, replacement of a treatment plant that meets increased wastewater load needs also to be economical, durable and more efficient. Failure to meet any of these goals may not be acceptable and would make the decision irrational. Given that there are always several acceptable alternative ways to design a treatment plant, the decision maker must identify several of them (an exhaustive search of all possible alternatives not always can or should be made). Consequently, the second criterion of rationality is that the decision maker must choose the best design from several alternative designs. The third criterion of rationality, according to Singh et al. (2001), is making the choice between alternatives in accordance with a rational evaluation process. The evaluation process might include considerations of costs (initial investment, operation and maintenance), the useful lifetime of the selected alternative and secondary effects, all of which (ideally) should be converted to a common measure of value. Most of these judgments are supported by existing or specially conducted measurements and modelling studies that quantify various components of the decision making process.

The challenges facing measurement and modelling of the water quantity and quality in urban processes should be contemplated in the wider perspective of risk-based decision support (McIntyre et al., 2003). Firstly, a high degree of measurement uncertainty is not necessarily an undesirable outcome, and undoubtedly is preferable to no indication of reliability at all. Secondly, measurement uncertainty should be viewed as a source of risk, as is traditional in other fields of engineering, and should be used to establish and achieve an acceptable failure probability in terms of water quality and quantity, rather than be used to criticize the measurement approach. Thirdly, it is worth noting that, in the context of decision support, we are not justified in investing resources in monitoring and estimating uncertainties unless the results will be instrumental in shaping the decisions that need to be made. The reader is referred also to Chapter 10 for guidance on the use of data in decision-making.

REFERENCES

AIAA. 1995. *Assessment of Wind Tunnel Data Uncertainty*. Reston, USA, American Institute of Aeronautics and Astronautics (AAIA).(AIAA S-071-1995).

ANSI/ASME. 1985. *Measurement Uncertainty: Part 1, Instrument and Apparatus*. Washington DC, American National Standards Institute (ANSI)/American Society of Mechanical Engineers (ASME). (ANSI/ASME PTC 19.I-1985).

ASME. 1998. *Test uncertainty: Instruments and Apparatus*. New York, American Society of Mechanical Engineers (ASME). (ASME PTC 19.I-1998).

Bertrand-Krajewski, J.-L. 2005. TSS concentration in sewers estimated from turbidity measurements by means of linear regression accounting for uncertainties in both variables. *Water Science and Technology*, Vol. 50, No. 11, pp. 81–88.

Bertrand-Krajewski, J.-L. and Bardin, J.-P. 2001. Estimation des incertitudes de mesure sur les débits et les charges polluantes en réseau d'assainissement: application au cas d'un bassin de retenue-décantation en réseau séparatif pluvial. *La Houille Blanche*, Vol. 6/7, pp. 99–108.

———. 2002. Uncertainties and representativity of measurements in stormwater storage tanks. Paper presented at the 9th International Conference on Urban Drainage, 8–13 September 2002, Portland, US.

Bertrand-Krajewski, J.-L., LaPlace, D., Joannis, C. and Chebbo, G. (2000). *Mesures en Hydrologie Urbaine et Assainissement*. Paris, Lavoisier.

BIPM. 1980. Recommendation ENC-1. *Report of the Working Group on the Statement of Uncertainties*. Paris, Bureau International des Poids et Mesures.

Coleman, H.W. and Steele, W.G. 1995. Engineering application of experimental uncertainty analysis. *AIAA Journal*, Vol. 33, No.10, pp. 1888–96.

ENV 13005. 1999. *Guide pour l'Expression de l'Incertitude de Mesure*. Paris, Association Française de Normalisation (AFNOR) (also available in English and German).

Fletcher, T.D. and Deletic, A. 2007. Observations Statistiques d'un Programme de Surveillance des Eaux de Ruissellement: Leçons pour l'Estimation de la Masse de Polluants [Statistical Observations of a Stormwater Monitoring Programme; Lessons for the Estimation of Pollutant Loads]. Paper presented at NOVATECH 2007: Conference on Sustainable Techniques and Strategies in Urban Water Management, 25–28 June 2007, Lyon, France.

Herschy, R.W. 2002. The uncertainty in a current meter measurement. *Flow Measurement and Instrumentation*, Vol. 13, No. 5, pp. 281–84.

ISO. 1993. *Guide to the Expression of Uncertainty in Measurement*. Geneva, International Organization for Standardization (ISO).

McIntyre, N.R, Wagener, T., Wheather, H.S and Yu, Z S. 2003. Uncertainty and risk in water quality modelling and management. *Journal of Hydroinformatics*, Vol. 5, No. 4, pp. 259–4.

Moffat, R.J. 1988. Describing the uncertainties in experimental results. *Experimental Thermal and Fluid Science*, Vol. 1, No. 1, pp. 3–17.

Singh, V.P., Strupczewski, W.G. and Weglarczyk, S. 2001 Uncertainty in environmental analysis. N.B. Harmancioglu, S.D. Ozkul, O. Fistikoglu and P. Geerders (eds) *Proceedings of the NATO Advance Research Workshop on Integrated Technologies for Environmental Monitoring and Information Production*. Dordrecht, Kluwer Academic Press, pp. 141–58.

Taylor, B.N. and Kuyatt, C.E. 1994. *Guidelines for Evaluating and Expressing the Uncertainty of NIST Measurement Results*. Gaithersburg, MD, USA, National Institute of Standards and Technology (NIST) (Technical Note 1297).

Chapter 7

Selecting monitoring equipment

D. Prodanović

Institute for Hydraulic and Environmental Engineering, Faculty of Civil Engineering, University of Belgrade, Bulevar Kralja Aleksandra 73, 11000 Belgrade, Serbia

7.1 INTRODUCTION

Integrated Urban Water Management should be based on true, measured data, with assessed uncertainties (as discussed in Chapter 6). Monitoring equipment, specially designed for the measurement of certain parameter(s), is used to obtain such data. A number of monitoring devices are available on the market and the task of optimal design and equipment selection is therefore not easy. Selection of monitoring equipment is typically made subsequent to the establishment of the goals and objectives of the monitoring programme (Chapter 3). The selected equipment has to not only fit the purpose, but the available budget, desired precision and accuracy and appropriate scale (e.g. a turbidity sensor used in a drinking water reservoir will need to be optimized to a different scale than, say, one used at the inlet of a wastewater treatment plant). It also has to cope with spatial variability of the measured variable (Chapter 5), to work within on-site environmental conditions, to work continuously between inspection and maintenance periods and to be able to be readily (and reliably) calibrated under field conditions. Relevant government laws and contract agreements should also be checked for possible selection constraints. Contract agreements for the purchase of measuring devices often dictate required measurement systems. These constraints may be in terms of accuracy, specific comparison of devices and procedures.

In order to assist the reader in the difficult job of selecting monitoring equipment, according to the principles given in the previous chapters, some basic guidance regarding sensors are given. Important sensor characteristics are listed, and the criteria used for the choice of monitoring equipment explained, considering the need to integrate different components of the urban water cycle.

7.2 DEFINITION OF TERMS AND HISTORICAL OVERVIEW

Monitoring can be defined as regular sampling and analysis of air, water, soil, wildlife, and other factors, to determine, for example, the concentration of contaminants and to track system behaviour and its response to management. Sampling is done using *monitoring equipment* devices (typically electrical) used for measurement of various parameters (after NDWR, 2000).

The primary role of the monitoring equipment is to convert some measured non-electrical quantity, such as water level, into an electrical quantity that can be stored and processed. This conversion is done within the element called a *transducer*, a device that converts one type of energy into another, or responds to a physical parameter (Labor Law Talk, 2005) for the purpose of measurement of a physical quantity or for information transfer (ANST, 2000).

Since the 1970s, the term *sensor* has been used widely instead of transducer. The sensor is a device that detects or senses a signal or physical condition. Most sensors are electrical or electronic, although other types exist. The sensors can either directly indicate the values (e.g. a mercury thermometer or electrical meter) or are paired with an indicator (perhaps indirectly through an analogue to digital converter, a computer and a display) so that the value sensed becomes readable.

In general, a history of the sensor development can be divided into two parts: one prior to the semiconductor era and the period where semiconductors have been used in order to improve the sensitivity, reduce the size and integrate the sensing and indicator functions. According to the level of integration, the sensor development can be grouped into four generations (Stanković, 1997):

(1) Raw sensors (resistive sensors, inductive displacement sensor, etc.)
(2) Sensors with built in amplification and temperature stabilization, with analogue output
(3) Modern hybrid sensors with analog and/or digital output, and the possibility to perform some data analysis (the direction of the data is still only from the sensor to the instrument)
(4) The latest generation of sensors with built-in two-way communication capability and additional instrumentation. The sensor itself knows its calibration data, transfer function, and so on, and is not just used to communicate between different sensors and the user, but to perform the more complex data analysis. The use of such a general purpose instrument permits easy expansion of the measuring chain.

Still used in many situations, typical monitoring equipment of older generation consists of the following components (Bertrand-Krajewski et al., 2000):

• Sensors, with or without an internal amplification
• Amplifier with a conditioner that will amplify and normalize a week signal from the sensor and also filter out the unwanted signals or average the measured signal
• Transmitter that will output standard current or voltage signal proportional to the measured quantity
• Display, with or without local registration unit
• Power supply (battery, solar panel, or mains connection).

A block diagram of the older instrument is shown on the left of the Figure 7.1. The basic characteristic of such a system is its passive role in monitoring: it can only collect and send data without any possibility to interact with environment. The more modern digital instruments (Figure 7.1, right) discussed in Section 7.3 below offer greater possibilities.

7.3 MODERN MONITORING EQUIPMENT

In the modern digital instruments (Figure 7.1, right), the microprocessor controls all functions of the instrument and since the sensor itself has the microprocessor too,

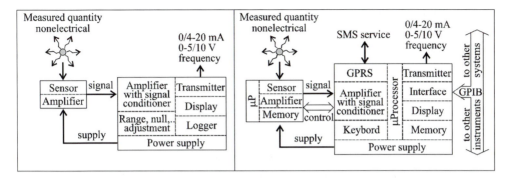

Figure 7.1 Diagramatic representation of monitoring equipment: old analogue instrumentation (left) versus new digital systems (right)

it can exchange the data and change the working parameters as needed. The data pre-processing can be done, and the instrument can communicate with the hierarchically dominant system, either by wire, fibre optics, or wirelessly. Such measuring systems are usually called *intelligent systems*, since they can *react and adapt to the measured variable*, in order to maintain the highest specified operating parameters. If turbidity of water is measured, for example, the operating range of a transparency sensor can be altered automatically if the concentration of suspended solids in water changes significantly, thus maintaining the maximum accuracy throughout the whole measured data.

Multi-parameter sensors are the state of the art in sensor development: the same sensor can now simultaneously measure more then one quantity, saving space and significantly reducing cost. For example, there are new sensor chips only 25 mm square with the capacity to measure temperature and five water quality parameters on their surface: free chlorine, monochloramine or dissolved oxygen (configurable), pH, redox and conductivity. The chip is manufactured on a ceramic substrate; it has no membranes, buffer solutions or glass bulbs, so a robust monitoring station can be constructed. When linked with a tiny low-power computing platform and wireless communication with aggressive power saving algorithms, a number of such stations will form a low-cost wireless sensor network for monitoring water supply and sewer systems (WSSS) (Stoianov et al., 2004). New developments in this field are happening all the time, and the reader is advised to seek out local guidance from both providers and users of such equipment.

7.3.1 Sensor characteristics important for equipment selection

In urban water related projects a number of different quantities are used. Those quantities that user can continuously monitor could be grouped as follows:

• Physical properties of water (temperature, density, viscosity, etc.)
• Quantities that describe current state of flow (level, pressure, velocity, etc.)
• Chemical parameters of water (pH, suspended solids, organic load, inorganic load, nitrates, nitrites, toxic substances, etc.)
• Biological content of water (total coliform bacteria, etc.)
• Climate (rainfall, air humidity, air temperature, solar radiation, etc.).

Apart from the listed quantities, data on a number of geometric quantities are also needed, such as the diameter or shape of a pipe, as mentioned in Chapter 4. Acquisition, accuracy assessment and handling of the geometric quantities, although being of equal importance, are not covered by this chapter.

For each measuring quantity, the user can find a whole range of sensors on the market, based on different working principles. Even sensors with the same working principle, if equipped with more or less sophisticated instrumentation, will produce varying results. A number of good textbooks (e.g. Stanković, 1997; Miller, 1983; Boros, 1985) and internet sites are available describing in detail individual sensor operation, range of usage, and the availability of sensors for measuring certain quantity (Muste, 2003). Those texts are oriented mostly towards either a certain group of users or types of measurements (laboratory measurement, sediment measurement, measurement in sewer systems). To avoid repetition of those details, only the basic characteristics common to all types of sensor and vital for good selection of equipment are listed here:

Accuracy/uncertainty and repeatability of the sensor: These are the two most important parameters (see details in Chapter 6). Basic accuracy is closely linked with the sensor's principle of operation while accompanying instrumentation can slightly improve or impair the operation (there are some examples where the manufacturer deliberately compromises accuracy in order to lower the market price). Repeatability is even more important than the accuracy: if the sensor is not accurate, it can be recalibrated, but only in the range of its repeatability.

Accuracy and repeatability of the whole measuring system: This is rarely assessed by the user, although the overall accuracy of the system is always lower then the accuracy of the sensor itself. The sensor has to be matched to field measuring conditions. For example, the ultrasound-based water-level measurement is widely used and is highly accurate, but only for clear and still water surfaces, without floating foam and debris, as commonly occurs in 'real' water systems. Similarly, the theoretically high accuracy of the ultrasound transit-time method of flow measurement in closed pipes will be largely reduced if the water is not homogeneous. Through final data validation (explained in Chapter 8) measures of the overall accuracy should be determined, and this uncertainty used in the reporting of any final output (Chapter 6).

Stability: Defined as constancy of repeated measurements, particularly over time, and under changing conditions, stability has two components: (a) long-term stability under constant working conditions, defined as a drift of the output signal per month or year, and (b) thermal stability, defined as a change of the output signal for a rapid temperature change of working environment. Thermal stability is important if the monitoring station must operate throughout the calendar year, especially in harsh conditions where variations in temperature can be up to 100°C. Specification of permitted time-drift and temperature-drift should be made when selecting sensors, based on allowable uncertainties specified in the objectives.

Resolution: Resolution refers to the capability of the measuring equipment to distinguish between small differences in an input signal (measured quantity). For analogue equipment, the resolution is theoretically infinite and is therefore practically equal to the noise level. Digital equipment is by its nature limited by its resolution. Contemporary digital equipment has minimized this problem using high-resolution converters (more than 16 bit) and auto-ranging options (automatic gain-adjustment). It is important to match the resolution of the sensor with all other components of the entire monitoring system (e.g. data logger, communication lines) to avoid degradation of the measured data.

Linearity: Some sensors are by their nature non-linear (for example, the rate of light absorption is nonlinear to the concentration of suspended solids). If the principle has good repeatability, the microprocessor can be used to linearize the output. But even being linearized, the non-linear system will not have constant relative error along the measuring range; the highest error will be in the range of maximum sensitivity.

Measuring range (or dynamic range): This is the range of change in a sensed quantity that can be measured with specified accuracy. For most sensors this range is 1:10, although a few can have a range of 1:100. The producers often specify an extended measuring range, where equipment can be used but with lower accuracy. For some sensors, exceeding the measuring range will cause permanent damage to the instrument (for example, differential pressure sensors are sensitive to overload, since the diaphragm can be damaged).

Dynamic response (slew rate, response time): This is a measure of how fast the sensor reacts to a sudden change of an input quantity. In general, three situations are possible:

(1) The input variable can have rapid changes in value, but the user is not interested in these, and the monitoring system needs only to record mean, averaged values. An example of such a case is measurement of daily pressure variations in pressurized pipes, where only mean values are needed and not peaks caused by transients. Filtration of the rapid transients can be done mechanically (using special housing constructions), electronically or combined.

(2) The user wants to measure rapid changes. Response time of the equipment is fundamental and it has to be matched with expected rate of changes in input variable. In some situations, using inverse Fourier transformations or convolutions in time domain, the frequency response of the equipment can be linearized. However, the user has to be aware of the large volumes of data generated by such equipment.

(3) The third situation is a mixture of two: when the user wants to measure mean values during certain periods (when river level is slowly changing, for example) and switch to fast response when certain conditions are fulfilled (e.g. as the flood wave is approaching). In the latest generation of monitoring equipment, the user can interact with control software to change the instrument's behaviour in different situations, and the monitoring stations themselves can communicate with others to exchange current data and information on their operating status (arrays of sensors) as shown in Chapter 24.

7.3.2 Criteria for the selection of monitoring equipment

Selecting the proper measurement device for a particular site or situation is not an easy task. A good knowledge of physical, chemical and biological processes at a measurement site is needed, as well as knowledge of different sensor's techniques applicable to those processes. Technical parameters are not the only criteria that should be considered. Other factors (ecological, financial, institutional, etc.) will often impose constraints that will limit the list of the possible monitoring equipment.

Throughout Chapters 3 to 5, the objectives and applications of the monitoring, the selection of variables, and consideration the temporal and spatial variability, were discussed. Bearing in mind the need to integrate the measurements in urban water systems in a way that different users can share the data, a checklist of items that a user should consider when selecting the appropriate monitoring equipment is given below. The list should be seen as a guide only (not compulsory instruction), and the user is also advised to consult other local sources and guidelines (e.g. USBR, 2001). The relative importance of each item in this list depends ultimately on project objectives.

Users of data: Identify and consider the needs of both current and potential users. Some possible uses of data include: (i) real time control (RTC), where online data are important. and users will not bother with data storage and post-processing; (ii) numerical simulation models, where a vast amount of short-term and long-term historical data are needed and the exact conditions of the system state are important; (iii) legal requirements for trade metrology, mostly regulated by government with the equipment regularly checked for accuracy specification; (iv) storage of historical data, performed by national or state institutions, for some (identified or unidentified) future need; and (v) system control and system management, the most demanding use, as data users need online data similar to short-term, mid-term and long-term historical data and also data mining and data extraction techniques.

Continuous monitoring versus ad hoc measurements for different purposes: Ad-hoc measurements could be a viable alternative since they are much cheaper and can cover a wider spectrum of measuring techniques, but with much more human involvement (and usually with bigger problems of sampling representativeness). A special kind of ad hoc measurements are diagnostic measurements, when higher accuracy equipment is used in order to quantify the values of certain parameters. Examples include the measurements of pump characteristics, natural frequencies for some construction, and the roughness factors for pipes.

Measuring accuracy: The user has to decide what is the required accuracy of measured data (based on the objectives and allowable uncertainty), and calculate how this will influence the whole project. Specification of the allowable uncertainty (and thus allowable inaccuracy) becomes a critical specification for monitoring equipment. As a rule of the thumb, the accuracy of equipment used should be at least three times better then the required. Typically, however, the price of monitoring equipment will be approximately exponentially related to its accuracy.

Possibilities to calibrate and recalibrate the equipment: To maintain rated accuracy, each piece of the measuring equipment has to be calibrated prior to its installation, and then periodically recalibrated (as suggested by the manufacturer). Also, the user has to verify the instrument after it has been installed at a measuring site (although the accuracy of available methods for such verification is generally lower). Ease of calibration, and the methods for verification of the equipment, should be key considerations in equipment selection. Modern measuring equipment should follow 'good laboratory practice' (GLP) standards (AGIT, 2003) and allow storage of the calibration results in the internal memory, forming a database where long-term behaviour of the equipment can be checked. If this is not the case, the user will need to keep a database of calibration results, in order to understand this behaviour.

Ability to check the whole monitoring station: Whenever possible, perform tests of the equipment by inputting a known volume or quantity of the measured variable. For example, in flow measuring devices in small sewers, one can drain the whole tank of known volume into the upstream manhole. Through later data analysis, volume added to the base flow can be deduced, providing an effective check of the equipment.

However, do not forget to note the event in the database as a test event, so that the results do not become mixed with the data from the true rainfall event.

Available resources: The following resources should be considered: (i) finances; (ii) time, since in most situations measurements are to be conducted at very short notice (maximum flow during floods, for example); (iii) available space for sensors and monitoring stations, especially for water quality measurements where a number of parameters are needed (the reason manufacturers have developed multi-parameter sensors); (iv) The level of education and skill of staff to operate the equipment; (v) power supply (battery, solar charger or regular mains supply) and the implications of this for service and reliability; and (vi) communication possibilities (telephone lines, GSM or GPRS, optical cable) and their implications for service and data downloading.

Available techniques for measurement: During the process of equipment selection, all available techniques should be considered. The most accurate technique is not always the most suitable; issues of robustness and system compatibility may be paramount. Monitoring equipment selection charts (Figure 7.2) are commonly available for most monitoring variables (USBR, 2001).

Sizing (working range): Monitoring systems should work within its optimal range of values of measured quantity. Choosing a device that can handle larger than necessary input values could result in elimination of measurement capability at lower (or higher) values, and vice versa. For practical reasons, it may be reasonable to establish different accuracy requirements for high and low values of measured quantity. To assess needed measuring range, it is convenient to use hydraulic modelling software (e.g. HEC-RAS (USACE, 2005), EPANET for water supply systems), or other models, (depending on the variables to be monitored) to predict likely ranges of behaviour. In the simulated environment, the number of needed measuring points and ranges for each sensor can be optimized using different criteria (Kapelan, 2002). Besides a static working range, the dynamic response of the equipment has to match expected variations of the input signal to avoid aliasing (deterioration of the signal due to insufficient sampling frequency).

Conditions at the measuring site: The selected device should not alter site hydraulic conditions so as to interfere with normal operation and maintenance. Also, the basic accuracy of the selected device should not deteriorate as a result of hydraulic conditions or the measuring site selected. Figure 7.3 provides an example of flow measurement where the flow conditions changed from rapid (supercritical) to tranquil (subcritical) and

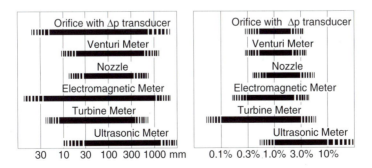

Figure 7.2 Suitability of pipe flow measurement devices as a function of diameter (left) and accuracy (right)

Source: Adapted from Radojković et al., 1989.

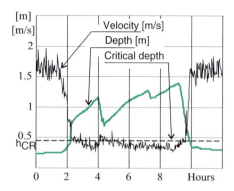

Figure 7.3 Change of flow regime due to varying downstream conditions

Source: SI, 2004.

then back to rapid, due to gate operation downstream (SI, 2004). A stream hydraulics simulation model such as the widely-available HEC-RAS (USACE, 2005) software can help the user to simulate and determine optimal position for flumes, weirs, or other controls.

Sensors with local storage: The current trend in production of new equipment is the addition of local storage capacity of each sensor. This will lower the risk of data loss: if supervisory equipment is out-of-order, the sensor will still work and collect the data; upon the re-establishment of communication, the sensor will report the locally stored data. However, the monitoring equipment now has to cope with possible ambiguity: as measured data are stored in the sensor and in the main database, data mismatch can occur. Appropriate data validation procedures should be implemented to address such possibilities (see Chapter 8).

Sensors with inbuilt communication capabilities: Nowadays sensors are often equipped with local wireless communication capabilities. In most cases this is not the primary method of sensor communication with other sensors in supervisory system (usually there is a centralized telemetry controller), but this allows the user to occasionally check the sensors' functionality and to obtain current readings through simple inquiry. Of course, data security then becomes an issue that has to be addressed, since the sensor readings are potentially can be intercepted (for example, using a mobile telephone).

Sensors with built in intelligence: In addition to local storage and communications capability, many sensors are able to: (a) analyse gathered data (e.g. find extreme values, statistical values, slope of data, etc.); (b) change their own function according to some programmed logic (e.g. when slope of input data is higher than a certain threshold value, the sampling rate should be increased; if this is happening too often, the threshold value should be changed and reported to the user); (c) apply the results of recent calibration that are stored in each sensor (which allows online change of the sensors, e.g. if pH sensor is replaced, the instrument will detect this and read the new calibration parameters and apply them to the measurements, writing this event into the log-file); and (d) report to the user if measurements are suspicious (through an SMS or other telemetry service). A number of sensors can thus be organized into several distributed monitoring stations, each communicating with each other. For example, if the automatic sampler has to take samples of clean water just before a cloud of suspended solids approaches the station, the upstream sensors will announce the coming event and the monitoring station can calculate the exact time when the first sample has to be taken. Based on previously measured concentration of solids, the sampler can be operated in a programmed regime.

Time synchronization: IUWM will require that measurements from a diverse range of monitoring equipment are integrated, to allow analysis of interactions to be undertaken. To allow easy data integration, it is necessary to have synchronized measurements with local time accurate to within one-third of the shortest measuring time increment, Δt. Nowadays, with the proliferation of the global satellite positioning system (GPS), an accurate synchronized clock signal should be used.

Robustness and sustainability of equipment: Monitoring equipment in general is expensive. The user has to consider the ability to use the same equipment at several locations, with different sampling criteria and working parameters. Highly dedicated, closed systems are better for permanent monitoring stations, but if equipment is to be shared among several users, it is better to have an open, user programmable system.

Coherency of all elements within monitoring system: All elements within the monitoring system (Figure 7.1) should match with their parameters: accuracy, resolution, and dynamic response. As a rule of the thumb, the sensor and conversion principles are the most critical parts of the chain, and so the rest of the equipment should have better parameters. A common mistake is to use the analogue 4 mA to 20 mA output (mostly with 1% accuracy) to transmit the measured signal to the main computer and then convert it into the digital form with a 0.01% accuracy converter.

Environmental limitations: Not all environments are sensor friendly and not all measuring techniques are environmentally friendly (e.g. some tracer techniques). The user has to analyse potential mutual impacts and to design protective measures. For example, the most expensive water quality probe will be useless if a floating plastic bag wraps around it, or if an aggressive water environment melts the cast aluminium housing. The measuring method can also influence the environment. For example, installing a weir or flume constricts the channel, slows upstream flow, and accelerates flow within the structure. These changes in the flow conditions can alter local channel erosion, local flooding, public safety, local aquatic habitat, and fish movement up and down the channel. For stream systems with high sediment loads, placement of a properly sized Parshall flume or low-head, sharp-crested or drop-box weir can help ensure that sediment flows through the flow control.

Operating conditions at measuring site: Conditions critical for proper operation and longevity of devices include suitability of power supply (stable source of supply), presence of vibrations (in vicinity of pumps, for example), direct sunlight (the temperature in a closed cabinet under direct sunlight can go up to 60°C or 70°C), high humidity (up to 100%), corrosive and aggressive environment (even in some clean water reservoirs), extremely low temperatures (most batteries will deplete in such conditions and in freezing water devices with moving parts in contact with water will stop functioning), and the concentration of debris (e.g. in rivers, especially during flood events). The potential for vandalism, especially in open field sites, has to be considered and proper housing selected.

Field/construction works: Installation will often be a limiting factor. For example, the flow in a sewer system can be easily and accurately measured using a critical depth flume. But the construction of such flume in an old, continuously operating sewer trunk is almost impossible (or, at least, not economically feasible).

Training of users and after sale service: The general trend in development of the measuring equipment is to hide its complexity from the user with 'intelligent' processors that will optimize the sensor operation according to local conditions. To keep the overall accuracy within the equipment's specified limits, however, it is essential to train the user for correct instrument usage (for example, the meaning of different parameters within the instrument, and the explanation of error messages, etc.). After-sale service and support is important: it may be better in some situations to select lower quality equipment with good, experienced local support, than to purchase the latest technology product that nobody in the surrounding area is capable of servicing or supporting.

7.4 SPECIFIC CONSIDERATIONS FOR DATA INTEGRATION WITHIN IUWM

What specific criteria for measuring equipment might apply to deliver a better integrated monitoring system across all components of the water cycle? Firstly, the underlying considerations discussed in the previous sections of this chapter remain important, and cannot be ignored. However, for data to be efficiently shared between different users, working across different water cycle components, the most important task is to *keep the data and monitoring system well documented*. This applies not only to the name and serial number of the sensors used, but also to descriptions of the measuring site including sketches and/or photographs, hydraulic and environmental conditions, equipment conditions, battery voltage and internal temperature. Importantly, the results of accuracy assessment for the sensors and for the whole system, history of sensor calibration, explanation of data storage format used, should all be rigorously documented. All such 'added' data that accompany the primary measured data are called *metadata*, and more details regarding the collection, storage and utilization of metadata are provided in Chapters 9 and 10.

During the design phase of the monitoring site, it is essential to handle the metadata with the same attention as the primary measured data. One part of the metadata will be created by the monitoring equipment itself. Contemporary measuring equipment in most cases will follow the GLP standard, and will store data about working conditions and calibration results. Another part of the metadata will be created during data validation phase (Chapter 8), while some of metadata has to be added manually, such as descriptions of site, observations and exceptions, and timing of additional tests. *The monitoring programme must define metadata that is compulsory, considering all possible future users of the data, across all potentially relevant water cycle components.*

7.5 SPECIFIC CONSIDERATIONS FOR DEVELOPING COUNTRIES

When selecting the monitoring equipment, the overall development of the region has to be considered too. Modern equipment means higher initial costs and the need for educated users to establish the network and to conduct regular inspections, recalibration and maintenance. The problem is emphasized if the measuring equipment is part of a development assistance programme: in most situations, the donated equipment will be of the latest generation, where the initial high cost is covered by a donor programme, but later operating and maintenance costs need to be covered by local users. The benefits from such programmes would be much greater if the equipment installed is manufactured by local companies and part of available resources are spent on the education of both local manufacturers and local users.

During the process of monitoring station optimization, the price/performance ratio should be thoroughly considered. In general, the ratio is not the same for countries with high standards where working labour is very expensive and for developing countries where 'human-powered' monitoring may be relatively less expensive. When new monitoring equipment has to be installed in developing regions, it is advisable to introduce the technology step by step: firstly the systems of older generation (which will often be relatively less expensive and usually robust), to allow the users to accept the technology, to develop skills and learn the basics. Over time (and depending on the need) new, more expensive smart systems can be installed. It is critical, however, that the selection of monitoring equipment matches local needs and context. Spending large amounts of money on unnecessarily complex and sophisticated equipment, without a clear need to do so, is likely to result in redundant and unused equipment, thus wasting already limited resources.

REFERENCES

AGIT. 2003. *Guidelines for the archiving of electronic raw data in a GLP environment*. Work Group on IT. www.glp.admin.ch/ (Accessed 02 July 2007.) (GLP-ArchElectRawData1_0.pdf)

ANST. 2000. *Glossary*. American National Standard for Telecommunications. www.atis.org (Accessed 02 July 2007.)

Bertrand-Krajewski, J.-L., LaPlace, D., Joannis, C. and Chebbo, G. 2000. *Mesures en hydrologie urbaine et assainissement*. Paris, Lavoisier.

Boros, A. 1985. *Electrical Measurements in Engineering*. Budapest, Hungarian Academy of Science.

Kapelan, Z. 2002. Calibration of WDS Hydraulic Models. Ph.D. thesis, Department of Engineering, University of Exeter, UK.

Labor Law Talk. 2005. *Dictionary*. encyclopedia.laborlawtalk.com (Accessed 02 July 2007.)

Miller, R.W. 1983. *Flow Measurement Engineering Handbook*. New York, McGraw-Hill Book Company.

Muste, M. 2003. *The Instrumentation Database*. Madrid, The Hydraulic Instrumentation Section (HIS), International Association of Hydraulic Engineering and Research (IAHR) www.iihr.uiowa.edu:88/instruments/home.jsp (Accessed 02 July 2007.)

NDWR. 2000. *Water Words Dictionary*. Nevada Division of Water Resources, Department of Conservation and Natural Resources. http://water.nv.gov/Water%20planning/dict-1/ww-index.cfm (Accessed 03 July 2007.)

Radojković, M., Obradović, D. and Maksimović, C. 1989. *Computers in communal hydraulic – analyze, design, measurement and management*. Belgrade, Serbia, Gradjevinska knjiga.

SI. 2004. *The report of flow measurements in sewer system Cukarica*. Belgrade, Serbia, Svet Instrumenta (SI) (In Serbian).

Stanković, D. 1997. *Physical-technical measurements*. Belgrade, University of Belgrade. (In Serbian.)

Stoianov, I., Whittle, A. Nachman, L., Kling, R. and Dellar, C. 2004. Monitoring water supply systems by deploying advances in wireless sensor networks. Paper delivered at Workshop on Innovation in Monitoring and Management of Ageing Infrastructure, Cambridge, UK.

USACE. 2005. HEC-RAS. US Army Corps of Engineers (USACE), Hydrologic Engineering Center. www.hec.usace.army.mil/software/hec-ras/hecras-download.html (Accessed 02 July 2007.)

USBR. 2001. *Water Measurement Manual*. Wasington DC, US Department of the Interior, Bureau of Reclamation (USBR). www.usbr.gov/pmts/hydraulics_lab/pubs/wmm/ (Accessed 02 July 2007.)

Chapter 8

Data validation: principles and implementation

J.-L. Bertrand-Krajewski[1] and M. Muste[2]

[1]Laboratoire LGCIE, INSA-Lyon, 34 avenue des Arts, F-69621 Villeurbanne CEDEX, France
[2]IIHR – Hydroscience and Engineering, University of Iowa, Iowa City, IA 52245, USA

8.1 INTRODUCTION

In situ measurement implies that the instrument sensors are physically located in the environment they are monitoring. These sensors collect time-series data that flow from the sensor to the data repository continuously, creating a data stream. Typically these sensors operate under harsh conditions, and the data they collect must be transmitted across various types of data communication networks; thus, the data can easily become corrupted through faults in the sensor or in data transmission. Undetected erroneous data can significantly affect the value of the collected data for applications. Critically important are the situations where the data are used for real-time forecasting, when there is limited time to verify the quality of the data. For this reason, robust and scientifically sound methods for detecting erroneous data before it is archived are necessary. Due to the vast quantity of data being collected, these methods must be automated in order for them to be practical. Since it is often difficult to determine whether an anomalous measurement has occurred due to a sensor or data transmission fault, or due to an unusual environmental system response, many fault detection techniques seek to identify anomalous measurements – measurements that do not fit the historical pattern of the data, but may not necessarily be caused by sensor or data transmission faults. The causes of these measurements can then be investigated to determine whether or not the measurement actually represents the environmental system state.

Given that the research and management of urban water systems is based mainly on the observation of systems, processes and phenomena, applied metrology plays a crucial role in determining the efficacy of these activities. *In situ* measurements of rainfall, flow rates and pollutant concentrations and loads are conducted under harsh and often difficult-to-control conditions, because the sensors are subjected to many functional, technical and operational constraints (continuous operation, the detrimental effects of wind, temperature, pressure, salinity, humidity, gases, the intrinsic variability, large range of values to be observed, fouling, influence of human and instrument errors, as well as technical failures, etc.).

Sensor manufacturers are continuously developing and enhancing instruments in order to account for some of the above-mentioned problems, by implementing various technologies (e.g. self-cleaning sensors, temperature compensation, auto-calibration, internal error checking, etc). Unfortunately, the reliability of these instruments remains frequently insufficient under the severe field conditions prevailing in urban water systems, especially in sewer systems, treatment plants and surface waters. The measurement results, even when collected with the most reliable instruments, are often defective and do not accurately represent the phenomena targeted by the monitoring programmes. They are subject to several problems like noise, missing values and outliers. Consequently, they must be systematically scrutinized, verified and validated before further use.

It is assumed herein that before initiating data validation, all the phases described above in this report have been carried out adequately (i.e. definition of objectives – Chapter 3, selection of monitoring sites and variables – Chapter 4, selection of appropriate spatial and temporal resolutions – Chapter 5, definition of allowable or acceptable uncertainties – Chapter 6, and appropriate selection of monitoring sensors and instruments – Chapter 7). It is also assumed that laboratory and/or *in situ* calibrations for the sensors and instruments as well as preliminary testing of the monitoring equipment have been carried out according to the best practices in metrology. Local guidelines should be consulted for this process.

Among the above considerations, two are of critical importance:

- Rigorous calibration of the sensors before installation and use. Calibration allows potential bias to be corrected, and uncertainties to be quantified (Bertrand-Krajewski and Bardin, 2001). A case study for the calibration process is provided in Chapter 6.
- *In situ* monitoring of the installed sensors and ancillary equipment. Regular surveillance is essential for avoiding degradation of the measurement conditions (e.g. clogging and/or progressive fouling of sensors by grease, solids and various other wastes). The risk of measuring non-representative values (e.g. pH sensors may measure the pH of the fouling layer around the electrode and not the pH of the effluent, Doppler velocity sensors may be blinded by cans, plastic bags or other debris transported by the flow) will be considerably reduced by a continuous and careful inspection and maintenance programme. Specific monitoring procedures and frequencies have to be established for each type of sensor. Periodic re-calibration of sensors must be made to prevent drifts and modifications of sensor response (e.g. piezometric flow level sensors should be checked and calibrated every 4 months to 6 months depending on their functioning conditions).

A rigorous data quality control and assurance can be obtained if these two critical activities are specified through written guidelines and implementation programmes.

8.2 BASIC PRINCIPLES OF DATA VALIDATION

The first basic rule to be applied for data validation can be summarized as follows: *The real targeted value of the measurand cannot be actually measured. Therefore, one should consider that measurements are wrong until there are sufficient and objective reasons to admit that the data are representative and reliable.*

Accordingly, *all* measured values and data should be submitted to the validation procedure. This seems to be a rather radical approach, but saves time and resources in the long term. While data validation is critically important for fluctuating variables such as flow, concentrations, and loads, it should be mentioned that 'static' variables, such as catchment area, imperviousness, structural description of urban water systems (length, diameter, material and roughness of pipes and sewers, characteristics of river reaches and sections), topographic data in maps and GIS, need also to be checked and validated. The following example is relevant in this regard:

> Rigorously calibrated flow level and flow velocity sensors were used for measuring flow rates and volumes during storm events for 6 months in a 1.8 m egg-shape sewer. Using independent rainfall measurements, the runoff coefficient of the catchment was evaluated: as expected, its value was lower than the imperviousness coefficient and the results were considered acceptable. However, during the conduct of a study on long term drinking water consumption and independent analysis of infiltration in the sewer system, it was found that the sewer cross-section was wrongly assumed: instead of a type T180 sewer (as indicated both in the municipal GIS and by the local sewer maintenance team), the sewer was of A180 type, leading to differences of up to 50% in flow rates for water levels lower than 0.2 m. *The lesson learned from this example is that even for data which may seem as if they can be taken for granted, double-checking is necessary.*

The user should be cautious with data that he/she did not measured directly. He/she should check, validate, and have records or tracks of the elements which certifies the validation of the data to be used. Unfortunately, checking and verification of this type of data is rarely included in validation protocols. While these 'redundant verifications' are not further referred to in the next paragraphs (because there are no specific methods associated with these verifications), *it is critical to devote an equal effort to assess and verify all the types of data, values and information that is used for/in monitoring programmes.*

It should be also emphasized here that the validation and its outcomes need to be recorded in appropriate log books, files or other similar type of documents. It is important to keep track of all actions, and to share the documents with other users in order to avoid duplication of effort, as well as to simply keep well documented data. This type of information, which is an integral part of the metadata, is crucial. Many existing and previous data cannot be used in a reliable manner because of the lack of information regarding their conditions of acquisition, quality and validity. This consideration is particularly important for integrated urban water management (IUWM), because data will need to be shared across diverse monitoring networks and among a diverse range of monitoring system operators. *The value of data will be greatly diminished if such data-quality related metadata are not rigorously maintained and published so that all users are aware of it.*

Usually, data validation is initiated by detecting measured values which are outside the uncertainty interval associated with the assumed or expected true value. Those abnormal values are called *outliers*. This *detection phase* involves two pre-requisites: (i) estimation of the maximum and actual allowable uncertainties, and (ii) estimation of the most likely true value based on previous knowledge, experience, or modelling results. In other words, data validation cannot be made independently of its measurement context.

Outliers are mainly generated by two sources:

- *the sensor or the instrument*, in which case the phenomenon to be measured is under normal conditions, but the sensor response is inaccurate due to major troubles.

This may be due to inappropriate sensor operation, drifts from the sensor and/or of calibration, clogging, fouling, adverse actions on sensors (such as maintenance, repair, etc.), human operating error (e.g. incorrect sensor programming), among other factors

- *the measurand (i.e. the phenomenon to be measured)*, in which case the sensor is functioning under normal conditions, but the phenomenon to be measured is disturbed due to unexpected events. For example, in sewer systems, flushing waves may dramatically change both flow and concentrations values. Moreover, upstream modifications, especially in 'looped systems', may affect significantly the phenomena compared to normal antecedent contexts.

Distinguishing between these two sources of anomalies is a difficult task (i.e. the difference between measured and expected values may be due either to the sensor or to the phenomenon or both). Therefore, *detection* only of abnormal values is not sufficient. A second phase, the *diagnostic*, should be carried out to identify the cause(s) of the observed differences. A careful diagnostic routine helps in formulating solutions for avoiding outliers in the measurements.

Data validation should be carried out at two levels:

- *local (or internal) validation*, whereby the validation data is obtained with one individual measurement at a point by one or more sensors. This analysis verifies the intrinsic validity of recorded data. This validation level is usually associated with the detection phase, but often the diagnostic phase is also needed
- *global (or external) validation*, that is, data validation carried out by taking measurements with several sensors, in different points, and/or comparing the data with models or alternative observations. For example, upstream and downstream data along a river reach or rainfall volumes and measurement volumes downstream a catchment can be compared for this purpose. This validation level is usually associated with the diagnostic phase, but detection aspects are also involved.

Data validation can also be made either offline or online with the latter typically used for real-time control systems. The following sections focus specifically on online validation.

After validation, the following data and information must be stored (Chapter 9):

- *as an initial raw data series* (this must *never* be discarded)
- *as a validated data series*, containing only the validated values, which will be used subsequently for monitoring, legal purposes, design, diagnostic studies, modelling, and other purposes, and.in some cases, a validity index (or data quality index) estimated for each value.

As previously discussed, the record of all validation actions must be stored in a separate file, including all decisions made (acceptance or rejection of values, replacement of values and how replacements have been made, context, diagnostic, etc.). This record will be very helpful in order to (i) progressively improve the quality of both diagnostic and data, (ii) document the process and instruments for other users of the data, and (iii) allow future interventions or interpretations of the data when new knowledge or techniques become available. These recorded elements are an integral part of the metadata and data quality assurance presented in Chapter 9.

Based on the above, the eight principles of data validation can be summarized as follows:

(1) Data validation shall be made *systematically* for all monitoring programmes (it is not an optional specification).
(2) *Always* check and validate *all* data from *all* types of measurements used in the monitoring programme (structural *and* functional data, static *and* dynamic).
(3) Validation should include *two phases*: (i) a detection phase, and (ii) a diagnostic phase.
(4) Validation should be made at both *local and global levels*.
(5) Validation cannot be made independently of its operational and environmental *context*.
(6) Validation is not pertinent without *rigorous and periodic calibration* of the instruments and *correct estimation of measurement uncertainties*.
(7) Validation can be made *off-line or online* depending on the monitoring purposes.
(8) Validation actions and results shall be *recorded* for memory, transmission, further diagnostics and reversibility.

8.3 VALIDATION METHODS

In contrast to uncertainty estimation for which standardized methods and approaches exist (see Chapter 6), there are no standard methods for data validation. Users have to define their own objectives, methodology and tools. This section attempts to provide some minimum basic specifications, applicable in many usual contexts, to carry out the validation process. These specifications are not exhaustive and, for specific conditions, sensors or contexts, more elaborated specifications (e.g. methods, algorithms) can be developed and applied. For example, one might use elaborated filters to smooth very noisy and disturbed signals, or early detection of failure (Brunet et al., 1990), and pre-validation between the transducers and the transmitter (Aumond and Joannis, 2006). Statistical methods such as 'principal components analysis' (PCA) can also be used when numerous redundant sensors are used (e.g. Mohamed-Faouzi et al., 2006).

Usually, data validation is carried out by inspecting short data records using simple calculations, analysis, and visualization tools. If the amount of data subjected to validation is voluminous, there is a need for development of automatic or semi-automatic routines for conducting the validation. Such routines are inherently better documented, eliminating the subjectivity of the validation process, and increasing its repeatability and transparency.

8.4 LOCAL LEVEL VALIDATION

Data validation implementation is illustrated here using the *Observatoire de Terrain en Hydrologie Urbaine*, or Field Observatory for Urban Hydrology (OTHU) Project in Lyon, France (Bertrand-Krajewski et al., 2000). The example summarizes the work reported by Mourad and Bertrand-Krajewski (2002; 2003). A fuller description of the OTHU Project is provided as a case study in Part III of this book (Chapter 23).

8.4.1 OTHU context and objectives

The OTHU Project was initiated in 2000, with specific efforts targeting metrological aspects over a 10-year monitoring programme. The OTHU Project entails continuous monitoring at five sewer system sites located in three experimental catchments. The sites were equipped with a variety of sensors (e.g. limnimeter, flow velocity sensor, pH, conductivity, temperature, turbidity, UV-probes, etc.) and with refrigerated automatic samplers. The typical time step for data acquisition is two minutes, resulting in the collection of a large amount of data. Given the extent of data produced, manual data validation is not possible and semi-automatic validation procedures, methods and tools have been developed and implemented.

8.4.2 Statistical and signal processing methods

A preliminary review of the conventional methods used in signal processing area, in particular statistical tests and application of decision theory, showed that there are several methods for data validation, detection of outliers and defects in data series. However, most of these direct methods are only applicable to random data and/or steady state processes. Non-steady state and partially auto-correlated process variables, which are typical in urban water systems (e.g. in sewer systems where large variability and fluctuations are observed during storm events), cannot be analysed with such methods. Moreover, their application strongly depends on hypotheses and assumptions regarding the process and measured data.

It is apparent that the theoretical formulation of these statistical and processing methods, as well as the required characteristics about the data, are difficult tasks for the field of urban hydrology for several reasons: (i) most techniques to detect outliers in statistical data samples are theoretically applicable if data are distributed according to pre-defined statistical laws that are not necessarily pertinent for urban hydrology data (Barnett and Lewis, 1990), (ii) filtering aims to entirely reconstitute the signal, but may arbitrarily modify values that were initially reliable and correct (Ragot et al., 1990; Bennis et al., 1997). Instead, verification of sensor redundancy (or alternatively their signals), as well as use of methods to detect changes or trends in system behaviour appear to be more suitable for the field of urban hydrology (see for example methods described by Gilbert, 1987; Brunet et al., 1990; Ragot et al., 1990). The pre-validation procedures used in the OTHU Project partially used these verification methods.

8.4.3 Data validation methods used in urban hydrology

There are many examples of data validation in the field of urban hydrology. Most of them concern the validation of rainfall data (Lupin, 1990; Soukatchoff, 1990; Auchet and Hammouda, 1995; Berthier et al., 1998; Jörgensen et al., 1998; Maul-Kötter and Einfalt, 1998) and flow data (Pilloy, 1989; Berthier et al., 1998; Auvray et al., 1999; Mpe A Guilikeng and Fotoohi, 2000; Wyss, 2000). In most of these cases, validation was carried out manually without or with very limited automatic assistance. Increasingly elaborated and automated procedures were used in more recent cases reported by

Piatyszek et al., 2000, Mourot et al., 2001, and Piatyszek and Joannis, 2001. Unfortunately, there are very few publications of data validation in hydrology providing sufficient details to allow the exact methodology to be followed. There is even less information regarding validation of measurements of pollutant concentrations (e.g. Joannis et al., 2006; Mohamed-Faouzi et al., 2006; Ciavatta et al., 2004; Pressl et al., 2004; Winkler et al., 2004).

8.4.4 Development of an automatic pre-validation method

Data validation establishes if a measurement is good or bad, i.e. valid or invalid for further uses. This decision is made using several criteria based on information available from prior measurements, other sources of knowledge and/or expertise. This information can be related to (i) the values themselves, (ii) the sensors (iii) the measurement environment and context or (iv) a combination of any of these elements. Therefore, the operator should successively (i) analyse the available information, (ii) extract the relevant information for a particular case, and (iii) adapt the information for the case of interest.

The automatic pre-validation method developed for the OTHU Project contains a set of seven parametric tests, each one corresponding to a validation criterion. Following the test, each data point in the series receives an elementary mark (i.e. *a data quality indicator*). The elementary marks used in the pre-validation phase are labelled as: 'A' for reliable values, 'B' for doubtful values and 'C' for defective, outlying or aberrant ones. Alternatively, Mpe A Guilikeng and Fotoohi (2000) use graduated marks ranging from 0 to 1 (or from 0 to 100%) to refine the resolution of the data quality estimation. It is considered, however, that the use of the graduated marks in hydrologic application is unnecessary because in many instances such quality indicators are difficult to interpret. For example, when referring to a measurement of mass or volume balances, how can one assess a measurement with a data validity indicator of 46%? Whatever application is envisioned in hydrology, that is, monitoring, design, calibration or verification of models and other operational purposes, it suffices if the data are declared valid or not (a binary decision). This is the reason why only three marks have been chosen for the OTHU Project's automatic pre-validation.

The final validation, carried out manually by the operator, should use only two marks: valid (A) or not valid (C). Missing or invalid data should be replaced by estimates obtained in a careful manner. Replacements should be (i) kept to the minimum possible, (ii) fully documented, (iii) reversible, and (iv) made only if absolutely necessary. Replacement values may give the false perception that a monitoring programme is in good order. The need to replace large amount of data is highly indicative of a defective system that should be subjected to a validation diagnostic. The overall measurement objective is not to fill the gaps in order to provide continuous data series that might appear complete and accurate, but to provide representative, honest, real and faithful data representing the observed phenomena and indicating the quality of the data acquisition process itself.

8.4.5 Validation criteria

The automatic parametric tests used for pre-validation consist of the following seven criteria.

8.4.5.1 Criterion 1: Status of the sensor

The sensor can be in one of the following modes: *on*, *off*, or *hold*. During this test the mode of the sensor connected to the data logger is checked (to determine if representative measurements are being collected, or if the sensor is on 'hold', and continues to collect measurements during maintenance, checking or verification). For this purpose, a special electronic box has been built that generates a specific 4 mA to 20 mA signal for each sensor during actions that affect or disturb measurements. The signal is enabled manually by the operator at the beginning of the action, usually maintenance and cleaning of the measurement devices and sensors, and disabled after re-establishment of normal operating conditions. The duration of abnormal functioning can be exactly detected and recorded: the corresponding values from the concerned sensor are marked C, otherwise the mark is A. An example for maintenance is given in Figure 8.1. Before maintenance, the box delivers a 4 mA signal. At the beginning of the periodic maintenance at 11.30 h, the operator switches the appropriate button on the box, which results in the output of 10 mA. The value in mA is different for each sensor under maintenance. At the end of the maintenance at 11.35, the operator switches again the button, and the output is back to 4 mA. During the automatic pre-validation, the period between 11.30 and 11.35 will be automatically detected and the invalid recorded data will be marked 'C'. While 'C' denotes invalid data, additional information should be recorded in a log-book and included in the metadata for further analysis (see Chapter 9). This method can also be used to indicate periods during which the sensor is not usable, i.e. during repair, replacement or any other kind of failure. Other applications of the method can be envisaged according to the local context and needs.

8.4.5.2 Criterion 2: Possible range for the measurement context

The measurement range is sensor- and location-specific and establishes the intervals within which measurements should be expected for a given measurement context. The range is defined by the sensor measuring range and/or by the physical conditions. For example, pH values must lie in the range between 0 and 14; the temperature in Celsius

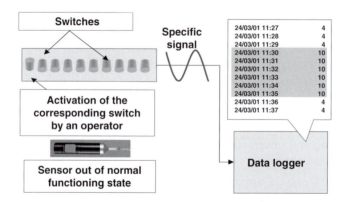

Figure 8.1 Use of electronic signal to indicate the state of the sensor. The change in output between 11.30 and 11.35 allows automatic detection of the 'in maintenance' state of the sensor during this period (see also colour Plate 9)

degrees cannot be negative in flowing wastewater; the water level after offset correction cannot be less than zero in a sewer pipe; and a water level sensor with a measuring range of 0 to 2 m will not give values out of this range. This very robust test allows detection of physically anomalous values. If a value is outside its possible physical range, it is marked 'C'. Otherwise, it is marked 'A'. An example is shown in Figure 8.2 for a flow velocity sensor. It can be noted in the figure that around 09:00 a.m., zero or negative velocities are recorded; hence they will be marked as invalid.

8.4.5.3 Criterion 3: Range of most frequently observed values

This is the measurement range within which a specific sensor records data at a specific site during normal flow and operating conditions. The limits of the range are set and adjusted gradually based on the range observed in previously acquired data and relevant information and knowledge collected from other similar measurement sites. For example, statistical frequencies of observed values may be used to define the local likely range as within the 95% or 99% confidence interval. The local likely range is less than or equal to the physical range defined for Criterion 2 above. For example, with a probability of 99%, the water temperature lies between 4°C and 20°C in a given domestic wastewater sewer and the pH is between 6 and 8. Any value outside the range of most frequently observed values is considered as doubtful and marked 'B', otherwise, it is marked 'A'.

8.4.5.4 Criterion 4: Maintenance periodicity

In order to ensure the reliability of measured values, all sensors should be periodically cleaned and maintained. The maintenance periodicity, T_e (in days), depends on both the sensor and its local functioning conditions. The date of the last maintenance, T_0, is recorded. Any value measured between T_0 and $T_0 + (T_e + 1)$ is marked A. If measured between $T_0 + (T_e + 1)$ and $T_0 + (2T_e + 1)$, it is marked B. The mark is C if the time elapsed since T_0 is beyond $T_0 + (2T_e + 1)$ days. Implementation of this criterion is illustrated in Figure 8.3.

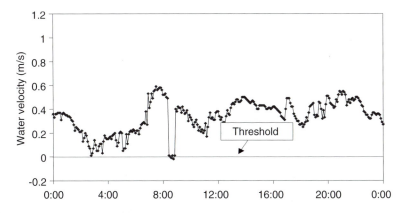

Figure 8.2 **Example of recorded flow velocities outside the defined possible range of the measurement (see also colour Plate 10)**

8.4.5.5 Criterion 5: Signal gradient

A sudden or erratic increase or decrease in measurements generate steep gradients that are not expected for the measured processes in the local environment and conditions, and therefore must be suspected as measurement system malfunctioning. Various methods were tested for detecting such behaviour, based on simple thresholds on absolute and relative variations, but the results were not entirely satisfactory. In order to illustrate this conclusion, consider the measurement of water level in a combined sewer. If the minimum and maximum values of the signal gradient are too close (narrow interval), false detection will occur during storm events when large but real signal gradients occur. On the contrary, if a larger interval is set, there is a risk that abnormal gradients during dry weather periods will not be detected. The relative gradient (i.e. the gradient divided by the absolute value of the signal) is also not satisfactory because, for low amplitude values, a small variation may lead to high values of the relative gradient. Thus, the choice of the appropriate interval is a difficult task. It is, however, possible to define simple gradient tests adapted to specific contexts, such as one threshold value for dry weather, another for wet weather. As mentioned, this approach needs to be carefully adapted for the specific measurement context.

A specific method combining a filtering algorithm and the detection of outliers by threshold assessment has been developed to circumvent these limitations. It is obvious that sudden changes in the behaviour of a system (usually due to a sensor fault or due to the presence of a non-representative phenomenon) generate high gradients. By filtering the signal using a moving average, high gradients are smoothed. The difference between the original signal and the filtered signal is a pertinent indicator of sudden changes and abnormally high gradients. Setting both threshold values for fault detection and the appropriate width of the window (e.g. from some minutes for water depth and flow velocity up to some days or weeks for rainfall volume) used in the averaging filter is always specific for each measurement site and each sensor used. These factors are determined by successive testing (trial and error) to take into account the background noise of the signal, the measurement uncertainties and the most frequently observed values. The aim of this iterative process is to reach an efficient detection sensitivity (i.e. all abnormal gradients should be detected) but without false detections

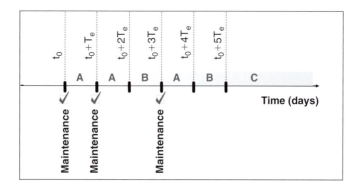

Figure 8.3 Recommended marking of maintenance operations data (see also colour Plate 11)

(i.e. no normal gradient should be marked as abnormal). Doubtful values are marked B, otherwise they are marked A.

Given x_i, the original signal, one calculates successively:

$yi = \Sigma_{j=-m}^{m} x_{i+j}/N$ the smoothed signal by central moving average on N successive values; and

$\varepsilon_i = x_i - y_i$ the residuals, which are usually normally distributed with a mean of zero.

The Page-Hinkley test (Ragot et al., 1990; see also the next section) is then applied to detect abnormal residues that correspond themselves to abnormal signal gradients. Figure 8.4 illustrates the results obtained with measurements of the water level in a sewer pipe. In order to facilitate the detection of such abnormalities, it is also possible to use the square of the residuals as a defect indicator. Other tests may also be used, such as the variance test applied to the residuals.

8.4.5.6 Criterion 6: Sensor redundancy

When redundant sensors are available at a given site, their values and the signal dynamics are compared in order to detect unusual data trends or gaps. This comparison has limited relevance if the sensors have not been rigorously calibrated and systematic errors have not been previously corrected for. If proper calibration has been done, both sensors should give the same values with a maximum acceptable gap given by their respective uncertainty confidence intervals. If the confidence intervals of both signals are not overlapping, the difference between the two signals is used as an indicator to detect diverging values. The analysis of the difference can be carried out by means of threshold values: when the difference between the two signals exceeds a predefined limit, the corresponding data are declared as doubtful and their values shall be later analysed manually by the operator. This approach is explained below.

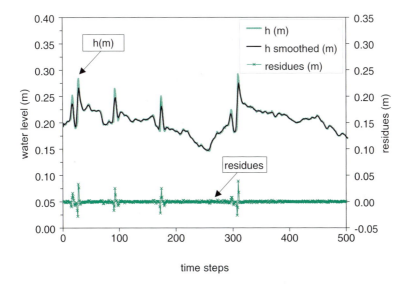

Figure 8.4 Example of the detection of abnormal gradients

When redundant data are available, the individual results and the uncertainty intervals (if available) should be compared with each other. Consider two results $\bar{A}$ and $\bar{B}$ and their respective standard uncertainties $u(A)$ and $u(B)$ corresponding to 95% confidence intervals (see Chapter 6): $\bar{A} \pm 2u(A) = \bar{A} \pm U_A$ and $\bar{B} \pm 2u(B) = \bar{B} \pm U_B$. Validation of the measurement occurs if

$$|E| < U \qquad \qquad 8.1$$

Where:

$$E = \bar{B} - \bar{A} \quad \text{and} \qquad \qquad 8.2$$

$$U = 2\sqrt{u(A)^2 + u(B)^2} = \sqrt{U_A^2 + U_B^2} \qquad \qquad 8.3$$

For example, assume that measured and benchmark results are available over a range of variation for the independent variable, X_i, as illustrated in Figure 8.5. According to Equation 8.1, for both measurements and both uncertainty intervals to be correct, the true value must lie in the region where the uncertainty intervals overlap. The larger the overlap, the more confidence we have in the validity of the measurements. As the difference between the two measurements increases, the overlap shrinks. When there is no overlap between uncertainty intervals, a problem clearly exists and at least one of the two measurements is wrong. Values should be marked B. Either uncertainty intervals have been grossly underestimated, an error exists in the measurements, or the true value is not constant. Investigations to identify bad readings, overlooked or underestimated bias or other problems are necessary to resolve the discrepancy.

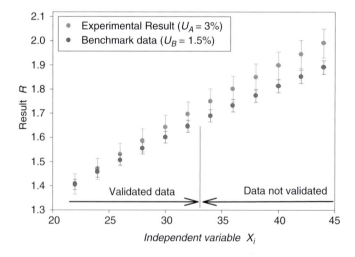

Figure 8.5 Validation of experimental results for redundant trail (see also colour Plate 12)

Other detection methods can be used, such as the Page-Hinkley algorithm (Brunet et al., 1990), as discussed below. The use of only two sensors can lead to the detection of doubtful values. However, in case no additional and contextual information is available, it may be very difficult to decide which value among the two should be rejected. From this point of view, three redundant sensors compose a more effective system (e.g. Menéndez Martinez et al., 2001) regarding flow measurements in drinking water networks.

Given $S1_i$, the values of the first sensor, and $S2_i$, those of the second one, the residuals $\varepsilon_i = S1_i - S2_i$ are calculated (see Figure 8.6). In case no defect is present, the ε_i values are normally distributed with a mean value $\mu_0 = 0$. In order to detect any variation of μ_0 with an amplitude greater than Δ, two tests are applied: the first one to detect increasing μ_0, the second one to detect decreasing μ_0. With λ the detection level and n the test memory size, one calculates:

$$\text{test 1: let } U_n = \sum_{k=1}^{n} \left(\varepsilon_k - \mu_0 - \frac{\Delta}{2} \right) \text{ with } U_0 = 0 \quad \text{and} \quad m_n = \min_{0 \le k \le n} U_k.$$

An abnormal positive gap is detected when $U_n - m_n > \lambda$.

$$\text{test 2: let } V_n = \sum_{k=1}^{n} \left(\varepsilon_k - \mu_0 + \frac{\Delta}{2} \right) \text{ with } V_0 = 0 \quad \text{and} \quad M_n = \max_{0 \le k \le n} V_k.$$

An abnormal negative gap is detected when $M_n - V_n > \lambda$.

The test is reinitialized after each detection; Δ and λ are to be set according to (i) the uncertainties of the two sensors, and (ii) non detection and false detection rates. According to Ragot et al. (1990), the initial value of λ can be set as $\lambda = 2h/p$ where $h = 2$ for normal distributions and p is the value of Δ expressed as the number standard deviations of the signal. The choice of appropriate values for Δ and λ is a key

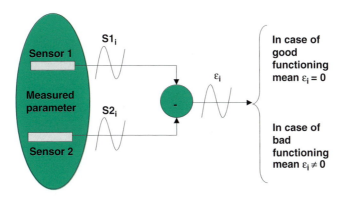

Figure 8.6 Sensor redundancy in a duplex system

factor to obtain a high performance level for the test. As an example, a high value of λ may miss the defects while a low value of λ may lead to detection of false defects. The values detected by the test are marked 'B'. Otherwise they are marked 'A'.

It should be noted here that the above test performs well only if the redundant sensors have similar dynamic responses. If one sensor has higher sensitivity, its data should be passed through a low pass filter prior to implementing the test. This aspect might be involved in the selection of the instrument for the monitoring programme (see Chapter 7).

8.4.5.7 Criterion 7: Signal/information redundancy

This method is in principle similar to the previous one, except the comparison is applied to data recordings of correlated variables. The provenance of the compared variables may be the same monitoring system (e.g. water level and flow velocity in a given sewer, rainfall intensity and flow rate, etc.) or alternative sources of information (e.g. measured conductivity and calculated flow rate in a sewer during storm events, measured velocity and velocity estimated from the water depth, etc.). The measurement uncertainties should of course be taken into account in all calculations. This test is also applicable for global validation when various signals are compared.

The measured signal $S1$ is compared to another equivalent signal $S2$ calculated by means of a model and other signals S_i (see Figure 8.7). The model can be of any type: simple correlation, multiple regression, empirical, statistical, conceptual or fully physically based models. When a model is used, its intrinsic response time and memory should be analysed first to avoid any misinterpretation and any bias in the application of the test. The model may be considered as a virtual sensor delivering the signal $S2$. (see example below in Section 8.4.7.4 Failure of a flow velocity sensor). All detected values are marked B, otherwise A.

Now consider the case where both sensor and signal redundancy tests are applied. If one value has been marked A for sensor redundancy and B for signal redundancy, the pre-validation global mark is B and further analysis is needed to make the final

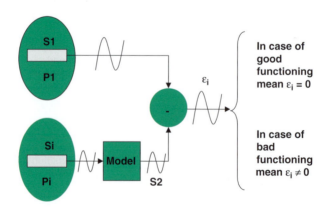

Figure 8.7 Signal redundancy in a duplex system

validation. Indeed, it may happen that both sensors give the same value (sensor redundancy) but that the phenomenon is not coherent with the expected value (signal redundancy). In other words, both sensors give coherent information, but this information does not correspond to the knowledge included in the model.

Reciprocally, if a value delivered by the Sensor 1 is marked B for sensor redundancy after comparison with the Sensor 2, and marked A for signal redundancy, and if all other elementary marks are A for the Sensor 1, the global mark B for the Sensor 1 can be replaced by A as the defect may be attributed to Sensor 2.

8.4.6 Pre-validation assessment

As the application of the above parametric tests depends on both the available information and the nature of the measured variables (e.g. rainfall intensity does not use the same validation procedures as those for water level), the seven tests can not be applied uniformly for all variables. When a test is not applied, the corresponding mark is 'N'.

Before applying the tests, the information required to set the test parameters should be available. A significant initial effort should be devoted to collect this information and to calibrate the tests specifically for each sensor and each location.

After the set of seven tests (or any appropriate sub-set) has been applied, a global mark for each value should be established: the global mark is equal to the lowest elementary mark given by one of the seven tests, thus:

- If all elementary marks are A, the global mark is A
- If at least one elementary mark is B, the global mark is B
- If at least one elementary mark is C, the global mark is C.

The ASCII files issuing from this automatic pre-validation procedure appear as shown in Figure 8.8.

The global marks are then used to automatically extract doubtful and false data (i.e. values with marks B and C) for final manual validation. The operator's tasks are considerably simplified, by concentrating their focus only on the data marked with B and C by the automatic pre-validation procedure. The operator can be assisted by automatic visualization tools offering a set of graphs for displaying synthetically the data and the information used for pre-validation. This tool should be associated with a knowledge database containing the characteristics of the most frequent defects and failures and their possible causes. In addition, some expert rules for analysing the marks attributed to each value should be formulated to facilitate the analysis and the interpretation of the

```
conductimetre yokogawa
mesures validées
Date-Heure          ;cond-µS/cm ;Limite inf ;Limite sup  ;notes            ;note globale
19/03/01 00:00      ;408.942    ;405.942    ;411.942      ;N A B N A N N    ;B
19/03/01 00:02      ;410.773    ;407.773    ;413.773      ;N A B N A N N    ;B
19/03/01 00:04      ;410.773    ;407.773    ;413.773      ;N A B N B N N    ;B
```

Figure 8.8 Example of ASCII file resulting from the automatic pre-validation procedure

detected defects and help the operator to meaningfully replace doubtful values when it is both feasible and necessary.

Prototype software named DAVE (Data Validation Engine) has been coded in Visual Basic® and implemented for validation implementation in the OTHU Project. Of course, other research and commercial software exist for this purpose, which are more or less flexible and adaptable to specific needs. Validation tools may also be developed and implemented at the request of specialized consulting companies. It is important to note that specifications of such software must be defined by (or at least in close collaboration with) the final user, to fully account for the specificity of the site, sensors, and monitoring objectives.

8.4.7 Examples of application

The above method and procedures have been tested and validated for two types of data: rainfall intensity and flow rate (water depth and flow velocity measured simultaneously in a sewer system). The efficiency of the method has been evaluated as satisfactory according to the available information and to the results obtained. Different types of defects and doubtful values were automatically detected.

8.4.7.1 Abnormal rainfall intensity

For rainfall intensities, local likely range and signal gradient tests facilitate detection of all abnormal values. An example is given in Figure 8.9, where abnormal brief and high intensities have been detected. A careful evaluation of the process led to the conclusion that the measured values are not reasonable for the measurement context. After comparison with other close rain gauges, it appeared this rain gauge was the only one which recorded precipitation with very high intensities. Consequently, the final mark was C.

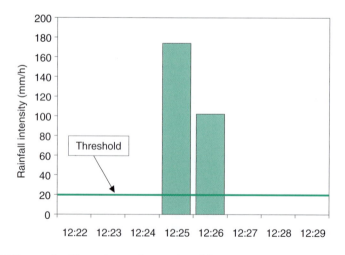

Figure 8.9 Rainfall recording illustrating an abnormal rainfall event

8.4.7.2 Erratic behaviour of a sensor

Figure 8.10 presents a case of erratic behaviour of a piezoresistive sensor. The zones at the top of the graph identify the periods where values are doubtful and receive the pre-validation global mark B. Firstly, the local likely range test has marked B all values higher than 0.96 m. Secondly, all other defects have been detected and marked B by the signal gradient test and by the signal redundancy test using the flow velocity as the correlated variable. This kind of defect is easily detected by the signal gradient test, where the original signal has been smoothed using moving average with a window of five values. This combination of various tests is very valuable to reinforce conclusions and marking. The signal shows large random fluctuations. The signal redundancy test reveals that the sensor is defective and that the measured values do not correspond to any real phenomenon. However, the lack of additional information about the context in this sewer during this period prevents any diagnostic of the causes of the defect.

8.4.7.3 Abnormal phenomenon

Figure 8.11 shows an unexpected cyclic behaviour of the water level in a sewer pipe. This phenomenon was due to the tests of a flushing gate made far upstream (approximately 500 m) of the monitoring site. However, the existence of these flushing experiments was unknown, both during the measurements and during the automatic pre-validation. Consequently, the signal gradient test marked B the values at the beginning of each flushing cycle. Some weeks later, when the origin of these cycles was eventually identified, the complete series has been manually validated and marked A by the operator.

8.4.7.4 Failure of a flow velocity sensor

In this example, a series of flow velocity data has been subjected to six tests (sensor redundancy has not been applied). For five of the tests, all marks were A. Only the signal redun-

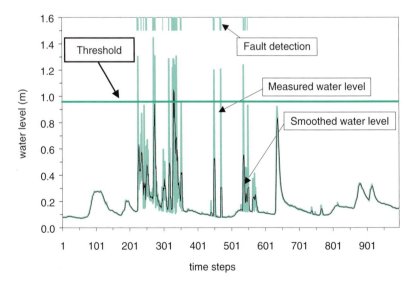

Figure 8.10 Hydrograph illustrating erratic behaviour of a piezoresistive sensor

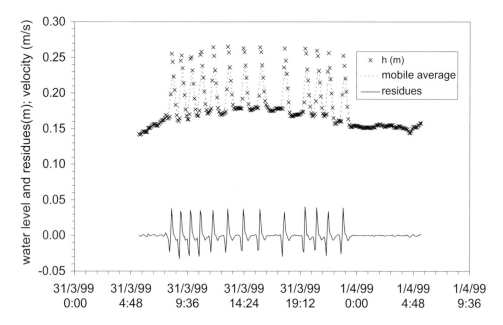

Figure 8.11 Record illustrating unexpected cyclic variation in the water level in a sewer pipe

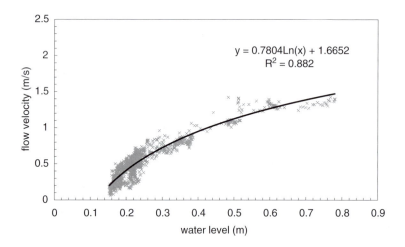

Figure 8.12 Empirical model with flow velocity as a function of the water level

dancy test indicated possible errors in the flow velocity signal when comparison was made with the water level signal. A previous analysis of water level and flow velocity data was used to establish a simple empirical model calculating the flow velocity as a function of the water level, when hydraulic conditions are normal and without any disturbance such as occurs with a backwater effect. This model is illustrated in Figure 8.12.

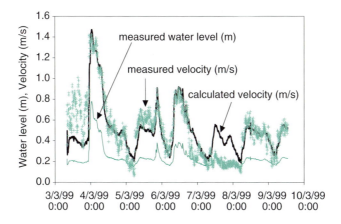

Figure 8.13 Hydrograph illustrating the failure of a water velocity sensor

Source: Data from Bernard, 2000.

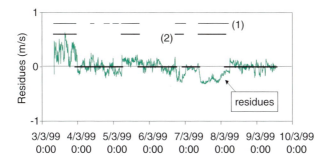

Figure 8.14 Detection of defects with two sets of parameters: (a) $\Delta = 0.1$ and $\lambda = 2$; (b) $\Delta = 0.1$ and $\lambda = 4$

The data series are represented on Figure 8.13. The residuals of the signal redundancy test are represented in Figure 8.14, with two sets of parameters Δ and λ. The standard deviation of the residuals is approximately 0.05. It appears that the differences between measured and calculated velocities, observable in Figure 8.13, are well detected by the test as shown by the corresponding residuals in Figure 8.14. Without any additional information, the pre-validation procedure would have been not able to find the cause of this defect. During the manual final validation, after comparison with other data series showing similar daily patterns, the operator has been able to conclude that the velocity sensor was defective.

8.4.7.5 Other types of problems

Many other, sometimes less obviously, doubtful and irregular values have also been detected by the pre-validation tests. An example is provided in Figure 8.15, where drifts are detected in the sensor or signal redundancy. Some defects have been interpreted in

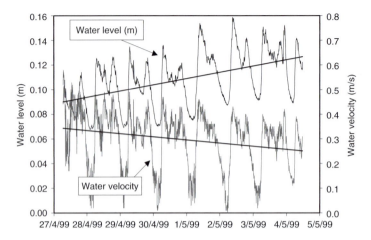

Figure 8.15 Hydrograph illustrating the drift of water level and flow velocity sensors (see also colour Plate 13)

the diagnostic phase and their causes have been elucidated and explained, but others remain unexplained due to a lack of information.

The methodology and the tests used for the automatic pre-validation of data in the OTHU Project are relatively simple. Nevertheless, they are efficiently detecting most defects, outliers, and irregular or doubtful values and gradients in voluminous data series. The strength of this methodology lies in the application of a large set of complementary parametric tests based on and calibrated with site and sensor specific information. The application of several tests also leads to a preliminary automatic diagnostic of the measurements, due to the fact that each test corresponds to a specific behaviour of the signal. It appears also that contextual information, such as the state of the sewer system, sewer maintenance and works, activities in the catchment, accidental spills, is absolutely necessary to improve the data validation process and the long-term reliability of the recorded values. In the initial phase, the parameters of the tests are set in such a way that they lead to a high rate of detection, even false detection. In subsequent refinement stages, the original validation parameters are progressively adjusted, reflecting the accumulation of knowledge about the specific monitoring system. Such a phased approach leads to attaining satisfactory and rational compromise in assessing valid, missing and false data.

8.5 GLOBAL LEVEL VALIDATION

In principle, global level validation is a kind of generalization of the signal redundancy test (i.e. Criterion 7 in the pre-validation). As an example, information at a given point in an urban sewer system is often highly correlated or even redundant with information obtained at other upstream or downstream points. Naturally, it is expected to observe time lags, data attenuation, amplification, and cumulative effects (mass balances) that can be predicted by numerical models.

The preference is for fast, simple, and robust models. Frequently, simple regression models are very convenient to identify, calibrate and update when new reliable data are available. As an example, Hamioud et al. (2005) have developed a method based on regression models and on 'fuzzy logic' to interpret the observed residuals. The method allows detection of the defective sensor among a set of sensors installed at different locations within an urban water system. A similar approach to the one described for the OTHU Project is presented by Hill and Minsker (2006). They used two strategies for detecting anomalous data: anomaly detection (AD) and anomaly detection and mitigation (ADAM). The AD strategy simply uses the previous measurements for future predictions, whether or not they were classified as anomalous, while the ADAM strategy replaces anomalous measurements with model predictions before making future predictions. The authors used four data-driven methods to create prediction models: naïve, clustering, perceptron, and artificial neural networks. Data-driven methods like these develop models using sets of training examples. Each example contains a feature set (i.e. the set of variables used to make the prediction) and a target output. A set of 30,000 training examples collected by a windspeed sensor were used by the authors to develop their validation models.

8.6 DATA VALIDATION AND PERFORMANCE INDICATORS

The increased use of real-time monitoring systems for supporting decision making process further reinforces the need for developing and implementing performance indicators for providing good quality data. In such systems there is little time to verify data quality, therefore methods for quickly and accurately identifying faults in the data collection process is vital. Many performance indicators may be calculated to evaluate the quality of the data and of their validation. The most obvious is the percentage of valid data. Sensor faults are determined by identifying anomalous measurements in the data stream, where anomalous measurements are operationally defined as measurements that fall outside of the bounds of an established prediction interval, usually either a 95% or a 99% prediction interval.

As an example, in France some municipalities have sub-contracted monitoring programmes. The payment of the sub-contractor is partly based on the percentage of valid data, or validity percentage. The minimum validity percentage may be set (e.g. to 90%). If the real percentage is higher than 90%, the sub-contractor is paid more; if it is less than 90%, the sub-contractor is paid less. Such an approach could be attractive, but the municipality should check its sub-contractor who will have a 'natural tendency' to have a high validity percentage. In other words, the validity percentage shall be submitted itself to a validation procedure controlled by the municipality. It is also important to monitor the validity percentage evolution over observational time (i.e. to determine if there is any increase, constancy or decreases of the monitoring programme quality). If the validity percentage does not change, there may still be interesting or important 'underlying' changes occurring. For example, the percentage of not valid data due to sensor failure may decrease, while the percentage of not valid data due to abnormal phenomenon increases.

Beyond the detection phase, the diagnostic phase should be monitored by means of performance indicators. Each performance indicator should be defined as the number or percentage of defects due to specific reasons, such as sensor failure; human errors in programming, maintenance; clogging, fouling, and other environmental problems; drift of calibration, and other abnormal phenomena. The percentage of unexplained defects is a very pertinent indicator for tracking progress in the diagnostic phase.

8.7 VALIDATION OF EXTERNAL LABORATORY ANALYSIS

When all aspects of monitoring programmes (from definition of objectives to information and report, including measurements themselves) are under the responsibility of one single organization, one may assume (and this must be checked) that ensuring the requested level of data reliability and quality along the whole chain of actions and metrology should be feasible if appropriate and necessary means are provided. However, frequently there some parts of a monitoring programme which are sub-contracted to external organizations. *In this case, it is crucial that all involved organizations use uniform rules for calibrating and maintaining the sensors, for evaluation of the uncertainties and validation of the data, and that this if overseen by the final user(s) of the data.*

This is particularly important for laboratories carrying out physical, chemical and biological analyses of samples of water, soil, sediments, fauna and flora, a large part of the activities and costs of urban water systems monitoring. According to the authors' own experience, and despite the fact that most laboratories are certified and/or accredited by various institutions, results are frequently given without the necessary information, such as uncertainties, exact dates of sample receipt, conditioning and analysis, and artefacts due to preservation, all information essential to the validation process. Frequently also, results are given only in printed form, instead of electronic files, which may contribute to error if and when the values have to be entered again for further use.

'Probably the most significant omission in requirements for analytical technologies is a requirement for analytical quality control (AQC). Increasingly AQC is being recognized as essential for data from monitoring programmes to be reliable and comparable' (Nixon et al., 1996, p. 47). The above observation was written in 1996 by Nixon et al. on requirements for water monitoring in Europe, and much progress has obviously been made since that time. Nevertheless, compared to other fields (such as food industry, metallurgy and chemical industries, etc.), measurements and analyses in urban water management have not reached the same level of quality and reliability. Quevauviller (2002; 2006) gives very valuable information and proposals (mainly regarding the use of certified reference materials to be used as internationally recognized standards) to improve the quality of analytical results. This topic is beyond the scope of this document, but organizations in charge of urban water systems monitoring should devote attention to these questions. Indeed, analyses of samples represent very high operation costs and it is absolutely crucial that the quality and reliability of these analyses should be guaranteed. It perhaps bears repeating that the quality of decisions made on the basis of monitoring data will only be as good as the data itself.

REFERENCES

Auchet, P. and Hammouda, A. 1995. Pluviométrie: validation et utilization des données pluviométriques à Nancy. *Techniques–Sciences–Méthodes*, Vol. 11, pp. 831–32.

Aumond, M. and Joannis, C. 2006. Mesure en continu de la turbidité sur un réseau séparatif d'eaux usées: mise en oeuvre et premiers résultats. *La Houille Blanche*, Vol. 4, pp. 121–28.

Auvray, J., Urvoy, Y., Mirandola, I., Morin, E., Renaud, M. and Bousquet, J.-P. 1999. Une expérience de diagnostic permanent sur le réseau d'assainissement de Vitry-le-François. *Techniques–Sciences–Méthodes*, Vol. 12, pp. 70–77.

Barnett, V. and Lewis, T. 1990. *Outliers in Statistical Data*, 3rd edn. New York, John Wiley and Sons.

Bennis, S., Berrada, F. and Kang, N. 1997. Improving single variable and multivariable techniques for estimating missing hydrological data. *Journal of Hydrology*, Vol. 191, No. 1–4, pp. 87–105.

Bernard, N. 2000. Simulation des impacts sur les eaux souterraines de l'infiltration des eaux pluviales en milieu urbanise – application au cas de l'aquifere de l'est lyonnais. Ph.D. thesis, INSA, Lyon, France.

Berthier, E., Auzizeau, J., Fasquel, M., Flahaut, B., Rouaud, J.-M. and Andrieu, H. 1998. Le suivi hydrologique de bassins versants expérimentaux en milieu urbanisé. *Bulletin de Liaison des Laboratoires des Ponts et Chaussées*, Vol. 218, pp. 59–75.

Bertrand-Krajewski, J.-L., Barraud, S. and Chocat, B. 2000. La mesure de l'impact environnemental des systèmes d'assainissement: exemple de l'observatoire de terrain en hydrologie urbaine *(OTHU)*. *Actes du Troisieme Congrès Universitaire de Génie Civil*, 27–28 June 2000, Association de Génie Civil Universitaire, Lyon, France, pp. 35–42.

Bertrand-Krajewski, J.-L. and Bardin, J.-P. 2001. Evaluation of uncertainties in urban hydrology: application to volumes and pollutant loads measured in a storage and settling tank. *Proceedings of the ICA 2001 Conference on Instrumentation, Control and Automation, 3–7 June 2001, Malmö, Sweden*. International Water Association, pp. 605–12.

Brunet, J., Labarrère, M., Jaume D., Rault A. and Vergé M. 1990. *Détection et diagnostic de pannes*. Paris, Hermès.

Ciavatta, S., Pastres, R., Lin, Z., Beck, M.B., Badetti, C. and Ferrari, G. 2004. Fault detection in a real time monitoring network for water quality in the lagoon of Venice. *Water Science and Technology*, Vol. 50, No. 11, pp. 51–58.

Gilbert, R.O. 1987. *Statistical Methods for Environmental Pollution Monitoring*. New York, Van Nostrand Reinhold Company.

Hamioud, F., Joannis, C. and Ragot, J. 2005. Fault diagnosis for validation of hydrometric data collected from sewer networks. Paper presented at the Tenth International Conference on Urban Drainage, 21–26 August 2005, Copenhagen, Denmark.

Hill, D.J. and Minsker, B. 2006. Automated fault detection: preparing real-time data for adaptive management. Paper presented at AWRA Summer Speciality Conference on Adaptive Management of Water Resources, 26–28 June 2006, American Water Resources Association, Missoula, Montana, USA.

Joannis, C., Aumond, M., Rufflé, S. and Cohen-Solal, F. 2006. Détection et diagnostic d'anomalies affectant des résultats de mesures en réseaux d'assainissement. *Actes du Séminaire GEMCEA Validation de Résultats de Mesure en Continu Issus de Réseaux de Surveillance*, 24 March 2006, Paris, France.

Jörgensen, H.K., Rosenörn, S., Madsen, H. and Mikkelsen, P.S. 1998. Quality control of rain data used for urban runoff systems. *Water Science and Technology*, Vol. 37, No. 11, pp. 113–20.

Lupin, S. 1990. Archivage et banque de données pluies: l'expérience de la COURLY. *Actes de la conférence Métrologie en Assainissement Pluvial Urbain*, 24–26 April 1990, Nancy, France.

Maul-Kötter, B. and Einfalt, T. 1998. Correction and preparation of continuous measured raingauge data: a standard method in North Rhine-Westphalia. *Water Science and Technology*, Vol. 37, No. 11, pp. 155–62.

Menéndez Martinez, A., Biscarri Trivino, F., Sanchez Gomez, A. B., Gomez Gutierrez, A. A. and Castaño Rubiano, M. 2001. Uncertainty and redundancy in flow metrology. *Actes de la Conférence A&E 2001 Automatique et Environnement*, 4–6 July 2001, Saint-Etienne, France.

Mohamed-Faouzi, H., Mourot, G. and Ragot, J. 2006. Diagnostic de fonctionnement de capteurs d'un réseau de surveillance de la qualité de l'air par analyse en composantes principales. *Actes du Séminaire GEMCEA Validation de Résultats de Mesure en Continu Issus de Réseaux de Surveillance*, 24 March 2006, Groupement pour l'Evaluation des Mesures en Continu dans les Eaux et en Assainissement, Paris, France.

Mourad, M. and Bertrand-Krajewski, J.-L. 2002. A method for automatic validation of long time series of data in urban hydrology. *Water Science and Technology*, Vol. 45, No. 4–5, pp. 263–270.

———. 2003. Pré-validation automatique de données environnementales en hydrologie urbaine. Paper presented at Colloque A&E 2001 Automatique et Environnement, 5–6 July 2001, Saint-Etienne, France.

Mourot, G., Maquin, D. and Ragot, J. 2001. Validation de mesures en pluviométrie. Paper presented at Conférence A&E 2001 Automatique et Environnement, 4–6 July 2001, Saint-Etienne, France.

Mpe A Guilikeng, A. and Fotoohi, F. 2000. Traitements de données pour la surveillance des réseaux d'assainissement: exemple de Dijon. *Actes de la Conférence SHF–GRAIE Autosurveillance et Mesures en Réseau d'Assainissement*, 5–6 December 2000, Lyon, France, pp. 225–32.

Nixon, S.C., Rees, Y.J., Gendebien, A. and Ashley, S.J. 1996. *Requirements for Water Monitoring*. Copenhagen, European Environment Agency (EEA).

Piatyszek, E., Voignier, P. and Graillot, D. 2000. Fault detection on a sewer network by a combination of a Kalman filter and a binary sequential probability ratio test. *Journal of Hydrology*, Vol. 230, pp. 258–68.

Piatyszek, E. and Joannis, C. 2001. Méthode de validation de données débitmétriques de temps sec en réseau d'assainissement. Paper presented to Conférence A&E 2001 Automatique et Environnement, 4–6 July 2001, Saint-Etienne, France.

Pilloy, J.-C. 1989. La débitmétrie en collecteurs, critique et validation de données. *Bulletin de Liaison des Laboratoires des Ponts et Chaussées*, Vol. 163, pp. 83–91.

Pressl, A., Winkler, S. and Gruber, G. 2004. In-line river monitoring – new challenges and opportunities. *Water Science and Technology*, Vol. 50, No. 11, pp. 67–72.

Quevauviller, P. 2002. *Matériaux de Référence pour l'Environnement*. Paris, Editions Tec et Doc.

———. 2006. *Métrologie en Chimie de l'Environnement*, 2nd edn. Paris, Editions Tec et Doc.

Ragot, J., Darouach, M., Maquin, D. and Bloch, G. 1990. *Validation de Données et Diagnostic*. Paris, Hermès.

Soukatchoff, V. 1990. Critique – validation des données pluies. *Actes de la conférence Métrologie en Assainissement Pluvial Urbain*, 24–26 April 1990, Nancy, France.

Winkler, S., Rieger, L., Saracevic, E., Pressl, A. and Gruber, G. 2004. Application of ion-selective sensors in water quality monitoring. *Water Science and Technology*, Vol. 50, No. 11, pp. 105–14.

Wyss, A. 2000. Traitement et validation des données de débits et de pluie pour le diagnostic des réseaux d'assainissement: exemple de Genève. *Actes de la Conférence SHF–GRAIE Autosurveillance et Mesures en Réseau d'Assainissement*', 5–6 December 2000, Societe Hydrotechnique de France (SHF) et Group de Recherche Rhône-Alpes sur les Infrastructures et l'Eau (GRAIE), Lyon, France. pp. 233–40.

Chapter 9

Data handling and storage

D. Prodanović

Institute for Hydraulic and Environmental Engineering, Faculty of Civil Engineering, University of Belgrade, Bulevar Kralja Aleksandra 73, 11000 Belgrade, Serbia

9.1 INTRODUCTION

Integrated urban water management is likely to involve the collection of vast amounts of data, from many different sources. To achieve true integration, the accumulated data must be able to be easily shared among different users. This chapter therefore discusses:

- portability of data formats
- data application for different models
- lifetime of data and storage systems
- data security.

Firstly, a short historical overview is given, followed by an explanation of typical data flow processes in integrated urban water management (IUWM) systems. Since the database management system is at the core of data handling and storage, a short introduction into database concepts and structure is given. A description of various inputs to databases, including an important explanation of metadata and its value, is presented, as well as possible outputs from such databases. The concept of an 'integrated database environment' is briefly explained, and guidelines for future users and creators of data handling and storage systems are presented.

9.2 HISTORICAL OVERVIEW

Data have been collected throughout history. Prior to the advent of computing technologies, data were physically stored on material such as stone (and later paper), providing relatively permanent storage. Drawbacks of such an approach were numerous: storage was very expensive; only a limited number of users could have access to those data; little information about possible accuracy of the data was available; and later processing of the data was difficult. However, there was one major advantage: stored data could be used even after a few millennia.

Computerization has fundamentally changed data handling and storage. The first computer applications were just copies of the old 'paper' procedures: data were mostly entered by hand and computers were used just as recording devices. Data and their formats were strictly function-specific and hence the number of users and applications of the data were limited. Longevity of the hardware and data formats was initially very short (usually less than a decade).

During the last few years, development in the field of computerization and communications has lead to a change in the paradigm of data use. Large databases are now linked with automatic monitoring equipment, computers are connected into networks and grids, diverse users and applications can share data and true integration of different water related systems is possible if not commonly practised. However, although the technology is largely mature, there are still some issues to be resolved, mostly regarding data lifespan and security.

9.3 DATA FLOW AND DATABASES

Guidance provided in this chapter is ordered to closely follow the typical data flow in water related data systems. Figure 9.1 graphically presents the data-flow process, from a source of the data (see Chapter 7) where monitoring equipment is used to measure a certain quantity and to add some basic description about measuring conditions (metadata or GLP, good laboratory practice), through Chapter 8 where measured data are validated and a measure of data quality or accuracy is added to the metadata (shown in Figure 9.1 as QA, quality assurance). This chapter discusses data storage and manipulation keeping in mind the final users and the data applications (discussed in Chapter 10).

As can be seen from Figure 9.1, the database management system (DBMS) is at the core of contemporary data handling and storage systems. The DBMS is a suite of elements consisting of the database and accompanying programmes for data input, output, control, data recovery, and so on. There are a number of commercially available DBMSs, made for different computer platforms. Regardless of the database structure, all DBMSs have the following characteristics (Prodanović, 1997):

- *Data independence*: User programmes that communicate with the database through the DBMS are independent of access strategy, physical structure and characteristics of the physical storage device. The DBMS acts like insulation between the programmes and database.

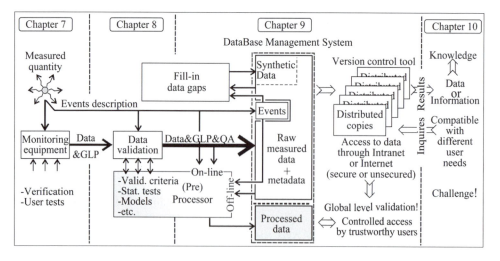

Figure 9.1   Various stages of the data flow process that leads to knowledge creation (relevant chapters in this book noted above; GLP = good laboratory practice, QA = quality assurance)

- *Data structuring*: Use of a large database would be a nightmare if the data were not structured in a meaningful way. The DBMS should take care of data structure and should be able to express the most complex relationships which may exist between the data.
- *Storage, validation and recovery*: The database is a 'shareable' resource and thus potentially vulnerable to errors (or even deliberate attack). The DBMS must validate the data before storing it and should be able to perform full recovery of the database. A mechanism for transaction tracking allows the DBMS to maintain the integrity of the database. Permanent storage of the data is the DBMS's primary role, so backup procedures and hardware redundancy are compulsory.
- *Redundancy control*: There are two types of redundancy. The first one is within acquired data when the same quantity is measured using two or more sensors, in order to assess the uncertainty during data validation phase (see Chapter 8, validation Criteria 6 and 7). The second is redundancy within the database itself, when several copies of the same data exist in different tables in order to improve database performance (speed and/or security). The DBMS can control only the second type of redundancy, by keeping all copies up-to-date.
- *User views and security*: In general, a DBMS is designed to be used by a number of users who may access the same database, and each user can work with certain parts of the data. However, when the database is used to store the raw measured data, it is not a good practice to allow access for all users. Although the users could be divided into different groups by the DBMS, and each users' group or individual user can have a number of security restrictions (username, password, ownership, right to read data, right to change data, etc.), it is much better practice to limit access to the main database to a few authorized users, and to have separate databases for all other data users. The DBMS will take care of making the distributed copies and keep the copies up-to-date. If the user discovers that some data within the main database has to be changed, he/she will need to send a request to an authorized user to complete the change.

The common structures used for databases are: hierarchical, network, relational and object-oriented. The first two structures were more popular during the 1970s and 1980s, while the relational database structure is dominant these days. A full object-oriented structure is possible but needs a lot of computing power, so an object-oriented approach applied over the relational database is more popular, mostly because it adds the 'object' flavour to the well established relational model.

Each of the four database structures has some benefits and drawbacks, depending on the type of data to be stored. Hierarchical and network structures are the best when a clear relation between stored elements exists (in some textbooks those structures are called 'navigational' structures). Those structures allow high-speed access to large volumes of the data, as well as easy update and inclusion of the new data.

The relational database structure is the most popular for general data storage. It uses the concept of linked two-dimensional tables. To store or to extract the data, queries are used; in most relational databases, the standard query language (SQL) is used. The main disadvantage of the relational structure is the time-consuming search along the tables, although indexing and creation of key-elements are used to accelerate the process. On the other hand, due to its flexibility, the relational databases are the best choice for general data storage systems.

The object-oriented (OO) approach is a contemporary trend in information technology. These systems are more modular, easier to maintain and easier to describe than the traditional, procedural systems. However, object orientation is often used without clear definition of what it actually provokes. One of the definitions of object orientation is that an entity of whatever complexity and structure can be represented exactly as one object, in contrast to other database structures where an entity has to be broken down to the lowest level. An OO database system will store and manipulate objects as they are, that is, with their state, behaviour and identity maintained, and not as a collection of their components. However, actual database implementation is not exactly as the definition might imply: each object is really stored as a number of basic entities, together with the topological relationship and methods used to manipulate the object; all the complexity is hidden from the user, in essence encapsulated. In general, the resulting OO database will be much bigger than a relational database, with a consequent reduction in speed.

9.4 INPUTS TO AND OUTPUTS FROM THE MAIN DATABASE

The main database has to handle numerous data inputs and output requests. The data inputs in most cases are in the form of data streams, coming from different communication channels using different hardware. To handle this, some kind of standardization is needed for data format and communication protocols.

Basically, there are five types of inputs into the main database:

Type I: *True raw data* as the result of some measurement. Such data could be:

- *the value of a certain quantity that is fixed in space (e.g. a point value, values along a line, or values in space within the fixed frame) and variable in time.* Inputs are mostly in the form of time series with constant or variable time steps, or sporadic measurements performed at certain instants in time. Examples are numerous, for example, flow rate in a pipe, air temperature at a meteorological station, water level in a borehole. Such data in most cases could be automatically imported, either through a direct link (e.g. GSM, radio link, optical link) or by manual file exchange.
- *the value of a certain quantity which may change in time and in space in a way that no fixed space frame can be used.* The volume of such data can be extremely large, depending on the spatial and temporal resolution (see Chapter 5). Some biological measurements fall into this category, such as algae blooms and fish behaviour and migration patterns.
- *positional data and attributes such as those about infrastructure systems, objects on the ground, cadastral data, soil characteristics, positions of known leaks from water pipes, borders of retention ponds.* Most of these data are best handled by a GIS (geographic information system) as part of database management system. However, it is important to keep in mind that existence of the features, their position and/or attribute values are not constant in time. For example, if a hydraulic model of a water network is made based on GIS data supplied by a water utility, and continuous monitoring of pressures and flow in the system is used to calibrate the model, it can happen that the network layout is not continuously up to date. During the calibration period, the maintenance crew from water utility may close or change certain pipes in the network.

Type II: *Metadata*. Metadata is literally data about true data. It is an essential part of a database, since metadata is critical for integration between different databases and users.

Metadata is a term that is understood in different ways by diverse professional communities that design, create, describe, preserve, and use information systems and resources (Gilliland-Swetland, 2004). Until the mid-1990s, 'metadata' was a term most prevalently used by communities involved with management and interoperability of geospatial data, where it referred to a suite of industry or disciplinary standards as well as additional internal and external documentation and other data necessary for the identification, representation, interoperability, technical management, performance, and use of data contained in an information system. The technical committee on Geographic Information/Geomatics (TC 211) has established the standard ISO 19115 'An International Metadata Standard for Geographic Information' that can be found at http://www.iso.org/ (Accessed 02 July 2007.).

More broadly, metadata is *the sum total of what one can say about any information object at any level of aggregation*. The information object is anything that can be addressed and manipulated by a human or a system as a discrete entity; it can be comprised of a single item, or it may be an aggregate of many items. In general all information objects, regardless of the physical or intellectual form they take, have three features: *content*, relating to what the object contains or is about, and intrinsic to the object; *context*, indicating the who, what, why, where, how aspects associated with the object's creation and extrinsic to the object, and *structure* relating to the formal set of associations within or among individual information objects and either intrinsic or extrinsic to the object.

Since the metadata become important in development of networked digital information systems, a broad concept of the metadata has been broken down into distinct categories that reflect key aspects of metadata functionality: administrative, descriptive, preservation, use, and technical metadata. However, in most large databases or data repositories, such as San Diego Supercomputing Center (www.sdsc.edu - Accessed 03 July 2007) the metadata must still be restricted to a specific format, such as the Dublin Core (http://dublincore.org - Accessed 02 July 2007), in order to make it usable.

Which categories of metadata are important for integrated urban water systems? The *administrative part* of the metadata takes care of acquisition information (and data accuracy), use rights and tracking of copies made, location information, version control, audit trails created by record-keeping system, among other tasks. The *descriptive part* describes or identifies information resources: cataloguing or indexing records, hyperlinked relations, annotations by users. The *preservation part* tracks actions taken to preserve physical and digital versions of resources (data refreshing and migration to new standards). The parts of the metadata that cope with *data usage* have to track the users and use of data, to select appropriate preview types. Finally, the *technical part* keeps a record of data formats, compression ratios, scaling routines, security issues (encryption keys), availability and response time for remote data sources, and so on.

When creating the database management system with metadata support, it is important to note that all input data, regardless of source, have a certain level of the uncertainty that is not constant over time. For example, daily readings from a turbidity sensor will have increasing uncertainty with the time elapsed since the last calibration (see Chapter 8). Similarly, older spatial data will have higher uncertainty, due to less accurate measurement equipment (Burrough, 1996).

As presented in Figure 9.1, the monitoring equipment and a separate data validation process should automatically create and add the metadata to the measured values. The procedures and metadata descriptions will vary with the type of measured variable. The minimum set of the metadata would include: type of monitoring equipment used, time and date of measurements (single measurement, or start of data collection), sampling conditions (sampling rate, start/stop conditions, alarm levels, etc.), sensor used, calibration data, environmental conditions (temperature, humidity, etc.), sampling place (from GPS data or manually entered), accuracy assessment, results of statistical tests (standard deviation, peak values, mean value trend, etc.) and results of other, well documented validation criteria. Since the metadata should help in data sharing among different users, it is essential to use well known standards wherever applicable (SEDAC, 2005). For example, the World Meteorological Organization's ISO 19115 guidelines (http://www.wmo.int/web/www/WDM/Metadata/documents.html - Accessed 02 July 2007) is a useful reference (GRDC, 2005). The use of specialized hydrological 'markup languages' can also facilitate easy integration of large hydrological databases, providing effective data sharing over the internet (Piasecki and Bermudez, 2003).

Type III: *Event descriptions*. These are generally separated from the regular flow of measured values. Strictly speaking, event descriptions could be considered as a part of the metadata and will be handled with the same metadata-handling routines. However, in most cases, event descriptions will be manually generated messages with a more loosely defined structure, such as photographs, sketches, sound records, or other data collections that are used to better describe the data and conditions during measurements. Under event descriptions one can place the explanation of tests conducted on a piece of system monitoring equipment, such as the exact date, time, and duration of a test and the volume of water poured into the siphon of a rain gauge, (data from which the calibration can be checked, and the volume of water added can be removed from the recorded time series of rainfall intensities), the description of measuring conditions that influence the calibration data (for example, bed silting in the critical depth flume), the changes in the surrounding environment (e.g. increased number of directly connected roofs to the sewer system within a catchment) or other important explanations necessary for later use of the measured data.

Type IV: *Synthetic data*. During the process of data acquisition, gaps in the recorded series can occur due to problems in communication lines, malfunction of monitoring equipment or other reasons. Depending on the measured quantity and possible data redundancy, it is possible to calculate the missing data and fill-in the gaps (see Chapter 8). For example, if three quantities are monitored, i.e. water level in a reservoir, the flow and the pressure in the outflow pipe near the reservoir, a very strong hydraulic relationship can be established among those quantities. Therefore, if the pressure logger runs out of battery charge, the missing part of the data can be easily calculated and stored within the database. Of course, the calculated (rather than directly measured) section of the time series should be marked, noting within the metadata that these are synthetic data. The calculation can be performed off-line either during the validation phase or later during database use (see Figure 9.1).

Type V: *Processed data*. Apart from the raw measured data, the input into the main Database can be both pre-processed and processed data. Processing can be done online, during raw data capture and online validation, or off-line (most commonly). *The imperative is to separate the raw data from the processed data.*

The pre-processed data can be described as another view on the same dataset, without new information in it. In most cases, pre-processing will consist of data conversion from the raw measurement units to scientific units using calibration data, removing trends, removing high frequency noise, or similar. Pre-processing can also include some kind of data compression, without any loss of data (with simple Run-Length-Encoding schemes, or more complicated schemes with sliding window-like LZ77 derivatives used in ZIP), or with loss of higher frequency components (like JPG algorithm for storage of images). Although the pre-processed data holds the same information as the raw data, it is *critically important that the user never remove the raw data from the database*. Safeguarding the raw data will allow the user to return and correct any errors observed at any later stage of data use.

The processed data will create new information out of the captured data. For example, using the frequency domain analysis of the pressure fluctuations and transients in pressurized pipe, the position and extent of leakage could be assessed. Or, total leakage from one district could be obtained by analysis of the difference between mean day flow and night flow. However, the role of the main database is the storage of the raw data, and it should not be overloaded with the processed data. The primary processing work should be done on distributed copies of the selected dataset. This reduces both the workload on the main database, and the risk of damage to the data.

Only a limited number of users should have access to the output side of the main database (in Figure 9.1 those users are called 'trustworthy users'). It is important to separate the main database, which holds the measured data and metadata, from numerous users with their unpredictable requests. The approach that should be applied is to have trustworthy users perform development and maintenance of the database, and have the output copies of selected datasets (or the whole database) accessible by other users. By limiting the public access to the main database, the security issues are much easier controlled.

Links from the main database to copies prepared for general usage should be *unidirectional*. If something is to be changed in the main database, it must first be verified by authorized users. These users must also ensure synchronization of the data between the main database and any derived copies. *Version control is important, since the users will work with different models, assumptions, or boundary conditions, and will produce different outputs using the same input data*. Also, version control has to cope with the possibility that working copies of the data from the main database could be different, since different selection criteria can be applied, or different data compaction algorithms used. For example, one copy of a rainfall dataset may have only daily mean values, while another may have only events with rainfall intensity higher than a certain threshold (see Chapter 8).

9.5 INTEGRATION OF DATABASES

The main database with its management system (as described and presented in Figure 9.1) is just one system for data handling and storage within the complex field of IUWM. In a complex water related organization several DBMSs will exist and will be linked.

The water utility system in Figure 9.2 provides one example of integration. A number of smaller DBMSs handle data and information related to water: LIMS is the Laboratory

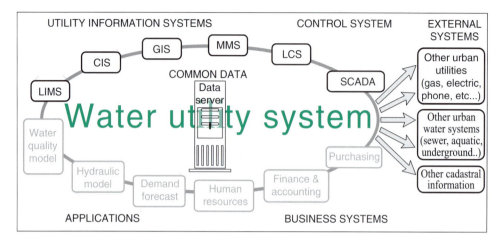

Figure 9.2   Example of data/information flow within the complex system of integrated urban water management (LIMS = Laboratory Information Management System, CIS = Customer Information System, GIS = Geographic Information System, MMS = Maintenance Management System, LCS = Leakage Control System and SCADA = Supervision, Control, data Acquisition and Data Analysis)

Information and Management System; CIS is the Customer Information System; GIS is the Geographic Information System; MMS is the Maintenance Management System; LCS is the Leakage Control System; and SCADA provides online network Supervision, Control, Data Acquisition and Data Analysis. All these systems are of vital importance for real-time functioning of the water utility, both technically and financially. Each system will collect data and will share subsets of those data with other systems. There should be, however, one common data repository system acting as a central supervisory unit to handle storage of the metadata (links and explanations of the type of the data and their available formats in the given DBMS), serve as a communication centre for links to other external systems, and handle the privileges and rights for access to specific data.

To achieve full integration the information system of the water utility should be linked with external systems, such as urban utility systems with spatial (positional) data regarding gas, electrical and telephone lines and cadastral systems where legal information about consumers and design consumptions (ie. the level of consumption on which the system design is based) can be found, and other water-related systems such as the sewer utility, underground water and surface water. Communication among those systems should all be passed through the central supervisory unit, which will process the requests, check access rights, consult its metadata database for the actual location of the searched data and ask the local DBMS to make a copy of the required data subset.

9.6 CONSIDERATIONS FOR DATABASE USAGE WITHIN AN IUWM FRAMEWORK

There is a considerable worldwide effort to integrate fundamental information and data in the field of water management. Numerous initiatives and projects have been conceived after recognition that data integration is urgently required if integrated urban water management is to be realized (Maurer, 2003, 2004).

Since data sharing among urban water components is imperative for the database to function in a sustainable long-term manner, its development has to be thoroughly analysed and investigated. The available standards for databases and metadata management (SEDAC, 2005) should be followed closely to facilitate data-sharing. In addition to the international standards, some considerations and recommendations that should be taken into account during the process of database creation can be summarized as follows:

- *Store the input data in raw format as received from monitoring equipment.* Store separately the used SI conversion factors and calibration data. Store the raw calibration data for each sensor and keep a record of previous calibrations to be able to analyse long-term stability of the sensor.
- *Collect adequate metadata.* Data in any database are useless without accompanying metadata (geographic position, sensor identification, responsible person, link to calibration data, assessed uncertainty, etc.). If possible, all potential future users of data should define their prerequisites for minimum and optimum metadata (Vyazilov et al., 2003).
- *Input data should be separated from information produced by (pre-) processing – never mix* the two (data/information version control). The raw data are unique, while the information developed from data processing can have different versions, depending on rules used in (pre-) processing. Metadata record the details of the various versions.
- *Redundancy in the input data is important, but it shouldn't be misused.* Redundant data will help in assessing data accuracy, in filling the gaps of missing data, and in crosschecking.
- *Hardware redundancy of the main database is compulsory.* A number of techniques exist, including disk mirroring, processor mirroring, and dual power supplies.
- *In the (pre)-processing stage, all measured time series should be placed on a common time line.* For example, if flow and pressure are monitored, and pressure readings were taken at 08.23.15, 08.28.29, 08.32.07, ... (i.e. with 5-minute sampling rate) and flow readings at 08.22.11, 08.32.09, 08.42.18, ... (i.e. with a 10-minute sampling rate), interpolate both readings to the same time interval, for example, 08.20, 08.30, 08.40, ... and so on. The interpolation is a part of pre-processing, so again, the raw data should be retained at all times. Note, however, that the results of such interpolation algorithms should be checked for data degradation, especially in the cases where gradients of the measured quantity are steep.
- *Data compression should be used, but with great care.* If the consequence of such data compression is a loss of information, the input raw data should not be compacted since data reduction will reduce overall accuracy. On the other hand, lossless compaction of the input raw data can be done, but it will reduce the reliability. If an error in the compacted data occurs, in most cases the whole time series will be lost. On the other hand, if an error occurs in a simple input raw data file, most of the data can usually be saved.
- *To achieve longevity of the main database, great care should be devoted to the selection of the software platform.* Often changes in software versions can lead to expensive database maintenance. Consider usage of open source platforms, since higher initial effort in database creation will generally produce lower running costs.

- *Use internationally established standards wherever possible*, for example, for storage of raw data (AGIT, 2003), for metadata (DCMI, 1999), for formating exported databases (GISHydro, 2005), for internal and external communication, for navigation within the database (SQL language) and so on.
- *Do not digitally pollute the database by storing unnecessary data.* The fact that current technology allows recording of high-resolution pictures and movies or time series acquisition with extremely fast sampling rates, does not mean that the database should be loaded with such data. If those data are redundant, contributing little to the primary purpose of the database, they will simply degrade its performance.
- *Portability of data format has to be maintained from the data source (monitoring equipment) to the data storage.* To achieve portability of the format means that the needs of all potential future users and models should be well considered from the outset. Later conversion of the data in the most cases is not straightforward.
- *Consider thoroughly the implications of temporal and spatial scales on data storage requirements* (see also Chapter 5). Data reduction techniques can be used, but this will increase the overall error, so careful selection of thresholds in data compaction should be done.
- *Database management system must address security issues.* Data security has two aspects:
  - data encryption, as secure data storage and control of access to data, and
  - prohibition on changing measured data.

  Only a small group of authorized users should have full access to all database resources. The main database should not be widely visible (through the internet, for example), so the main computer should not have an internet protocol (IP) address. One-way communication from the main computer to publicly accessible copies of the database should be used.
- *Appropriate cross-checks should be put in place to deal with possible discrepancies in data stored at different locations.* If intelligent instrumentation for data acquisition is used (see Chapter 7), the raw measured data will be stored at two locations, namely, in temporary memory within monitoring equipment and in permanent storage within the main database.
- *Performance indicators for data-storage systems should be established and continuously monitored.* Indicators could be, for instance, on the rate of database enlargement, number of megabytes per asset, number of database users per month, and so on. Such indicators can be used for maintenance and growth predictions.
- *Database maintenance and cleanup should be done on a regular basis.* There are two important issues that should be considered:
  - Which data should be kept and for how long, and which data should be thrown away? Clearly, within the database, some raw data sets should be kept 'as is' permanently, and other raw data sets are needed only for a limited period of time (they are disposable after completion of certain projects) and then they will be compacted, or kept only as statistical values (averages, extremes, etc.), or completely deleted. The procedure for regular database inspection has to be established in which every year (or every second year) the performance indicators on data usage should be calculated for data sets marked as 'disposable' and appropriate action taken.

- Long term data storage or the persistence of used data storage. Here, three points are to be resolved:
  - o How long will the storage media hold the information? What conditions and what kind of maintenance is needed to preserve the recorded data on media (humidity, temperature, rewinding or refreshing of tapes)? In general, magnetic media can be used for a few decades, optical media for 50 years to 100 years. If data are needed for centuries, then appropriate migration plans are needed.
  - o Changes in data storage standards and maintenance compatibility with previous standards? The time scale here is much shorter than the problem of longevity of the storage media. Again, an appropriate migration plan should be developed.
  - o Database management software is constantly changing and improving. To keep pace with this, use software solutions which are as standard as possible, preferably with open-source components where possible and well supported.

REFERENCES

AGIT. 2003. *Guidelines for the Archiving of Electronic Raw Data in a GLP Environment.* Working Group on IT. www.glp.admin.ch/ (Accessed 02 July 2007.) (GLP-ArchElectRaw Data1_0.pdf).

Burrough, P.A. 1996. Natural objects with indeterminate boundaries. P.A. Burrough and A.U. Frank (eds), *Geographic Objects with Indeterminate Boundaries.* Bristol, UK, Taylor and Francis.

DCMI. 1999. *Dublin Core Metadata Element Set*, Version 1.1, Reference Description. Dublin Core Metadata Initiative. dublincore.org/documents/1999/07/02/dces/ (Accessed 02 July 2007.).

Gilliland-Swetland, A.J. 2004. *Introduction to metadata – setting the stage.* Los Angeles, USA, J.P. Getty Trust. http://www.slis.kent.edu/~mzeng/metadata/Gilland.pdf (Accessed 03 July 2007.).

GISHydro. 2005. *Hydrologic modeling using GIS.* Austin, TX, US, Center for Research in Water Resources, University of Texas at Austin. www.ce.utexas.edu/prof/maidment/gishydro/home.html (Accessed 02 July 2007.).

GRDC. 2005. Metadata. Global Runoff Data Centre. http://grdc.bafg.de/servlet/is/2377/ (Accessed 02 July 2007.).

Maurer, T. 2003. Intergovernmental arrangements and problems of data sharing. Paper presented at Monitoring Tailor-made IV Conference on Information to support Sustainable Water Management: From Local to Global Levels. St. Michielsgestel, The Netherlands. grdc.bafg.de/servlet/is/3997/ (Accessed 02 July 2007.).

———. 2004. Transboundary and transdisciplinary environmental data and information integration – an essential prerequisite to sustainably manage the Earth System. Paper presented at Online Conference on INDUSTRY IDS-Water Europe 2004 www.idswater.com (Accessed 02 July 2007.).

Piasecki, M. and Bermudez, L. 2003. Hydroml: conceptual development of a hydrologic markup language. Paper presented at the XXX IAHR Congress, 24–29 August 2003, Thessaloniki, Greece.

Prodanović, D. 1997. Introduction to Geographical Databases, Development and Maintenance of Database for Urban Infrastructures, Database Matching with Simulation Models, and Data Structures for Physically Based Models. E. Cabrera and Č. Maksimović (eds), *Sistemas de informacion geografica (GIS) aplicados a redes hidraulicas.* Valencia, Spain, Grupo Mecanica de Fluidos, Universidad Politecnica de Valencia.

SEDAC. 2005. Metadata. Socioeconomic Data and Application Center. sedac.ciesin.org/metadata (Accessed 02 July 2007.).

Vyazilov E., Mikhailov, N., Ibragimova, V. and Puzova, N. 2003. Metadata as tools for integration of environmental data and information production. N.B. Harmancioglu, S.D., Ozkul, O. Fistikoglu and P.G. Fistikoglu (eds), *Integrated Technologies For Environmental Monitoring and Information Production.*

Chapter 10

Use of data to create information and knowledge

D. Prodanović

Institute for Hydraulic and Environmental Engineering, Faculty of Civil Engineering, University of Belgrade, Bulevar Kralja Aleksandra 73, 11000 Belgrade, Serbia

10.1 INTRODUCTION

Collected data constitute a basic source of information. They form a window that we use to look at our environment (Singh et al., 2003). Data are used for a large number of activities (see Chapters 1 and 3), such as for assessment of a system's performance, calibration and verification of Integrated Urban Water Management (IUWM) models, real-time control of IUWM systems. The portability, presence of metadata, and their reliability are critical to the process of data translation into required knowledge. The steps and procedures used for turning data into information, and then finally into knowledge are explained in this chapter.

10.2 DEFINITION OF TERMS

Data are classically defined as the basic building blocks of human knowledge and consist of separate, uncorrelated raw facts (IBM DB2, 2003). *Information* is data endowed with relevance and purpose. Using relationships among the original facts (raw data) a meaningful context is given to data. *Knowledge* is a step further from information. Knowledge is created only when human minds incorporate (accept) and act on information through *decisions*. Information can be created from the data using different, mostly computerized techniques. In the process of knowledge creation, however, technology can only help humans to select appropriate information, but *human beings must convert the information into knowledge*.

The classic definition of data implies that the raw datum by itself delivers no benefits to the final user. Knowledge is needed to decide on certain action (Figure 10.1). A massive dataset may even be a distraction if not processed into information. Also, information does not lead to the decision-making process until people learn it and accept it.

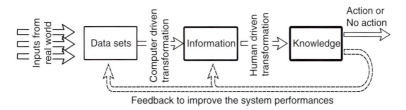

Figure 10.1 Data must be converted into knowledge in order to be useful

According to some authors (Santos and Rodrigues, 2003; Thearling, 2007) the knowledge discovery process presented in Figure 10.1 consists of a seven-step sequence (Han and Kamber, 2001; Babović et al., 2002):

(1) *Data cleaning* (as explained in Chapter 8), to remove noise from data and inconsistent data sets;
(2) *Data integration*, to combine different sources of data;
(3) *Data selection*, to retrieve relevant data for analysis: an appropriate data sampling strategy has to be defined;
(4) *Data transformation*, to process data into a form suitable for data mining, through dimensional reduction using aggregation operations;
(5) *Data mining*, to identify patterns (relationships, events or trends, which may reveal both regularities and exceptions among data) and enable model selection;
(6) *Pattern evaluation and interpretation*, to identify interesting patterns representing the knowledge;
(7) *Knowledge representation and usage*, to represent the gathered knowledge to the user and its use in decision making.

All these steps assume that the data about the real world are already acquired and stored within available databases, actually the most costly and technically complex task as it involves conversion of data from our 'analogue' world into a digital representation. Some authors call those preparatory steps 'data consolidation' (Thearling, 2007).

Figure 10.2 extends the process of data transformation into usable knowledge given in Figure 10.1. It shows the feedback process used to improve the overall system

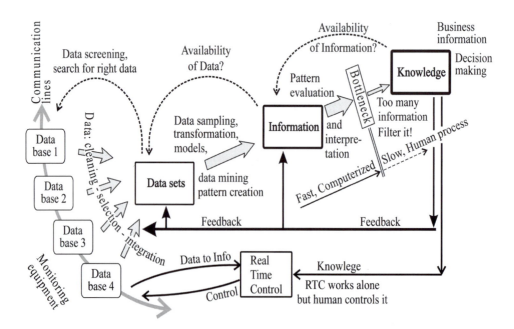

Figure 10.2 Conversion of collected data to knowledge as an interactive process

performance at each step. Two important considerations are underlined on the upper left part of Figure 10.2:

- *Data screening*: How can we know *what data are available* and *where they are*? In complex systems, like urban water, a number of organizations are in the business of data collection. Existence, accessibility and availability of such data become an important 'source of information' in itself and are key components of making and keeping those data useful.
- *Searching for the right data*: How can we know that the *right data* were collected and stored from the *right place* and at the *right point in time*? For knowledge creation, *no information* is much better than *bad information*. The single most elusive problem associated with transformation of data to knowledge is identifying errors while gathering accurate information. That is why data validation (Chapter 8) is so important and why metadata associated with accuracy assessment must be stored together with raw data.

The steps from data capture and data cleaning through pattern evaluation are nowadays carried out using computers. Large quantities of data can be easily processed and vast amounts of information created. However, the path from information to knowledge still has one important 'bottleneck', human beings. Humans have to be able to accept all those patterns and information in order to compile it into knowledge. This bottleneck we must be aware of when using automatic filtering of irrelevant patterns and identification of interesting information. Often the capacity of computers to produce patterns and information exceeds the capacity of human beings to interpret them.

The conversion of data to knowledge, as presented in Figure 10.2, implies that the users of data are managers and politicians that will use expertise of scientists in knowledge creation, to inform business and political decisions. But in many water-related organizations, most data are still measured in order to control and manage the system in real time. Real-time control (RTC) devices are designed to act upon a certain state of input parameters and measured data. In this way, RTC systems can be seen as 'data to knowledge converters', with more or less artificial intelligence (Figure 10.2), however, RTC systems are still controlled and programmed by humans.

10.3 FROM DATA TO INFORMATION

While data is simply the product of collection, information is defined by the content of the data, which is meaningful to the user and relevant to the question of problem being addressed. Data exists from the moment it is captured and stored, while information exists only if it is useful to some audience or decision makers in the most general and inclusive sense.

The database management system therefore has to be designed to support information retrieval. Portability of stored data and the presence of metadata are prerequisite, as described in Chapter 9. As different urban water systems are integrated, portability of the data becomes the biggest challenge, since different data users with different needs will have access to local databases.

A scheme for information retrieval in an integrated environment is presented in Figure 10.3. The data user (presented as 'Workstation') will ask the 'common data

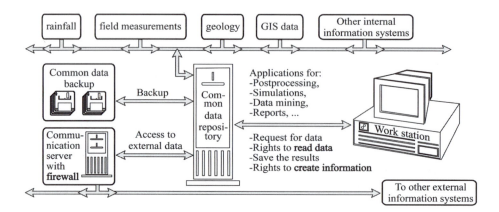

Figure 10.3 Schematic of an integrated database and data-processing system

repository' (the database with metadata about the raw data and storage locations) what data are available and where they are. Automated searching and sorting tools have been developed to keep track of where data are stored (e.g. SDSC, 2004). While searching for data to be analysed, care should be taken on systems with **Data-Rich information-Poor Syndrome (DRIPS)** (Maksimović, 1999).

An authorized user will extract data and then perform different techniques to develop information from this data. This may include calibration of simulation models, comparison with calculated or historical data, as well as, say, simply visualizing the data. Any derived datasets should be accompanied by metadata that describes exactly how it was derived.

Any information extracted from the data should have the following properties (Fedra, 2003; Harmancioglu, 2003):

- It should be *timely in relation to the dynamics of the problem to be addressed*. For example, when a pipe bursts, information on what valves to close in order to cut-off the flow is needed urgently. On the other hand, information on population growth, useful to predict future water needs is required over longer time frames.
- It should be *available when it is needed*. For example, predictions need to be made in time to react to an event.
- It should *maintain the accuracy of the data used to create it*. Proper creation and maintenance of metadata is necessary to ensure that only accurate data are used.
- It should be *accurate and precise in the frame of the information requirement*. For water supply, for example, certain precision will typically be a legal requirement, if it to be used for billing. On the other hand, excess precision will unnecessarily consume storage.
- It should be *easy to understand*. The information is just a step toward its acceptance by humans and subsequent conversion into knowledge. Understanding is prerequisite to this process.
- It should have a *format expected by and adequate* for the audience and users.
- Its *context should allow and facilitate interpretation*. Information should not be ambiguous; it must have a unique meaning.

- It should be *easily accessible* (free or not), that is, *inexpensive in relation to the implied costs of the analysed problem*. For example, if a water utility manager needs data on total consumed electrical energy in the previous year for rough assessment of power needed for the current year, two sources of data could be used: the first one, much faster and cheaper, will be by summing all consumers' electricity bills. The second one could be by taking readings from all supervision, control, data acquisition and data analysis (SCADA) systems, sorting the consumption according to different types of electrical devices and types of consumers, checking of electrical tariffs. The second approach is much more elaborate, will give better insight, but is also much more time consuming and more expensive.

The user has a number of information technologies available to process data (Shaimardanov et al., 2003). These technologies, none mutually exclusive and often combined, include:

- GIS, a set of tools for data retrieval, selection, manipulation and preview
- statistical analysis for re-processing of data, data aggregation and subsampling
- data mining for the automated search for certain patterns within data or for certain theoretical relations
- simulation models used alone or within data mining
- the internet for data acquisition, searching for data and knowledge, the dissemination of obtained information, and the distribution of the data mining work load to grid computers
- *Object orientation* as an encapsulation of the above methods.

10.3.1 Geographic Information Systems

Geographic Information Systems (GIS) facilitate the capture, storage, retrieval, analysis and display of geographic and spatial data (Burrough, 1993). The basic working forms are maps, that is, all data within GIS are spatially (geo-) referenced. GIS works generally with very large volumes of data and uses complex concepts that describe geometry of objects and relationships between them. The spatial objects have attributes or properties that could be a function of time but are independent from the location in space. The space in GIS is defined through layers and the relation between objects, so users can easily locate them. A GIS has two major roles: one is data capture and storage (see Chapter 9), and the other is data integration, analysis, integration with external models, as well as presentation and dissemination.

The GIS handles inquiries from various groups of users with different views of the same spatial data and with different processing needs (Prodanović, 1997). The key difference between GIS and other automatic cartography systems (AM/FM systems) or CAD programmes is their ability to integrate geo-referenced data from a number of layers (and different sources) and to create new information out of existing data. Successful GIS usage depends on:

- data availability (at the appropriate scale for the problem)
- adequate concept of internal data organization

- existence of metadata to allow integration of the available data
- a decision model for users that will integrate the gathered data, subsample them according to some criteria, transform them using the selected model, and create new information or patterns
- criteria for model evaluation, where the GIS visualization functions will help in presentation of model outputs (in time and in space), thus increasing the speed of information to knowledge transfer.

According to some authors, GISs were originally developed as an operational tool: to manage vast amounts of spatial data. Now GISs are shifting from being simple operational tools to being strategic decision support systems, incorporating more powerful analytical techniques (Sholten and LoCascio, 1997) as a result of a number of developments:

- visualization that has evolved considerably: three-dimensional virtual reality functions and multimedia offering greatly improved information evaluation by users
- communication possibilities that allow easy sharing of data with closed proprietary systems now transformed into widely accepted open GIS (Raper, 1997)
- advanced spatial data analysis methods based on neural networks, genetic algorithms, fuzzy data concepts, interpolation and extrapolation of data
- hardware progress relative to price, particularly for GPS.

10.3.2 Statistical analysis

The extraction of statistical information is probably the very first thing each data user will do when faced with a new dataset. For example, when the concentration of dissolved oxygen (DO) is monitored in a small creek, the user may calculate statistics such as the mean, median, minimum and maximum. Depending on the main database design (Chapter 9), some primary statistical values could even be stored as metadata in the (pre-) processing stage, thus speeding up the possible search by remote users for relevant raw data (e.g. search for records where the mean DO concentration is less than 5 mg/L).

In general, there are three statistical concepts that are used in data analysis (Singh et al., 2003) regardless of the dimensionality of data:

(1) extraction of aggregate characteristics, that is, calculation of mean values, along time or space, specifically the arithmetic mean, median, mode, harmonic mean and geometric mean (with the calculation of mean values reducing dimensionality of the data by one degree);
(2) extraction of variations of individual values from aggregate properties, including the calculation of deviation (mean and standard), variance, coefficient of variation, skewness; and
(3) change of the time/space domain into a frequency/time or wavelet domain. Standard methods of time series analysis include the analysis of frequency distribution of individual values (empirical or theoretical), usage of 'Fourier transform' to extract the dominant frequency components using periodic sine and cosine functions, or the wavelet transform with optimized mother functions.

All three concepts could be used either as data transformation tools to process the data into new information, or as pre-processing tools to reduce dimensionality of the problem.

Based upon the extracted statistical data, it is important to maintain an active feedback with the main database from which the raw data were extracted, as previously discussed. The feedback should provide information to the main database on:

- monitoring equipment, with respect to selected time and spatial resolution (for example, the sampling rate is too slow for the monitored variable, or the sensor range used is too wide), assessed accuracy (more accurate measurements are needed) and availability of metadata from monitoring equipment (requests for more specific metadata)
- new requests for (pre-) processing of the raw data and general work with the metadata
- sampling criteria used to extract the data from the main database and to create the remote database accessible by users.

10.3.3 Simulation models

Simulation models are *the representation of physical laws or processes expressed in terms of mathematical symbols and expressions* (i.e. equations). Such models are used as a basis for computer programmes where the effect of changing certain variables on the output result can be examined, for example, the analysis of the effect of daily variation in water consumption on water delivery system.

As the raw data are a representation of the physical world's current state through sampled fragments, the simulation model is used to represent the continuous complex interactions between different variables and processes. The simulation model should respond to inputs as it would in the real world, allowing the user *to interpolate the fragments of measured data into a continuous series*. Such models can also be used for past and future prediction (assuming constancy), and thus become very useful for making decisions about possible management actions.

Simulation models are a powerful and commonly accepted technique for extracting information from available data. The data are used during several stages, including:

- *Model creation*: The concept of the real world simulation is heavily driven by data availability. The old concepts of well-established models are now changing (Maksimović, 1999) towards data-driven physically based models (Drécourt and Madsen, 2001; Katopodes, 2003). With the integration of the data related to all aspects of urban water cycle, it is possible to develop a complex integrated model that will allow full interaction of several systems. A good example is presented within the UNESCO IHP VI Project 3.5.3 'Urban Groundwater Interactions' where three urban water components are linked: water supply, sewer system (channels and trunks), and underground water.
- *Model calibration*: Sufficient amounts of measured user-selected data are needed to calibrate the model. The calibration phase of the simulation model can, in some instances, result in changes in already accepted model concepts.

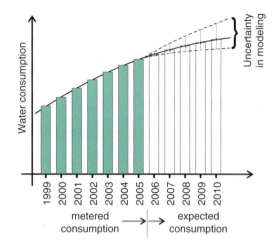

Figure 10.4 Extrapolation of water consumption using a simulation model with a specified level of
uncertainty

- *Model verification*: The data used in verification is different from data used during
the calibration phase. Model verification can be also done continuously using newly
measured data, so changes in the real world that are not implied by the model can be
discovered. Through the model verification, a measure of model uncertainty can
be established. There are two types of uncertainties that should be addressed (see
Chapter 6), specifically, first, how accurately is the true world represented and what
are the limits of model? What are the errors that the model will produce using
accurate input data, within specified model limits (i.e. when used for interpolations)
and outside the limits (if used for extrapolations)? The answers to these questions are
directly related to the model concept (whether it is a physical model, conceptual
model, simple parametric model, etc.) and data sets used for model calibration.
Secondly, the propagation of data uncertainty through the model and interaction
with the model's own uncertainty (see Figure 10.4) should be addressed.
- *Model usage*: Once the model is calibrated, the user can 'play' with it, trying different
scenarios to learn what might be the response in the real world. An important part
of each simulation model is the presentation of the results, converting the information
into usable knowledge.

The interactive use of models provides a 'feedback loop', helping to improve understanding of the system. In turn, the user with this new knowledge may determine to
collect better data to further improve the model.

Simulation models can also be integrated into bigger, more complex models of the
urban water system. An example of several simulation models integrated into the larger
conceptualization of the water utility system is presented in Figure 10.5. The demand
forecast model, for example, will take samples of data from the Customer Information
System (CIS), locations from the GIS, current flow and reservoir levels measurements
from the SCADA, and meteorological data from external system. The demand forecast

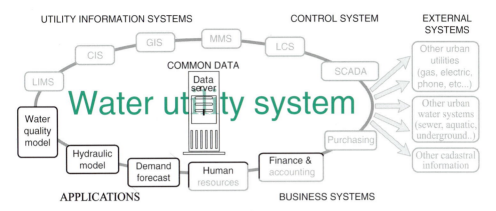

UTILITY INFORMATION SYSTEMS   CONTROL SYSTEM   EXTERNAL
                                                             SYSTEMS

Figure 10.5 Example of applications of an integrated use of simulation models as part of an Integrated Urban Water Management system (LIMS = Laboratory Information Management System, CIS = Customer Information System, GIS = Geographic Information System, MMS = Maintenance Management System, LCS = Leakage Control System and SCADA = Supervision, Control, data Acquisition and Data Analysis).

model is linked with a hydraulic model of water transport through the network, which will use also GIS data and data from the Leakage Control System (LCS). The hydraulic model is closely coupled with the water quality model and calibrated using the results of the Laboratory Information Management System (LIMS). The final goal of all these integrated models is to predict the quality parameters at the customer's connection and to suggest the appropriate actions to maintain those parameters within allowable limits.

A well-established and calibrated model can be used to extract reliable information even from incomplete data sets. If the pressure sensor in SCADA is out of order and its signal is missing, the simulation model can be used to compute a dummy number to fill in the gap within the database, as shown in Figure 9.1. Of course, this entry has to be marked within the metadata as 'simulated'.

Sometimes, the simulation model can be used to reduce the cost of equipment by monitoring a variable indirectly using some other easily measured quantity. For example, turbidity is often used to monitor total suspended solids (TSS) using simple correlation models (Fletcher and Deletić, 2007). If this is the case, it has to be recorded within metadata, since it affects knowledge derived from the TSS readings.

10.3.4 Data mining

The main drawback of using a simulation model for information extraction is that the user must have a good previous knowledge to prepare the model and to prepare the input data for calibration and model usage. When faced with a number of available data series within an integrated urban water database environment, analytical tools need to include intelligent reasoning in computerized data analysis. The general name for all such tools is 'data mining', the automated analysis of large or complex data sets in order to discover significant patterns or trends that would otherwise go unrecognized (Savic et al., 1999), or, according to Kurt Thearling, the extraction of hidden predictive information from databases (Thearling, 2007).

The term 'data mining' appears in the literature under a multitude of names, which includes knowledge discovery in databases, data or information harvesting, data archaeology, functional dependency analysis, knowledge extraction, and data pattern analysis (Savić et al., 1999). Recent improvements have seen advances in the use of data mining for environmental numerical and non-numerical data, including pattern recognition in spatial data (Wachowicz, 2000).

Data mining can be conducted as:

- *Unsupervized learning* (Roiger and Geatz, 2002) or *unidirected* or *pure data mining* (Savić et al., 1999) or *data-driven mining* (Babović et al., 2002): A data mining method that is left relatively unconstrained to build models and discover patterns in the data, free of prejudices (hypotheses) from the user. It is thus a true discovery process and is used usually for classification and clustering.
- *Supervised learning* or *directed data mining* or *theory-driven mining*: The user builds a 'learner model' or concept definition based on existing knowledge and understanding of physical processes. Data mining is used to train the model by comparing the known input/output relations. The model is then used to determine the outcome for new input instances.
- *Detection of anomalous data and patterns*: The user applies previous data mining results to analyse anomalous patterns and unusual data elements, that is, those that do not conform to the general patterns found.
- *Hypothesis testing and refinement*: The user presents a hypothesis to the system for evaluation and, if the evidence for it is not strong, seeks to refine it.

Within all the above learning categories, two main data-mining tasks can be identified (Santos and Rodrigues, 2003; Savić et al., 1999), namely,

- *prediction*, a task of deduction using the data to make prediction, incorporating classification, regression and time series analysis; and
- *discovery* or *description*, task of general data characterization, which may include deviation detection, database segmentation, clustering, associations, rules, summarization, visualization and text mining.

The process of data mining starts with data screening, cleaning and integration from different sources. Particularly in large, integrated databases where the data come from many different sources, there will likely be errors. With the assistance provided by metadata, screening systems will identify anomalous data. In most cases some data transformation will be used to prepare datasets before model running. Care should be taken during data transformation not to mask the features that carry the most important information.

Selection of training and validation datasets using appropriate sampling strategies is the next step. In order to achieve robustness and generalization, data mining is commonly done on a split dataset. The training set is used to develop the model and to evaluate fitness of the learned model, while the validation dataset is used to calculate the overall error between the modelled and target output. It is important to include a sufficient number of parameters (data fields) that may have some relevance to the problem being

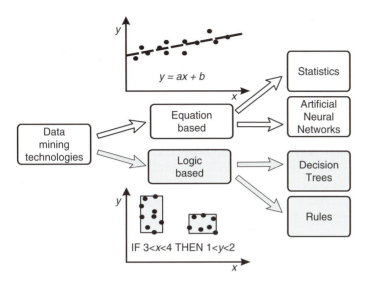

Figure 10.6 Main technologies for data mining

Source: Savić et al., 1999.

studied. The data mining system will then discover which ones are the most useful and what is the relationship among the parameters. Omitting a highly relevant parameter from analysis will cause deterioration in prediction performance of the system.

The final phase, knowledge discovery and encoding, involves running the system, validating the patterns discovered and finally encoding the results of data mining in software that can be used for prediction or classification purposes in future.

Data mining technologies can be classified into two major groups: equation based and logic based. The difference between these two approaches can be seen in Figure 10.6. The equation approach is carried out mostly on numerical data, using *statistics* and *artificial neural networks*. Typical of the statistical approaches is regression analysis (Figure 10.6). It works well for less complex sets of data, such as straight line or simple nonlinear smooth surfaces. However, transparency of the method (ability of humans to understand the equation) decreases with the complexity of the equation employed. On the other hand, the greatest advantage of artificial neural networks over other modelling techniques is their capability to model complex, non-linear processes without having to assume the form of the relationship between input and output variables.

The alternative approach, the logical approach, usually employs the conditional operators IF/THEN to represent the knowledge. The logical approach (as in Figure 10.6) is best at dealing with sharp-bounded properties of objects, although some fuzziness can be added in the condition operators. The logical approach can deal with both numeric and non-numeric data. Decision trees and rule induction are two of the most used techniques. Though similar, they differ in the way they discover information and more importantly in terms of their behaviour regarding new data items. The decision tree will use the simplest form of IF/THEN statements to represent the information and discover rules. Rule induction will use conditional relationships or so-called conditional logic (for example, 'IF it is raining, THEN it is cloudy') combined with associations between data fields or association logic (for example, the observed fact that in 80% of cases when fault x occurs then fault y is also encountered).

Data mining is still a relatively young discipline with wide and diverse application. Because of the relative slowness of data mining algorithms, the working sample of the main database has to be small, well sampled and well-prepared. The user must consider if the data mining approach is the right one, or if instead a simple data query (using SQL) would more efficiently solve the problem (Roiger and Geatz, 2002). This depends on the type of questions user wants to answer, and the type of knowledge he/she wants to discover:

- *shallow knowledge*, simple summaries (e.g. averages) or aggregates (totals) of an attribute over a selected set of cases, in which SQL is probably the most appropriate tool
- *multidimensional knowledge*, information about the frequent occurrence of values of different attributes (known as 'Association Analysis') or simple associations in large databases
- *hidden knowledge*, about patterns or relationships that are not obvious, which the user can't guess prior to data mining
- *deep knowledge*, about underlying patterns and relationships that can only be discovered using prior scientific or meta-knowledge. This is the current research frontier for data mining.

10.3.5 Self-organizing maps

A self-organizing map (SOM) can be considered as an equation-based data mining technique, which uses an artificial neural network algorithm in the unsupervised learning category. SOM is a data visualization technique introduced by Kohonen (2001), to reduce dimensionality of data through use of the self-organizing neural networks. The need for such a technique is obvious, since humans have limited possibilities in the visualization of complex, high volume and multi-dimensional data sets. SOMs reduce dimensionality by producing a map which plots and thus displays the similarities of the data by grouping similar data items together.

Many fields of science have adopted the SOM as a standard analytical tool: statistics, signal processing, control theory, financial analysis, experimental physics, chemistry and medicine. The SOM solves difficult high-dimensional and nonlinear problems such as feature extraction (WEBSOM, 2005) and classification of images (PicSOM, 2001), acoustic patterns, adaptive control of robots, demodulation and error-tolerant transmission of signals in telecommunications. SOM tools are directly compatible with GIS environments where they can be used either for pure visualization purposes or together with principal components analysis for spatial identifying relationships and time series analysis. The SOM will assist integrated urban water management by making the interpretation of multi-source and multi-domain data easier for a range of urban water cycle scientists and managers.

10.3.6 Internet and grid systems

Advances in internet infrastructure (e.g. TCP/IP protocol, HTTP communication protocol used for retrieval of documents written in HTML and XML languages), allow the rapid exchange of data and information between users. Expansion of wireless networks allows environmental monitoring stations to be simultaneously the WEB server. The user can access it from the internet and obtain status and current readings online.

Currently, the most challenging issue regarding an integrated water environment driven by internet proliferation is *web-based modelling* (Islam and Piasecki, 2004). This is the new paradigm shift in hydrodynamic modelling as well as in numerical modelling. To access the data and simulation model, the user (client) requires only a network connection, an intelligent browser and installation of required plug-ins, or client version of proprietary software. From the server side, in addition to the database management system and a high volume of disk space, a distributed version of the numerical model is needed, to share the data and tasks with one or more clients at the same time. Because of the potentially large amount of necessary data transfers between the client and server, a *client-side-request* and *server-side-simulation* approach should be used. Model-View-Controller (MVC) architecture can be used for this kind of system (Kurniawan, 2002). This separates the *simulation model* or business logic from the *model views* or presentation logics (Islam and Piasecki, 2004).

The key issue in development of such a web-based simulation environment is standardization. The development of a web markup language specific to the hydrologic community, such as HYDROML (Piasecki and Bermudez, 2003), should standardize data description and thus facilitate storage, querying, analysis, retrieval and exchange among data holding sites and end-users. The markup language differs from simple metadata. As with any language, it is constructed from two components; the first is the set of *grammar rules* (syntactic structure) and the second is the *dictionary* (semantics) to provide the 'words'. The latter is derived from the metadata and associate standards, while the former is provided through the use of the Extensible Markup Language, XML (W3C-XML, 2002). While quite a number of markup languages are currently being developed and used in a variety of areas, two stand out as perhaps closest to hydrologic sciences, namely, the Geographic Markup Language (GML, 2002) developed by ISO (norm 19136) with the OpenGIS community and the Earth Science Markup Language (ESML, 2002) originating from an effort to incorporate data elements from the earth observation community.

The further extension of web-based simulation leads toward *grid systems* and *grid computing*. A grid system is 'an ambitious and exciting global effort to develop an environment in which individual users can access computers, databases and experimental facilities simply and transparently, without having to consider where those facilities are located' (RealityGrid, 2001; Joseph and Fellenstein, 2004). It offers a model for solving massive computational problems by employing the unused resources (CPU cycles and disk storage) of a large numbers of disparate, often desktop, computers treated as a virtual cluster embedded in a distributed telecommunications infrastructure. The focus is on the ability to support computation across administrative domains, which is different from traditional computer clusters or traditional, distributed computing.

In spite of rapid developments in internet-based applications, there are still a number of considerations that need to be resolved. Some of the most important are security, trust and portability.

Security of web-based simulations and grid systems remains an issue that must be better addressed by the industry and scientific community. In trusted environments, within a department or within an open research community for example, where applications and data are not in a mission-critical or proprietary state, security is less of an issue. Beyond those safe harbours, there is a call for strong security measures, and the industry is working to overcome both real and psychological firewalls. Currently, several working groups within the Global Grid Forum are working on standardization of the many already existing security solutions.

Another important question is the degree of trust that a user can place in results and views obtained from a web-based simulation model. What form of accreditation of data is needed, and how will data uncertainty be communicated to the users? There is also a question of information availability: how to be sure that today's accessible pages and sites with relevant data and literature will be accessible tomorrow? Saving local files is a solution, but adds to storage requirements and data duplication.

Another consideration concerns portability of data. Now it is common to share data over the internet between people working on the same project, but questions still remain about how others can use the data. Data formats and accompanying metadata are tightly linked with programmes and applications that will use the data. Conversion from one format to another is sometimes painful (for example, conversion between rainfall-runoff models, say, from Hydroworks to Mouse or Simpson) or sometimes not possible without additional work (if concepts of programmes were not the same, such as in the conversion from a node-oriented model to a link-oriented one).

10.3.7 Object orientation

The Object Oriented (OO) concept is based on a real world concept: the fundamental construct is the *object* that combines both *data structure* and *behaviour* (Fedra, 2003). The OO concept was first developed as a database management system, but since then it has evolved into the mature concept capable of integrating a wide range of information technologies. Most programming languages are now object oriented, and GIS also works well within the object-oriented umbrella. In addition, databases with sharable resources and integrated metadata are becoming object oriented, and even numerical simulation models are being developed using object-oriented approaches.

In the OO system, the object is the basic element. Objects of the same structure and behaviour are grouped within classes. The basic properties of objects are (Prodanović, 1997):

- *Abstraction*, a mental process that extracts the essential aspects of the object that distinguishes it from all other kinds of objects for a particular purpose. For example, the complex real world object is reduced to the rectangular shape named 'Poly1' to represent a certain cover type, the percentage of imperviousness, specific wastewater flow, and so on. There are four basic concepts of abstraction: *classification*, *generalization*, *association* and *aggregation*.
- *Inheritance*, a mechanism permitting the development of new classes by modifying the existing one. The object 'Poly1' will have all characteristics of classes and subclasses (for example, the 'Cover' subclass used to describe the terrain cover) it was derived from. This is the primary way of minimizing data redundancy and of breaking the complex, real-world objects into manageable modules.

- *Encapsulation* or *content hiding*, which means that everything within the object is private to that object. The only way to work with its contents is through the object interface, the built-in *object operators*, *methods* and *rules*. The area of 'Poly1' is stored within this object. If the 'Cover' classes used to derive the 'Poly1' have a function that will report the polygon area, the user can ask the 'Poly1' for its area. Otherwise, access to that information is not possible.
- *Polymorphism*, meaning that the same object can respond differently, depending on the type of operation the user asks it to perform, and the current state of the object. The total area operator in the 'Cover' class will look at the 'Poly1' object to check if it is a background type of area or not, so as whether to include the area into the calculation or not.
- *Message passing*, meaning that the only way to work with an object is to pass a message to the object and to wait for the object's response. If the built-in methods in the object recognize the received message as a valid operation, they will react; otherwise they will ignore the message. The message system is very important, and it is employed for user communication with the objects, as well as internal communication among the objects.
- *Persistence*, meaning that an object will live as long as the (authorized) user decides so. Anything that refers to the object will then also delete references to it.

From the given list of an object's basic properties in OO systems, it can be seen that OO can be easily applied to distributed databases and the distributed concept of computing and modelling. A large database that covers all needs of a big water supply company can be separated into a number of smaller units (see Figure 10.3). All those units could be spread around the company, sitting on different computers within appropriate departments and using different operational systems. An intranet or full internet link can be used to connect the computers and databases. The object orientation approach is thus very helpful in achieving database integration and sharing.

There are a number of examples of the OO approach in resolving water resource management problems. Fedra and Jamieson (1996) have described and used three spatially referenced object types: river basin objects, network objects and scenarios. Those objects have functions that can obtain or update their current state and report the state to clients, and a number of classes used to derive the objects. Havnø et al. (2002) gave an excellent example of OO code architecture development on the Caloosahatchee Basin in central Florida. Another large project (WaterWare, 2005) came out as a result of EUREKA EU487. It is an integrated model-based information and decision support system for water resources management, developed and applied on the River Thames (England), the Lerma-Chapala Basin (Mexico), the West Bank and Gaza (Palestine), the Kelantan River (Malaysia) and the Yangtze River (China). Also, applications around the Mediterranean in the EU sponsored projects SMART and OPTIMA included river basins in Cyprus, Turkey, Lebanon, Jordan, Palestine, Egypt, Tunisia and Morocco. The OO approach is widely used also in Cooperative Research Centre for Catchment Hydrology (http://www.catchment.crc.org.au/-Accessed 02 July 2007) for development of a catchment modelling toolkit (http://www.toolkit.net.au/ Accessed 02 July 2007). These examples provide useful 'starting points' to identify what may be possible for a specific local application.

10.4 FROM INFORMATION TO KNOWLEDGE

Using all the described techniques in the previous subsection, the user can generate a vast amount of information. However, the information will become knowledge only after the user is able to process and understand it. Knowledge itself, however, is not the final goal; the ultimate goal is to use that knowledge to make decisions.

10.4.1 Resolving data bottlenecks

Generally, it is the transfer of information into knowledge that provides the bottleneck in the overall flow from data to knowledge. The 'transfer rate' of information into knowledge is limited, and in fact, the large rate of information production can cause a phenomenon known as 'information infarct' due to information overload (Maurer, 2003, 2004).

The problem of a user's limited capacity in processing acquired information is even more challenging if this is to be performed in a limited time frame, for instance, in order to undertake action in the case of emergency or disaster. In order to make significant progress towards understanding more complex integrated environments and to undertake the right actions at the right time, it is crucial to improve handling of the ever-growing amount of information. Some measures that could be considered are (Maurer, 2003):

- *Accelerating information transfer rates.*
- *Homogenization and standardization of information representation.* It is much easier for the user to work with standardized diagrams and tables, than to have to re-learn each time table headers and diagram axes.
- *Improvement of selection mechanisms for targeted information retrieval* to provide only the required information and thus reduce overload. Technology can help in selection and ordering of the most relevant information (a good example is the Google web search engine, which in the most cases will offer the most relevant information within first 5 to 10 listings).
- *Improvement of aggregation schemes to summarize information.* The pile of information can be reduced if usable aggregate information is presented to the user. Then, if users want deeper knowledge, they can further interrogate the summarized information.
- *Improvement of disaggregating schemes and interface definition* to facilitate sharing of the work-load among a number of project participants, thus decreasing the amount of information required by a single individual. Since more users will be involved in knowledge creation, the process will be quicker. Also, each user can select the most appropriate type of information according to need.

These proposed measures call for the organization of better coordinated structures with a high degree of complexity, both in a technical and an administrative sense. Examples in the technical domain include libraries, meta-databases, standardization efforts, generic concepts, expert and decision support systems. Similarly, in the human domain, this relates to improved coordination of organizations and programmes at the international, national, regional and local levels. Often developments in the technical field trigger change in organizational structures within society. Sharing of the work-load on a global scale is thus needed to cope with complex tasks that are to be solved. In the global knowledge society this leads toward establishment of balanced and stan-dardized education systems throughout the world. Chapter 11 provides some further discussion of the institutional requirements to support integrated data collection and use for urban water management.

10.4.2 Knowledge application

The true value of knowledge is only in its use. As shown in Figure 10.2, there are two main outcomes from knowledge creation: the feedback to various parts of the system and the action (or the decision *not* to take action) that is undertaken as a result of this knowledge.

Feedback towards the start of data-information chain will in general improve the performance of the whole system. Optimization of monitoring network, sampling site position, sampling frequency and methods used for data evaluation, based on early results and previous knowledge, can significantly reduce flow of the data, but improve its quality. Improvements to models can help to understand existing data.

The feedback will be also applied to autonomous real-time controllers (RTCs), devices that mimic data to knowledge transformation by taking some actions based on obtained data. Through the process of learning about the whole system and behaviour of each RTC, the controlling strategy can be changed and adapted to optimize the sys-tems performance.

Actions that the user will undertake based on knowledge will be the result of a *decision-making process*. Computers aid this process through decision support systems (DSS). The objective of DSSs for integrated urban water management is to improve planning and operational decision making processes by providing useful and scientifically sound information in a dedicated form to the actors involved in these processes, including public officials, planners and scientists, various interest groups, major water users and possibly even the general public.

A DSS will support and facilitate the process of assessing the possible consequences of measures and actions, before making a proper selection from the available alterna-tives. The ultimate objective is to ensure sufficient and sustainable water resources, thus contributing to the maximization of some (rather hypothetical) social welfare function.

Decision making involves a choice between alternatives. The DSS should help the user to analyse the alternatives and to rank them according to a number of selected cri-teria by which they can be compared. These criteria are checked against the objectives and constraints involving possible trade-offs between conflicting objectives. The con-straints are to be checked also, if no alternative can meet them.

Approaches in DSS span a wide range of conceptual levels, such as (Fedra, 2003):

- *information systems*, to provide information about the present state of a system permitting forecasts based on the observed trends
- *scenario analysis*, to support the exploration of numerous 'what if?' questions
- *comparative evaluation*, to assess different scenarios using performance indicators established (preferably according to some local, national or international standard)

in at least two scenarios (with graphical display of data allowing easy comparison in most cases)

- *optimization*, to reach a consensus. Since each scenario is described by more than one performance variable, direct comparison does not necessarily lead to a clear ranking. This can be resolved by introduction of a preference structure that defines the trade-offs between objectives. Numerous optimization techniques are then used, either directly, or more often with a discrete multicriteria approach, that will seek an efficient strategy to satisfy all the actors and stakeholders involved in the water resource and environmental management decision processes.

REFERENCES

Babović, V., Drécourt, J.P., Keijzer, M. and Hansen, P.F. 2002. A data mining approach to modelling of water supply assets. *Urban Water*, No. 4, pp. 401–14

Burrough, P.A. 1993. *Principles of Geographical Information Systems for Land Resources Assessment*. Oxford, Clarendon Press.

Drécourt, J-P. and Madsen, H. 2001. Role of domain knowledge in data-driven modeling. Paper presented at Fourth DHI Software Conference, 6–8 June 2001, Helsingør, Denmark.

ESML. 2002. Earth Science Markup Language. http://esml.itsc.uah.edu/ (Accessed 02 July 2007.)

Fedra, K. 2003. From data management to decision support system. N. B. Harmancioglu, S. D. Ozkul, O. Fistikoglu and P. Geerders (eds), *Integrated Technologies for Environmental Monitoring and Information Production*. Dordrecht, Kluwer Academic Publishers, pp. 395–410 (NATO Science Series, IV: Earth and Environmental Sciences, Vol. 23).

Fedra, K. and Jamieson, D.G. 1996. An object oriented approach to model integration: a river basin information system example. K. Kovar and H.P. Nachtnebel (eds), *HydroGIS"96: Application of Geographic Information systems in Hydrology and Water Resource Management*. Wallingford, UK, IAHS (IAHS Publication No. 235).

Fletcher, T.D. and Deletić, A. 2007. Observations Statistiques d'un Programme de Surveillance des Eaux de Ruissellement; Leçons pour l'Estimation de la Masse de Polluants [Statistical Observations of a Stormwater Monitoring Programme; Lessons for the Estimation of Pollutant Loads]. Paper presented at the NOVATECH 2007: Sixth International Conference on Sustainable Techniques and Strategies in Urban Water Management, 25–28 June 2007, Lyon, France.

GML. 2002. Geographical Markup Language. www.opengis.net/gml/ (Accessed 03 July 2007.).

Han, J. and Kamber, M. 2001. *Data Mining: Concepts and Techniques*. San Diego, CA, Academic Press.

Harmancioglu, N.B. 2003. Integrated data management: Where are we headed? N.B. Harmancioglu, S.D. Ozkul, O. Fistikoglu and P. Geerders (eds), *Integrated Technologies for Environmental Monitoring and Information Production*. Dordrecht, Kluwer Academic Publishers, pp. 3–16 (NATO Science Series, IV: Earth and Environmental Sciences, Vol. 23).

Havnø, K., Sørensen, H.R. and Gregersen, J.B. 2002. Integrated water resources modelling and object oriented code architecture. Copenhagen, DHI Water and Environment. http://www.dhisoftware.com/Bangkok2002/Proceedings/Papers%20Bangkok/BA%20029/code-architecture.doc (Accessed 02 July 2007.)

IBM. Data mining software built into the cyberinfrastructure system. www-306.ibm.com/software/data/iminer/ (Accessed 02 July 2007.)

IBM DB2. 2003. *Embedded Analytics in IBM DB2: Universal Database for Information on Demand*. IBM Corporation. ftp.software.ibm.com/software/data/pubs/papers/embeddedanalytics.pdf (Accessed 02 July 2007.)

Islam, A.S. and Piasecki, M. 2004. A strategy for web-based modeling of hydrodynamic processes. Paper presented at Seventeenth ASCE Engineering Mechanics Conference, 13–16 June 2004, University of Delaware, Newark.

Joseph, J. and Fellenstein, C. 2004. *Introduction to Grid Computing*. Prentice Hall.

Katopodes, N.D. 2003. Adaptive control of flow and mass transport by multi-sensor arrays. Paper presented at Thirtieth IAHR Congress, 24–29 August, Thessaloniki, Greece.

Kohonen, T. 2001. *Self-organizing Maps*, 3rd edn. Berlin, Springer (Springer Series in Information Sciences, Vol. 30).

Kurniawan, B. 2002. *Java for the Web with Servlets, JSP, and EJB*. Indianapolis, Ind., New Riders Publishing.

Maksimović, Č. 1999. *Y2K2C Project Initiative – Mission Statement*.

Maurer, T. 2003. Intergovernmental arrangements and problems of data sharing. Paper presented at Monitoring Tailor-Made IV: Conference on Information to Support Sustainable Water Management: From Local to Global Levels. St. Michielsgestel, The Netherlands

———. 2004. Transboundary and transdisciplinary environmental data and information integration – an essential prerequisite to sustainably manage the Earth System. Online Conference on INDUSTRY IDS – Water Europe 2004 [http://www.idswater.com (Accessed 02 July 2007.)].

Piasecki, M. and Bermudez, L. 2003. HYDROML: Conceptual development of a hydrologic markup language. Paper presented at the XXX IAHR Congress, 24–29 August, Thessaloniki, Greece.

PicSOM. 2001. Methods and systems for content-based image retrieval. Helsinki, Laboratory of Computer and Information Science, Helsinki University of Technology. http://www.cis.hut.fi/projects/cbir/ (Accessed 03 July 2007.)

Prodanović D. 1997. Introduction to geographical databases, Development and maintenance of database for urban infrastructures, Database matching with simulation models, and Data structures for physically based models. E. Cabrera, and Č. Maksimović (eds), *Sistemas de informacion geografica (GIS) aplicados a redes hidraulicas* Valencia, Spain, Grupo Mecanica de Fluidos, Universidad Politecnica de Valencia.

Raper, J. 1997. Geographic information on the web. Paper presented at ESF GISDATA Final Conference on Geographic Information Research at the Millennium, 13–17 September 1997, Le Bischenberg, France http://shef.ac.uk/uni/academic/D-H/gis/raper.html (Page now deleted.)

RealityGrid. 2001. Engineering and Physical Sciences Research Council, UK. http://www.realitygrid.org/index.shtml (Accessed 03 July 2007.)

Roiger, J.R. and Geatz, M. 2002. *Data Mining: A Tutorial Based Primer*. Addison-Wesley Publishing.

Santos, M.A. and Rodrigues, A. 2003. Information technology and environmental data management. N. B. Harmancioglu, S. D. Ozkul, O. Fistikoglu, and P. Geerders (eds), *Integrated Technologies for Environmental Monitoring and Information Production*. Dordrecht, Kluwer Academic Publishers, pp. 39–52. (NATO Science Series, IV Earth and Environmental Sciences,Vol. 23)

Savić, D.A., Davidson, J.W. and Davis, R.B. 1999. Data mining and knowledge discovery for the water industry. D. A. Savic and G. A. Walters (eds), *Water Industry Systems: Modelling and Optimisation Applications,* Vol. 2, Baldock, UK, Research Studies Press, pp. 155–64.

SDSC – San Diego Supercomputing Center. 2004. http://www.sdsc.edu/ (Accessed 03 July 2007.) and http://www.sdsc.edu/srb/Pappres/Pappres.html. (Accessed 02 July 2007.)

Shaimardanov, V.M., Mikhailov, N.N. and Vorontsov, A.A. 2003. Perspective decisions and examples on the access and exchange of data and information products using web and XML applications. N. B. Harmancioglu, S. D. Ozkul, O. Fistikoglu and P. Geerders P. (eds), *Integrated Technologies for Environmental Monitoring and Information Production*. Dordrecht, Kluwer Academic Publishers, pp. 435–48. (NATO Science Series, IV Earth and Environmental Sciences, Vol. 23).

Sholten, H.J. and LoCascio, A. 1997. GIS application research: history, trends and development. Paper presented to ESF GISDATA Final Conference on Geographic Information Research at the Millenium, 13–17 September 1997, Le Bischenberg, France http://shef.ac.uk/uni/academic/D-H/gis/key3.html (Page now deleted.)

Singh, V.P., Strupczewski, W.G. and Weglarczyk, S. 2003. Uncertainty in environmental analysis. N.B. Harmancioglu, S.D. Ozkul, O. Fistikoglu, and P. Geerders (eds), *Integrated Technologies for Environmental Monitoring and Information Production*. Dordrecht, Kluwer Academic Publishers, pp. 141–58. (NATO Science Series, IV: Earth and Environmental Sciences, Vol. 23)

Thearling, K. 2007. Data Mining Tutorial. http://www.thearling.com/dmintro/dmintro.html (Accessed 03 July 2007.)

W3C-XML. 2002. World Wide Web Consortium (W3C). http://www.w3.org (Accessed 02 July 2007.)

Wachowicz, M. 2000. How can knowledge discover methods uncover spatial-temporal patterns in environmental data? Paper presented at SPIE Aerosense 2000 Conference on Data Mining and Knowledge Discovery: Theory, Tools and Technology II, 24–28 April 2000, Orlando, Florida, USA.

WaterWare. 2005. A Water Resources Management Information System. http://www.ess.co.at/WATERWARE/ (Accessed 02 July 2007.)

WEBSOM. 2005. Self-Organizing Maps for Internet Exploration http://websom.hut.fi/websom/ (Accessed 02 July 2007.)

Chapter 11

Social and institutional considerations

R.R. Brown

School of Geography and Environmental Science, Monash University, Melbourne 3800, Australia

11.1 INTRODUCTION

This chapter outlines the key social and institutional considerations that should be addressed when designing and administering an IUWM data collection and management programme. These considerations come from an assessment of the social and institutional insights in a range of specialist fields including policy design, implementation research, institutional analysis, community indicators, corporate sustainability, socio-technical studies, urban planning, environmental governance, science communication and environmental reporting.

This chapter presents some of the important overarching 'principles for practice' that an urban water manager should follow when designing and administering an IUWM data collection and management plan. These socio-institutional principles include:

- *Leadership and commitment*: focuses on the science-policy interface through the synthesis of bio-physical IUWM data to enable the setting of measurable policy targets and benchmarks
- *Public participation*: recognizes the need for local knowledge, due to the significant uncertainties with scientific data and the many value-based decisions that need to be made within IUWM
- *Transparency and accountability*: works towards advancing ongoing corporate and public reporting of IUWM information over time to enable an environment of increased transparency, trust and accountability between and among stakeholders
- *Coordinated data access and sharing*: there are often numerous stakeholders and organizations involved with IUWM, who depend on information from each other
- *Evaluation and action learning*: focuses on the need to facilitate ongoing and adaptive change, to improve the application of data for learning.

The scope of each of these principles in relation to the social and institutional considerations for IUWM data collection and management is presented in Section 11.3. While it is beyond the scope of this chapter to present a comprehensive review of each social and institutional consideration that has been addressed and raised across this wide array of specialist fields, guidance is given to key references that can be drawn upon in this broad arena of research and practice.

11.2 THE IMPORTANCE OF SOCIAL AND INSTITUTIONAL FACTORS

Integrated Urban Water Management is still largely in its infancy, and many governments, organizations and communities are still operating within the traditional urban water management approach. This is despite widespread recognition of the current inefficient use of resources, continuing waterway degradation and delay in adoption of IUWM caused by ongoing investments in traditional approaches. Therefore, the socio-institutional context is an essential consideration to all aspects of IUWM.

Several commentators have attempted to explain the observed slow pace of change in urban water management, within an Australian context. For example, Hatton MacDonald and Dyack's (2004) review of 'institutional impediments' to water conservation and re-use found that the 'overarching' institutional impediment is a lack of coordination of the policies and regulations that govern water conservation and reuse. Brown (2005) and Wong (2006) highlight the fragmented administrative framework in which urban water management is implemented, suggesting this perpetuates a lack of attention to institutional learning within the urban water sector. The national environmental industry lobby has also identified a 'lack of trust' and 'inappropriate risk transfers' between stakeholder organizations, as key factors retarding the implementation of IUWM across Australia (The Barton Group, 2005). Many of these issues identified across the Australian context are also common in other places.

Overall there is an increasing and diverse group of international commentators highlighting the problem of institutional inertia, and its significance to the observed slow pace of progress towards IUWM (see for example, Lundqvist et al., 2001; Vlachos and Braga, 2001; Hatton MacDonald and Dyack., 2004; Wong, 2006; Brown et al., 2006). A number of these issues are encapsulated by Serageldin's (1995) identification of the 'silo effect', which describes the separation of responsibilities among organizations, and their inability or unwillingness to consider their mandate relative to those of other organizations, is highly relevant to IUWM. This is often expressed as 'vertical fragmentation' between levels of government and 'horizontal fragmentation' across levels of government. In Mitchell's (2005) review of the results of integrated water resource management efforts over the last 30 years, he suggests that aspiring to remove these boundary effects through structural re-organization often proves 'futile' and that focusing on enabling institutional learning and improving coordination between stakeholders is required.

The social and related behavioural aspects of IUWM, while increasingly being highlighted and investigated, are still underdeveloped in this field (see Chapter 22 for guidance on the collection of social and institutional data). A common starting point in the broader integrated environmental management literature is the wide acceptance of the need for *a broader range of inputs into data collection and decision-making processes*. For example, from the perspective of a national-level government agency, Fisher (2000, p. 68) states:

> Good environmental and natural resource management depends fundamentally on understanding human behaviour. While it is essential that we understand the biophysical aspects of a natural resource or ecosystem, environmental management at the government level largely deals with the consequences of human interactions

with the environment. However until recently, government agencies have tended to focus more on the natural science concepts (and to some extent economic concepts) underpinning sustainability and the social and behavioural aspects have been largely ignored.

This overall context poses a significant and complex challenge to urban water managers with the goal of institutionalizing the widespread application of IUWM. Consequently, the social and institutional considerations associated with IUWM data collection and management plans must focus on strategically enabling both social and institutional learning to enhance the long- term viability and commitment to IUWM.

With IUWM necessitating the involvement of a wide range of stakeholders, there is a need not only for the integration and presentation of different disciplinary data sets, but also sophisticated communication of this data and information that enables meaningful and common learning across multiple scientific and policy networks, and communities.

11.3 KEY PRINCIPLES

This section provides an overview of five overarching 'principles for practice' when designing and administering an IUWM data collection and management plan. Reference is given to key publications which underpin this overview. These five principles for practice should be considered as a 'package', because they are inter-dependent, and when practised together, provide the best opportunity for advancing the practice of IUWM. Therefore the principles for practice should be equally considered throughout the IUWM process. For example, if 'public participation' is being actively facilitated within the catchment, then an urban water manager should ask the following questions. Does this public participation activity:

- assist with the setting and review of science-policy targets associated with the '*Leadership and commitment*' principle?
- inform the design of public reporting processes associated with the '*Transparency and accountability*' principle?
- both review and inform the available science, and other knowledge, of the urban water problem associated with the '*Coordinated data access and sharing*' principle?
- include the public in the evaluation of the IUWM programme and ongoing learning over time associated with the '*Evaluation and action learning*' principle?

11.3.1 Leadership and commitment

Leadership and commitment are central issues to advancing the implementation of IUWM. There are a number of important insights that can inform initiatives to develop organizational leadership and commitment, whether it is driven by communities, scientists, water authorities, or a top-down regulatory intervention.

Mullen and Allison's (1999) comparative analysis of four different implementation models, including top-down, coordinated top-down, authority driven and locally driven citizen-led bottom-up, revealed three key factors for ensuring leadership and commitment to environmental management programmes and polices. These included the extent of stakeholder involvement, the availability of social capital and the presence of a real or perceived water resource concerns or problems.

These insights are significant considerations in moving towards an integrated management framework for data collection and decision implementation. No longer can the implementation of IUWM related decisions be confined to a single organization or discipline. This imposes a need for interaction and coordination between different players across functional sectors, government and non-government groupings and the public. Margerum (2001, pp. 422–23) suggests that addressing organizational commitment is one of the key criteria for ensuring implementation success, because:

> Organizational stakeholders are particularly important in implementation, because they are most likely to have the resources and capacity to carry out implementation actions. However, organizations also operate under complex structures and hierarchies that make participation in a consensus-building process difficult. Furthermore, once consensus is reached with organizations, it can be difficult for the information, policy direction, and actions to filter through the organizational structure to produce changes.

In the ideal situation, the urban water manager would be working towards a series of IUWM goals built around a clear 'vision' (Figure 11.1). This vision, built on significant community engagement, sets out what the most desirable sustainable urban water future would be for the particular region or catchment, given the local social, ecological and economic values. This process should also accept *that there will always be a lack of all the necessary information*, and should progressively address this knowledge gap through a *commitment to ongoing data collection and to learning*. This addresses an important issue of organizations and individuals being reluctant to make what seem like 'bold' decisions as they are not confident that they are the 'right' decisions. Therefore

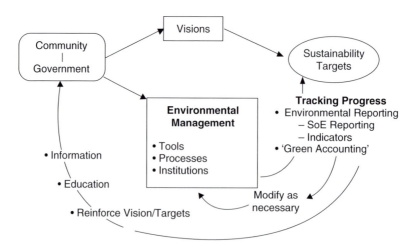

Figure 11.1 Envisaging sustainable urban water futures and tracking progress towards this goal
Source: Adapted from Harding, 1996.

processes such as these are important for enabling risk-sharing among stakeholder organizations, communities, scientists and politicians.

An essential step in this process is the *setting and monitoring of well designed IUWM targets*. These targets need to be critically informed by the best available data and science, which are then translated into policies for implementation. For example, in Australia in metropolitan Melbourne, there is a powerful target of reducing the nitrogen load in urban stormwater runoff to the receiving Port Phillip Bay by 500 t/yr. While this target was informed by a five-year, multi-million-dollar, scientific study on the health of the bay, the outcome of the science was translated into a strong and simple policy target that can be mandated though government regulatory instruments, and easily reported to politicians. It is currently adopted by the local water utility and municipalities and integrated into land development assessment tools.

Monitoring progress towards these IUWM targets should focus on '*feedback loops*' as also shown in Figure 11.1. This offers a means to continued engagement of all stakeholders in achieving the desired outcomes, as well as a basis for ongoing social and institutional learning. The feedback loops are also reinforced across the other principles for practice areas including 'transparency and accountability', 'coordinated data access and sharing' and 'evaluation and action learning'.

In developing measures for tracking progress towards IUWM the urban water manager may consider using one or more composite indices, or a range of separate indicators which together cover the areas of meaning for the achievement of sustainable urban water environments. However, there is significant complexity associated with the design and use of indicators (Lawrence, 1997). Most people are familiar with economic indices such as the Gross Domestic Product (GDP), which measures the value of all goods and services produced within a nation. Sustainability professionals have also become accustomed to environmental indices such as the daily pollution index which provides a composite measure of a number of different air pollutants ranked against a standard for air quality. Recently with the advent of *State of the Environment Reporting* there has been considerable activity in developing indicators for use in these reports.

Indicators and indices are a means of communicating information to enable us to easily appreciate the current state and trends of some attribute of importance to us. Indicators need to be designed for a particular purpose and user group, and used for that purpose only. Such composite indices are used because they are seen as providing a simpler, and hence more powerful and understandable message to average users. For example, a message from a composite water quality index could tell us that water quality today is – good, medium or 'poor' and consequently a swimming location may be made closed to the public if it is 'poor'. This is often seen as having greater meaning to morning newspaper readers or those listening to the news on television or radio than a string of separate indicators covering say, nitrogen heavy metals, suspended solids, phosphorus and *Escherichia coli*, among other factors.

Community and government jointly develop visions of a future sustainable urban water environment and use science and other relevant inputs to determine targets to achieve this vision. A carefully designed mix of IUWM tools influences progress towards achievement of these sustainability targets. Progress towards these targets is 'tracked' using stakeholder reporting and accounting systems. These in turn provide essential information to refine IUWM tools, inform the public and reinforce commitment to IUWM targets.

11.3.2 Public participation

This principle recognizes that there will always be significant uncertainties with scientific data and that there are many value-based decisions that need to be made in advancing IUWM. Particularly over the last three decades, the practice of 'public participation' has increasingly been identified as essential to not only understanding the social and behavioural aspects as (Fisher, 2000), but also enabling the successful implementation of integrated environmental management programmes.

11.3.2.1 What is public participation?

Public participation is essentially a disposition towards planning and decision-making, recognizing that power, control and political commitment are central issues. One helpful participation typology that has been put forward by Scales (1997) and others such as Shand and Arnberg (1996) is of the continuum of participation techniques. This continuum (Figure 11.2) offers different participation techniques, given the level of political commitment for 'power-sharing' with the community. See also Bishop and Davis (2002) for an overview of a broad range of other public participation typologies.

The left-hand side of the continuum represents minimal political commitment towards public participation, and therefore minimal power held by the public to influence decision-making and outcomes. Moving towards the right side of the continuum increases the political commitment to partner with communities in influencing the outcome.

Using *information gathering* as an example (third from the left on the continuum), the social sciences have developed specific techniques, such as telephone interviews, questionnaires, opinion surveys, face-to-face structured interviews and survey workshops. What is important to note for this level of participation is that the terms of reference for the information gathering has 'already' been decided before exposure to the community. This means that the community has not been involved in determining the scope of the relevant information needed to address the IUWM issue, and hence this participation activity is located in the left half of the continuum. By contrast, if the 'power-sharing and participatory decision-making' activity (sixth from left on the continuum) was employed, the community would be collaboratively involved *in framing the IUWM problem, deciding what information was needed, and how best to present and communicate this information.*

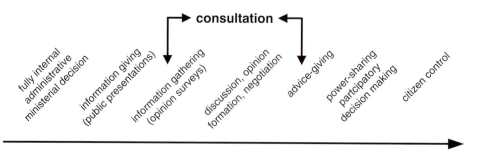

Figure 11.2 Continuum of public participation methods in government programmes

11.3.2.2 Why is public participation important to IUWM data collection and use?

The arguments presented for public participation, while not precisely presented within the IUWM literature, can be broadly grouped into three key areas: (a) facilitating learning and building new social norms and practices; (b) improving the process of expertise; and (c) advancing social justice and democratic rights of citizens.

Environmental problems, such as urban water problems, are as much political and social issues as they are scientific. Alone traditional urban water experts and physical scientists are not well equiped to address the social-political dimension. Therefore, the community can play an important role in the IUWM data collection and management processes by providing other forms of knowledge and 'values'. Much of the environmental literature today calls these ways of knowing and meaning, local and indigenous knowledge. It is these forms of knowledge and their input into the IUWM decision-making processes that clearly improve how critical knowledge gaps are addressed, as well as provide an important opportunity for influencing local 'values'.

Another argument in the sustainability discourse for public participation is to address social justice (principle of sustainability) and improve the social capital (knowledge, skills and ownership) vis-à-vis local, urban water problems and solutions so that local communities can have some say over decisions that impact their day-to-day lives and expectations on quality of life. Sometimes this is discussed as actively building citizenship potential so that people can come together as citizens and deliberate about their common affairs. Advancing the democratic rights of citizens through public participation is achieved by facilitating a process of deliberation (i.e. talking and being heard). This is considered to lead to shaping and sharing understandings and meanings of environmental issues within the community and between the community and other stakeholders.

Public participation in this sense is concerned with working towards building social integration and communication around issues of common concern and interest. Participation, within this broad argument, also seeks to maintain commitment to social and democratic justice by being concerned and aware of social equity, access to information and rights to participate and being heard and ultimately influencing decision-making processes.

11.3.3 Transparency and accountability

This principle directs ongoing corporate and public reporting of IUWM information over time to enable increased transparency, trust and accountability between and among stakeholders and the community. The decision of what to report and to whom, is therefore important. The IUWM information reporting process should be used to develop a common conceptual framework of the IUWM issues and possible solutions amongst the numerous stakeholders. This is an important basis for developing an environment of trust and collaboration.

An important aspect of transparent communication involves the appropriate tailoring of information to various stakeholders. They are likely to include stakeholder groups such as government agencies, local community groups, industry groups, environmental groups, professional associations, the media and others. Given the wide audience involved in IUWM, communication regarding data collection and analysis, an urban water manager must ask the questions of: To whom am I communicating? What is the most appropriate communication technique? In what format should the information be presented?

The process of communicating IUWM data is important and needs to be considered as essential for decision-making. It is important to recognize that concepts of trust and legitimacy should *not* be treated as issues that need to be addressed only after the data has been collected and analysed. To be effective these concepts need to be fostered over time from setting the scope of the data collection programme through to the overall communication strategy associated with the IUWM process. This communication strategy needs to be clear, transparent and tailored to all concerned. Clarity and transparency is a way of dispelling stakeholder concerns about hidden agendas, so that a more trusting dialogue between different and contending stakeholders is formed, even if there is a general lack of agreed position on IUWM issues.

A highly successful case study and example of the 'transparency and accountability' principle is the 'Healthy Waterways Partnership' Annual Report Card series by the Moreton Bay Waterways and Catchments Partnership (MBWCP, 2006). The annual Ecosystem Health Report Card presents an easy-to-understand snapshot of the health of the region's freshwater and estuarine/marine environments, providing 'A'(excellent) to 'F' (failing) ratings for nineteen estuaries, eighteen freshwater catchments and the region's receiving waterway, Moreton Bay. The Report Card is available online at http://www.healthywaterways.org/index.html (Accessed 02 July 2007.).

This Report Card is considered an important regional tool for communicating with politicians, local and state government agencies, industry and the community. It provides a scientifically rigorous progress account of the current ecological status and what needs to be improved in terms of waterway health across the region. The Report Card has been presented each year since 1999, providing insights into issues affecting waterways and the effectiveness of investments in waterway and catchment management during this time. The 2006 Report Card is the culmination of scientific monitoring of 381 freshwater and estuarine/marine sites during the period of July 2005 to June 2006. The local university plays an important role in supporting the rigorous scientific approach to measuring waterway health using a broad range of biological, physical and chemical indicators.

This monitoring and public reporting programme informs the development of IUWM actions across the region to protect and conserve areas of high ecological value, manage point sources through wastewater reuse and discharge standards, manage rural diffuse pollution sources by achieving good land management practices and addressing stream and gully erosion, and manage urban diffuse sources. Importantly this activity drives action within the region because it essentially assists in creating both a public and political accountability for waterway health.

11.3.4 Coordinated data access and sharing

There are numerous stakeholders that not only need IUWM related information, but can also contribute important and useful data. The more integrated the management approach, the greater the need for effective data sharing. To enable this, it is essential that there is a well facilitated and coordinated data management and access system. This system should allow for equitable data access across all stakeholders, and promote the timely and effective sharing of data sets. This will not only allow for wide contribution and application, but enhance inter-organizational relationships and hopefully commitment to IUWM (Johnson and Walker, 2000).

It is important that the key stakeholders work together to facilitate a dedicated funding source to support the ongoing management and visual presentation of all IUWM information, as well as establish a system of 'rules' for data sharing and updating among stakeholders (see also Chapters 9 and 10). This platform for coordinated data would include IUWM information across the social, ecological and economic dimensions. For example, it should allow biophysical scientists to access local waterway health information, through to a sustainability analyst wanting to conduct a triple bottom line assessment.

11.3.5 Evaluation and action learning

Throughout the literature on integrated forms of resource management there is a general consensus regarding the paucity of ongoing data collection, analysis and evaluation. This principle focuses on addressing this issue, as it significantly impedes social and institutional learning over time, and consequently the facilitation of adaptive management techniques. Evaluation is an essential component of any data collection and management programme, and the *action learning philosophy* is increasingly being recognized as essential to improving implementation rates. Evaluation should be interactive, iterative, and meaningful, and directed towards influencing what emerges in an ongoing positive and constructive way for advancing IUWM.

Many evaluation processes involve a linear inquiry beginning with well-defined questions, data gathering or field research, analysis of the data, generating conclusions and forming decisions which are often presented as recommendations for action. The 'action learning' (or 'adaptive management') approach explicitly recognizes that the IUWM issue involves uncertainty, and often conflicting objectives. Action learning, instead of being a one-off linear inquiry that starts with questions and ends with answers, is a cyclic process that both begins and ends with action. It continuously incorporates research findings, as intermediate conclusions, to refine the next cycle of the inquiry process. This is not to say that clear definition of objectives is unimportant (see Chapter 3), but it reinforces the need to revise objectives as understanding of the system grows.

The action learning framework requires that key stakeholders are collaboratively involved in designing the questions to be addressed throughout the cyclic process. The idea is that as different aspects about an IUWM issue are investigated and revealed, this informs and clarifies the purpose of the next data gathering stage and so on. Part of the iteration is associated with enabling participatory learning, which involves key players learning and evolving their perspective. There are also benefits of this approach in terms of positively increasing stakeholder commitment to, and ownership of, the outcomes of the IUWM process.

With the action learning process being cyclical, the separation between data gathering and data analyses is therefore a nominal one only. These two activities necessarily occur interactively throughout the action learning process. Given this, it could appear difficult for an urban water manager to know where to start. Therefore, it is very important in the initial stages to be able to conduct a preliminary assessment of the evaluation data needs. Evaluation needs should be clearly established from the outset of data collection processes, and interactively involve the key stakeholders in both the initial framing of the data gathering needs as well as providing data for the needs assessment.

According to Bellamy et al. (2001), evaluation has a range of purposes including improving programme management, improving transparency and accountability, reducing risk and uncertainty, fostering learning and improving process. Furthermore, Bellamy et al. (2001, p. 407) also highlight that the practice of evaluation across these multiple purposes within fields such as IUWM:

> ..... are critical elements of successful policy development and implementation. The challenge to create policy processes, institutional arrangements and natural resource management practices that contribute towards achieving sustainable and equitable resource use outcomes requires rigorous evaluation as part of the change process. Evaluation is fundamental to identifying change, supporting an adaptive approach that is flexible enough to meet the challenge of change, and enabling progressive learning at individual, community, institutional and policy levels.

Evaluation and action learning is important for enabling individual, community, organizational and institutional learning about both the IUWM problem and the relative effectiveness of various approaches designed to address it. This will not only assist with ensuring programme efficiency, but importantly contribute to the development of a critical knowledge data bank that will improve the credibility of IUWM decision processes with key stakeholders. With the increased number of disciplinary inputs and organizational players with differing interests involved in the IUWM process, the practice of evaluation therefore needs to address and inform multiple needs.

REFERENCES

Bellamy J.A., Walker, D.H., McDonald, G.T. and Syme, G.J. 2001. A systems approach to the evaluation of natural resource management initiatives. *Journal of Environmental Management*. Vol. 63, No. 4, pp. 407–23.

Bishop, P. and Davis, G. 2002. Mapping public participation in policy choices. *Australian Journal of Public Administration*. Vol. 61, No. 1, pp. 64–75.

Brown, R. 2005. Impediments to integrated urban stormwater management: the need for institutional reform, *Environmental Management*. Vol. 36, No. 3, pp. 455–68.

Brown, R.R., Sharp, L. and Ashley, R.M. 2006. Implementation impediments to institutionalizing the practice of sustainable urban water management. *Water Science and Technology*, Vol. 54, No. 6–7, pp. 415–22.

Fisher, M. 2000. Putting people in the picture 1: social sciences for natural resource management. *Australian Journal of Environmental Management*. Vol. 7, No. 2, pp. 68–69.

Harding, R. 1996. Introduction and 'macro' briefing paper. Australian Academy of Science Fenner Conference on the Environment, '*Tracking Progress: Linking Environment and Economy through Indicators and Accounting Systems*', 30 September–3 October, 1996, Institute of Environmental Studies, University of New South Wales, pp. 1–10 and 16–19.

Hatton MacDonald, D. and Dyack, B. 2004. *Exploring the Institutional Impediments to Conservation and Water Reuse – National Issues, Report for the Australian Water Conservation and Reuse Research Program*. Canberra, CSIRO/Australian Water Association.

Johnson, A. and Walker, D. 2000. Science, communication and stakeholder participation for integrated natural resource management. *Australian Journal of Environmental Management.* Vol. 7, No. 2, pp. 82–90.

Lawrence, G. 1997. Indicators for sustainable development. F. Dodds (ed.) *The Way Forward. Beyond Agenda 21.* London, Earthscan Publications, pp. 179–89.

Lundqvist, J., Turton, A. and Narain, S. 2001. Social, institutional and regulatory issues. Č. Maksimović and J.A. Tejada-Guilbert (eds), *Frontiers in Urban Water Management: Deadlock or Hope.* Cornwall, UK, IWA Publishing, pp. 344–98.

Margerum, R.D. 2001. Organisational commitment to integrated and collaborative management: matching strategies to constraints. *Environmental Management,* Vol. 28, No. 4, pp. 421–31.

MBWCP, 2006. Moreton Bay Waterways and Catchment Partnership, *Annual Report 2005–06,* Brisbane, Australia. Available at http://www.healthywaterways.org/index.html (Accessed 02 July 2007.).

Mullen, M.W. and Allison, B.E. 1999. Stakeholder involvement and social capital: keys to watershed management success in Alabama. *Journal of the American Water Resources Association.* Vol. 35, No. 3, pp. 655–62.

Mitchell, B. 2005. Integrated water resource management, institutional arrangements, and land-use planning. *Environment and Planning A.* Vol. 37, pp. 1335–52.

Scales, I. 1997. *Consultation and people with a disability.* Disability Council of New South Wales, Sydney. Available at http://www.disabilitycouncil.nsw.gov.au/archive/97/scales.html (Accessed 02 July 2007.).

Serageldin, I. 1995. *Toward Sustainable Management of Water Resources,* World Bank, Washington DC, USA.

Shand, D. and Arnberg, M. 1996. *Responsive Government: Service Quality Initiatives.* Paris, Organization for Economic Cooperation and Development (OECD).

The Barton Group. 2005. *Australian Water Industry Roadmap: A Strategic Blueprint for Sustainable Water Industry Development.* Report of The Barton Group, Coalition of Australian Environment Industry Leaders. Available at www.bartongroup.org.au (Accessed 02 July 2007.).

Vlachos, F. and Braga, G. 2001. The challenge of urban water management. Č. Maksimović and J.A. Tejada-Guilbert (eds), *Frontiers in Urban Water Management: Deadlock or Hope.* Cornwall, IWA Publishing, pp. 1–36.

Wong, T.H.F. 2006. Introduction. T.H.F. Wong (ed.), *Australian Runoff Quality: A Guide to Water Sensitive Urban Design.* Canberra, Engineers Australia. pp. 1–8.

Chapter 12

Financial considerations

T.D. Fletcher[1] and J.-L. Bertrand-Krajewski[2]

[1]Department of Civil Engineering (Institute for Sustainable Water Resources), Building 60, Monash, Monash University, Melbourne 3800, Australia
[2]Laboratoire LGCIE, INSA-Lyon, 34 avenue des Arts, F-69621, Villeurbanne CEDEX, France

12.1 INTRODUCTION

The design and implementation of monitoring programmes clearly depend on financial constraints. This chapter outlines the principles of matching monitoring activities to available budgets, and identifies some strategies which may be used to obtain the maximum value of data, for the minimum expenditure.

12.2 PRINCIPLES FOR BALANCING FINANCIAL CONSTRAINTS AND DATA REQUIREMENTS

It is critical that financial constraints be identified *at the start* of the monitoring programme, so that trade-offs between costs and outcomes can be made in an explicit fashion. Ideally, the point of intervention of financial constraints should occur at the time that monitoring objectives are being identified. If this does not occur, there is the danger that stakeholders of the monitoring programme will have unrealistic expectations about the objectives that can be met, and may not realize the compromises that have been made in terms of the methodology, to meet the budget constraints. In such a case, uncertainty in the data (see Chapter 6) may mean that few if any of the original objectives are really met.

Clearly, however, there is a need to review the relationship between objectives and costs, once the monitoring methodology is identified and specified, as the methods chosen will finally determine the costs (Figure 12.1). If the budget is inadequate to meet the proposed methodology, then the *objectives* should be reviewed. Revising objectives will then lead to a revised methodology that, after enough iterations, can be achieved with the available budget. This approach also avoids the tendency for developing a 'shopping list' of parameters to measure, without a clear link back to the questions that can be answered by measuring these parameters (Bertrand-Krajewski et al., 2000). Alternatively, if some objectives cannot be met with available resources, and if these objectives are critical, additional resources will need to be obtained.

It must be emphasized that financial constraints should not be considered as the ultimate criteria. When monitoring objectives are critical, increasing available resources is more important than decreasing the scope of the objectives. The hidden or delayed costs of missing data and inadequate measurements necessary for the efficient management of urban water systems should be explicitly accounted for. Monitoring is too frequently among the first aspects to be sacrificed, as its present and short-term costs are accounted for more easily than its later and mid-term benefits. Cost saving on monitoring early on

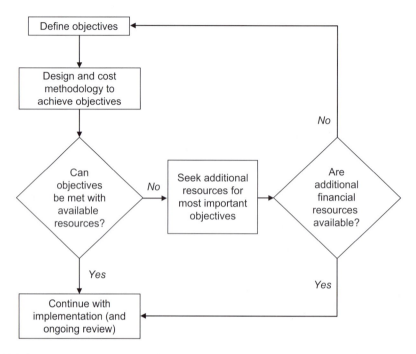

Figure 12.1 Interactions among various financial considerations in the design and implementation of monitoring programmes

in a project may provide false economy, resulting in much higher costs later due to inefficient operation and management.

In many situations, initially there will not be adequate information to determine the methodology necessary to meet the objectives. For example, if the objective of a monitoring programme is to determine the difference in nitrate concentration between two adjacent catchments of differing land use, then a pilot study may be used initially to understand the variability within each catchment and thus to determine the required number of samples to take from each catchment [see Zar (1999) for a discussion of the determination of required sample size using power tests]. In the absence of local pilot studies, it may be possible to extrapolate from other studies of the relevant data, although this should be undertaken with caution (i.e. with careful consideration as to whether there are differences that may preclude comparison between the studies).

Chapter 6 provides a detailed discussion of the analysis of uncertainty, and its role in design and implementation of monitoring programmes. Considerations of uncertainty are important, since only once uncertainty has been quantified can the required methodology to meet the objectives be confidently specified. Defining the allowable level of uncertainty is a key aspect in setting of the monitoring programme. The most important is not to reach the minimum possible uncertainty, which may be incredibly expensive, but to ensure that appropriate resources are spent to obtain the level of uncertainty which is coherent and compatible with the future use and application of the data.

The level of resources allocated to a monitoring programme should be proportional to the consequences of the issue to be monitored (and managed). To be able to do this, however, depends on having some baseline understanding of how the system of interest operates; this understanding itself may not be possible without baseline monitoring data. A pilot study (as discussed above) to determine uncertainties could be used to compare the potential error in a given monitoring design with the economic, environmental and social consequences of decisions made based on errors of the identified magnitude. For example, design of a water quality treatment plant based on inadequate understanding of variation in inlet water quality will have consequences far beyond the additional cost of monitoring to decrease uncertainty in predictions of variability in concentrations of key pollutants. A typical example: many operators usually hesitate to allocate money for monitoring programmes, but do not negotiate when very expensive capital infrastructure costs are proposed. All too frequently, it is considered that money can be saved for monitoring (typically interpreted as soft infrastructure) while spending money for investments (hard infrastructure and actual construction) is usually considered as unavoidable. This can occur even when the monitoring cost, as a proportion of the infrastructure investment, is minimal. A more balanced view is needed to account for the benefits expected from investments based on a better and reliable knowledge obtained from monitoring programmes. On the other hand, monitoring programmes should be clearly, rationally and economically defined with a clear anticipation of the application of the results. Monitoring programmes are not only a cost, but the only way (in association with modelling) to design and operate urban water systems in a correct, efficient and sustainable manner.

12.3 STRATEGIES FOR MAXIMIZING DATA VALUE WITH LIMITED RESOURCES

There are often opportunities for increasing the efficiency of expenditure on monitoring programmes. Some opportunities are identified below.

12.3.1 Interrogation of existing data sources

The most cost-effective way of collecting data may be to utilize existing data resources. Where there have not been significant changes in system operation, or land use, for example, historical data may provide the necessary level of detail to meet the monitoring objectives. Yet, the use of ancient, historical or existing data should be made very carefully. Existing data typically lack clear information regarding context: *who made the measurements? for what purposes? with what methods, protocols, and uncertainties?* Experience has shown that there is frequently a large gap between the amount of existing data and its genuinely usable fraction. Additionally all existing data must be checked and validated before being used for new objectives.

12.3.2 New technologies

Traditionally, new monitoring technologies have been considered to be expensive. As identified in Chapter 7, however, there is now a wide range of monitoring technologies that are relatively inexpensive to purchase, install and operate. Secondly, technologies such as remote sensing (e.g. satellite imagery) may provide access to data at very low cost. The advent of wireless networks and smart sensors has dramatically cut installation complexity (and thus costs), and allows more flexible deployment of systems within and between catchments. The relative cost of technological solutions will depend, however, on availability of local supply and support for the technology and the cost of labour. In some situations, it may be cheaper (and possibly preferable for other reasons) to employ manual sampling techniques rather than rely on technology. This decision can only be made with an understanding of local circumstances.

12.3.3 Sharing of monitoring networks

The nature of IUWM is such that there is a high probability of more than one agency needing the same data. Given this, every effort should be made to establish a collaborative approach to data collection, to exploit economies of scale, and to reduce or eliminate duplication. This is also a good opportunity for increasing available resources to share costs between partners and institutions.

12.3.4 Integration of monitoring and models

The integration of monitoring and modelling approaches may provide an appropriate compromise where resources are limited. For example, consider the situation where a proposed monitoring programme will examine a number of phenomena, including flow, runoff water quality and the response of aquatic ecosystems. In this case, some phenomena may be better understood (e.g. rainfall-runoff relationships) than others (e.g. water quality and ecological response). A lower intensity of rainfall and flow monitoring may be undertaken (supplemented by model predictions, calibrated against monitored data), with resources allocated to higher resolution measurement of water quality and ecosystem health indicators. Such decisions should be made in a fully transparent manner, however, with explicit assessment of the uncertainties (refer to Chapter 6).

12.3.5 Composite sampling

The spatial and temporal resolution of sampling will have a major influence of monitoring costs. Selecting an appropriate resolution is thus an important consideration (refer to Chapter 5), and depends on the initial objectives of the monitoring programme. For example, consider the situation where the objective is to quantify the pollutant loads generated from a catchment. In this case, the temporal variation within a storm may be unimportant, and flow-weighted composite samples may be used to determine the mean concentration for each monitored storm event. However, if the objectives require knowledge of variation in loads within the catchment (to allow prioritization of treatment strategies, for example), then spatial replication will be needed and a pilot study would help to determine the extent and density of sampling required.

REFERENCES

Bertrand-Krajewski, J.-L., Barraud, S. and Chocat, B. 2000. Need for improved methodologies and measurements for sustainable management of urban water systems. *Environmental Impact Assessment Review*, Vol. 20, No. 3, pp. 323–31.
Zar, J. H. 1999. *Biostatistical analysis*, 4th edn. Upper Saddle River, NJ, US, Prentice Hall.

Consideration and integration of specific urban water cycle components

Chapter 13

Monitoring to understand urban water cycle interactions

T.D. Fletcher and V.G. Mitchell

Department of Civil Engineering (Institute for Sustainable Water Resources), Building 60, Monash, Monash University, Melbourne 3800, Australia

13.1 INTRODUCTION

Part I of this book discussed the guiding principles critical for the development of a high quality urban water system monitoring programme, having clearly defined the programme's objectives, key variables to be monitored and an acceptable level of uncertainty. Part II shifts focus to discuss specific considerations that relate to the major elements of the urban water system, covering biophysical components, such as urban meteorology, water supply, stormwater and wastewater systems, as well as aquatic ecosystems. There is also a discussion of the data required to understand social and institutional components of the urban water system, including human health.

The purpose of this chapter is to provide an overview of monitoring requirements to identify, quantify and understand the interactions between each of the urban water cycle components. Accordingly, this chapter draws on and synthesizes the more detailed and specific information provided in Chapters 14 to 22. Many of these principles and considerations raised here are further highlighted in the two case studies provided in Part III of this book.

Each chapter in Part II examines the specific considerations on how to monitor individual components of the urban water cycle, discussing the potential interactions between each urban water component and other aspects of the urban water cycle and, more importantly, approaches to ensuring that these interactions are monitored effectively. Each of Chapters 14 to 22 can therefore be read as stand-alone documents. However, by definition, the interactions between them necessitate consideration of the information provided in the accompanying chapters. For example, to understand how to monitor the impacts of combined sewer overflows on aquatic ecosystems, an understanding of the functioning of combined sewers (Chapter 18) and aquatic ecosystems (Chapter 20) is required. Appropriate cross-references are therefore provided in each chapter.

There are a number of considerations critical to developing an integrated monitoring programme in support of integrated urban water management. Firstly, attention should be given to the monitoring programme as early as possible in the life of the relevant urban water infrastructure. For new infrastructure, for example, the monitoring programme should actually be designed *at the same time* as the infrastructure itself is being designed. This allows monitoring equipment to be installed at the appropriate locations, so that representative data can be collected.

Secondly, monitoring should be, as discussed in Part I, based on a sound conceptual model of the potential interactions between different components of the urban water

Table 13.1 Monitoring interactions between urban climate, water supply and other components of the urban water cycle

Component	Urban climate		Water supply	
	Potential interactions	Monitoring considerations	Potential interactions	Monitoring considerations
Water supply	Precipitation, temperature and evapotranspiration influence demand, particularly for outdoor uses.	Collect demand data contiguously with climate data (or arrange data-sharing between meteorological and water agencies). Ensure matching spatial scales are used (particularly if there is a significant rainfall gradient).		
	Extreme events such as severe freezing, or floods, may damage urban water supply infrastructure.	An efficient data and forecast sharing protocol with the meteorological agency will allow preventative actions to be scheduled.		
	Temperature and rainfall will affect pathogen growth and transport.	Models of pathogen response to climate will depend on compatible data series of each. Attention should be paid to the appropriate temporal and spatial scales. Such a model can then be used for pathogen growth predictions and risk management.		
Wastewater combined sewers	Infiltration into wastewater from surrounding soils will depend on soil moisture, which in turn will depend on evapotranspiration and rainfall.	Consideration to annual and long-term climate cycles needs long-term climate data at appropriate locations and scales, matched to groundwater data. The groundwater	Improper wastewater disposal may pollute drinking water supplies, resulting in reduced water availability and/or increased treatment cost.	Monitor quality of intake, paying particular attention to parameters that indicate pollution from wastewater (e.g. pathogens, oxygen demand). Ensure that water quality

Component			
	and wastewater monitoring should be designed with consideration for existing climate data. Rainfall patterns will determine the frequency of combined sewer overflows.		monitoring downstream of wastewater disposal is in place with data made available to water supply agency.
	Spatial variability in rainfall extremes needs to be considered. Historical records or stochastic modelling may be needed. Additional climate stations may be needed to capture the spatial and temporal variability.	Increased consumption of drinking water means larger flow rates in wastewater system.	Ideally, real-time demand data will allow prediction of treatment plant inputs. Forecasts based on historical data will generally be adequate. Demand data should be available to wastewater treatment operators.
		Reduction in drinking water consumption will reduce dilution and affect treatment processes. Leakages may cause wastewater to get sucked into water supply pipes, or vice versa.	Sewer leakage should be regularly measured using direct or indirect methods. Mass-balance methods can be used to identify if there is leakage from one system to another.
Stormwater	Stormwater pollutant concentrations will depend on rainfall intensity, as well as antecedent dry weather period.	Stormwater may impact drinking water quality, especially in mixed-land use catchments	Regular ongoing monitoring of stormwater quality is required, rather than ad hoc responsive monitoring. Stormwater quality data can be used to specify water treatment requirements.
	Local rainfall data is needed at high temporal resolution (e.g. at 1 minute intervals). Data needs to be near centroid of catchment.	Stormwater may provide an alternative water supply, for potable or non-potable purposes (e.g. rainwater tanks, large-scale stormwater harvesting).	Monitor stormwater quality at inlet, in storage, and at distribution points. Data will need to be compared with climate data of appropriate temporal resolution.

(Continued)

Table 13.1 (Continued)

Component	Urban climate		Water supply	
	Potential interactions	Monitoring considerations	Potential interactions	Monitoring considerations
			Potential interaction of stormwater and water supply networks, due to leaks.	As for wastewater, leak detection using mass-balance methods or direct monitoring should be regularly undertaken.
Groundwater	Groundwater levels depend on recharge rates which in turn depend on soil moisture. Soil moisture is governed by rainfall and evapotranspiration. Groundwater pollutant concentrations will depend partly on evapotranspiration and subsequent soluble ion concentrations.	Consideration of annual and long-term climate cycles needs long-term climate data at appropriate locations and scales, matched to groundwater data. Groundwater monitoring systems should be designed considering existing climate data. As above	Leakage from the water supply network will increase groundwater recharge and potentially increase the groundwater table.	Monitoring groundwater quality can be used to detect water supply leakage, in conjunction with groundwater level monitoring, and leak detection of the water supply network.
Aquatic ecosystems and urban streams	A wide range of indicators of ecosystem function and health are dependent on climate.	A conceptual model of interactions between climate and ecosystem should be used to design an integrated monitoring system.	The discharge of polluted and cold water from mains or reservoirs can degrade aquatic ecosystems.	Monitoring of ecosystem indicators (e.g. macro-invertebrate community assemblage) should be integrated with water quality data; ensuring data sharing between reservoir operator and agency for environmental management.

	Hydraulic stresses on lotic aquatic ecosystems will depend on rainfall.	Ensure that rainfall data can be provided at appropriate spatial scale and location (either by data-sharing or installation of specific gauges).	Extraction of water from waterways will affect flow regime.	Integrate measurement of (i) extractions, (ii) flow regime and (iii) ecosystem health indicators.
Human health	Temperature and rainfall will affect water quality of inflows to water supplies, and in storage. Long dry periods may lead to toxic algal blooms.	Ensure efficient data-sharing protocol between meteorological agency, water supplier and health agencies. Ideally, forecast data should be used to support predictive models for human health risk assessment.	Cross-connections or leaks into the water supply distribution network could result in contamination, causing human disease or long-term health effects.	Leak detection (direct or indirect methods) data should be used in conjunction with (i) monitoring water quality in the water supply system and (ii) analysis of human health data (e.g. rates of gastroenteritis).
Society and institutions	Societal attitudes and behaviour will determine community responses to climate-induced impacts such as drought and floods, with consequences for infrastructure requirements.	Social profiling should be undertaken as part of the assessment of risks and management strategies for droughts, floods, and other climate-induced water-related disasters.	Community preferences will determine the required (a) security of supply and (b) water quality.	Regular community profiling should be undertaken to understand community preferences and attitudes.

Table 13.2 Monitoring interactions between wastewater (and combined sewers) and other components of the urban water cycle

Component	Wastewater (and combined sewers)		Stormwater	
	Potential interactions	Monitoring considerations	Potential interactions	Monitoring considerations
Stormwater	Cross-connections between wastewater will result in contamination of stormwater. In separate systems, stormwater inputs can affect efficiency wastewater treatment. In combined systems, the proportion of stormwater and wastewater will affect treatment plant performance.	Undertake leak detection of wastewater using direct or indirect methods. Monitor stormwater quality for pollutants indicative of wastewater inputs (pathogens, oxygen-demanding substances, etc). Implement an integrated monitoring programme for the two streams with compatible timing and location of measurements. Use mass-balance analysis (based on simultaneous measurements of wastewater and stormwater).		
Groundwater	Groundwater may leak into wastewater or vice versa. Groundwater depletion via wastewater system may cause damage to overlying structures.	Monitoring groundwater quality can be used to detect for wastewater leakage, in conjunction with groundwater level monitoring, and leak detection of the wastewater network.	Exfiltration from stormwater sewers recharges groundwater and raises its level and may cause geotechnical problems by decreasing soil stability. The effect on groundwater quality depends on the quality of stormwater.	Exfiltration is difficult to detect and may need detailed mass-balance monitoring of stormwater, linked to monitoring of groundwater quantity and quality. Tracer measurements may be required, but will be difficult.

Aquatic ecosystems & urban streams	Wastewater and combined sewer systems influence aquatic ecosystems via (i) continuous discharge of (usually treated) effluent, or (ii) overflows of untreated water.	(Integrated monitoring of wastewater discharge (quality and quantity) and ecosystem health indicators (i) needs data-sharing between relevant agencies and (ii) high temporal-resolution monitoring to characterize the overflow behaviour, matched with upstream and downstream ecological monitoring. Long-term ecological monitoring may also be required to detect trends related to wastewater discharge frequency and amount.	Increased peak flows will redefine channel size and shape and isolate floodplains.	A dense network of strategically located bores will help to identify possible leakage points. Integrated programme of hydrologic/hydraulic monitoring (discharge rate, velocity, shear stress, etc.) with geomorphological data (pool and riffle development, substrate classification, channel dimensions, etc.). Integrated monitoring of (i) stormwater inputs and (ii) receiving water quality, to determine storm water contribution and assimilation capacity of waterway. Data will need to be combined with sampling of sediment quality, to determine long-term accumulation effects. Ensure that ecological monitoring is linked with hydrologic and water quality monitoring of receiving waters; base programme on strong conceptual model of interactions.
			Water quality of urban streams will be greatly affected.	
			Aquatic life in lentic and lotic receiving waters will be affected by hydrologic and hydraulic impacts, as well as water quality.	
Human health	Cross connections may result in wastewater inputs to drinking water, posing a threat to human health.	Leak detection (direct or indirect methods) data should be used in conjunction with (i) monitoring water quality in the water supply system and (ii) analysis of human health data (e.g. rates of gastroenteritis).	Polluted stormwater could lead to direct pollution of drinking water, particularly where water is extracted from downstream surface waters.	Integrate monitoring of drinking water quality (at inlet, storage and distribution) with monitoring of potential sources of contamination (including stormwater, and other potential inputs).

(Continued)

Table 13.2 (Continued)

Component	Wastewater (and combined sewers)		Stormwater	
	Potential interactions	Monitoring considerations	Potential interactions	Monitoring considerations
	Wastewater discharge to waterways may pose a threat, through pathogens, or bioaccumulation of micropollutants and toxicants in human food sources (e.g. fish).	Ensure a coordinated monitoring plan of (i) leakages and overflows, (ii) water quality in waterways and (iii) contamination levels in sediments and aquatic organisms, such as fish.	Poor quality stormwater poses a risk to humans where downstream surface waters are used for recreational or fishing purposes. The risk is greatest where high levels of pathogens are present (usually due to contamination from wastewater).	Design a risk-based monitoring programme for waterway water quality and pathogen monitoring, based on stormwater inputs (e.g. sampling triggered by storm events, etc). Sampling should be of sufficient temporal resolution to understand the variation in water quality and pathogen loads, not just event mean concentrations.
Society and institutions	Trends and variability in water use will affect wastewater production. Community attitudes will also determine the required level of treatment of wastewater, and the manner of its disposal. Similarly, community attitudes will determine what frequency of combined sewer overflow is acceptable.	Wastewater management should be based on ongoing surveys to determine community preferences.	Values placed on receiving waters by communities will influence how stormwater is managed (i.e. the degree to which environmental values of receiving waters are taken into account). Similarly, community values regarding property, will determine the acceptable impact of flooding, and thus the importance of flood management.	Ongoing surveys of community preferences regarding natural waterway values, and the impacts of stormwater, should be undertaken. Trends in preferences may be as important as the 'current state' of preference. Similarly, socio-economic data, such as land-use and property valuations, can be useful for prioritizing flood management programmes.
	Recycling of wastewater will depend on the ability of the wastewater agency to be involved in water supply, and on the constraints or facilitation provided by other institutional stakeholders.	Before embarking on a project to develop wastewater recycling, information should be gathered on the institutional constraints and opportunities, including institutional capacity, regulatory restrictions, etc.	The responsibilities for management of stormwater may be shared between multiple agencies, with potential for gaps and inconsistencies.	Mapping of responsibilities, as well as surveys of institutions (through interviews, etc.) should be used to assess the capacity (within and between organizations) for stormwater management.

Table 13.3 Monitoring interactions between groundwater and other components of the urban water cycle

Component	Potential interactions	Considerations for monitoring interactions
Aquatic ecosystems and urban streams	Groundwater may discharge into urban streams affecting their quantity and quality or recharge from them, in which case the streams affect the quality of groundwater. Groundwater levels will also affect flow regimes in surface waters.	Integrate ecosystem health monitoring with water quality and hydrological monitoring of groundwater (and surface waters). The integrated monitoring system should be developed using a conceptual model of the potential interactions between groundwater and aquatic ecosystem function and composition.
Human health	Leakage from wastewater and stormwater (including in combined sewers) into groundwater pose potential risks to human health, where the groundwater (i) is used for water supply, or (ii) leaks into the water supply system.	Undertake regular groundwater monitoring with (i) locations based on model of potential points of contamination to drinking water and (ii) frequency based on understanding of temporal dynamics and inputs of pollutants to groundwater. Pilot monitoring may be required to understand these processes before the long-term monitoring programme is established.
	Long-term contamination of groundwater (e.g. from heavy metals) will cause chronic health problems, where ground-water is extracted for water supply.	Integrate monitoring of groundwater quality (and sources of contamination) with human health data. For example, analyse relationships between groundwater quality and instances of (or trends in) certain conditions known to be caused by particular pollutants.
Society and institutions	Limited community understanding of interactions between surface water and groundwater may inhibit attempts to achieve more integrated management.	Determining the understanding of the interactions of various urban water system components through community surveys, interviews, and other means will underpin management strategies.
	Responsibility for groundwater may or may not rest with the institutions who manage other aspects of the water system which affect groundwater quantity and quality. Institutional constraints may limit the degree to which the respective agencies collaborate.	Mapping intra- and inter-organisational impediments and constraints to integrated management can be useful for identifying policy interventions or the need for restructuring organisational responsibilities.

system. Interactions between urban water cycle components should be identified, based on a conceptual model of the system, and an integrated monitoring plan – based on that model – should be developed. Even when the only interest of a particular agency rests with, for example, drinking water quality, it is evident that many other aspects of the urban water system may impact water quality, which in turn may impact on wastewater volumes and quality. Isolated monitoring of individual aspects of the urban water system is unlikely to deliver a robust management strategy that can detect or predict changes and facilitate response to them.

13.2 MONITORING URBAN WATER CYCLE INTERACTIONS

Examples of interactions between each of the urban water cycle components are provided in tables in each of Chapters 14 to 22. A selection of these examples is provided in the tables (Tables 13.1 – 13.5), in order to illustrate possible approaches to monitoring these interactions.

Chapter 14

Urban meteorology

F.H.S Chiew[1], A.W. Seed[2], Y. Zhang[3] and T. Adams[4]

[1] CSIRO, Land and Water, Canberra 2601, Australia
[2] Bureau of Meteorology Research Centre, Melbourne 3000, Australia
[3] National Risk Management Research Laboratory, US Environmental Protection Agency, 26 W. Martin Luther King, Cincinnati, OH 45268, USA
[4] Ohio River Forecast Center, National Weather Service, 1901 South State Route 134, Wilmington, OH, 45177, USA

14.1 INTRODUCTION

The socio-economic consequences of climatic conditions in urban areas are amplified by the altered hydrologic cycle that comes with urbanization and conversion of land use to impervious areas. These conditions put unique demands on the field of meteorology, and the need to incorporate meteorology into urban water management decision-making. This chapter outlines the applications of urban meteorology within the integrated urban water management (IUWM) framework, and provides some specific considerations with regard to the collection and use of meteorological data in the urban context. The user is also referred to local guidelines for specific information on measuring meteorological variables and selecting measurement equipment, for example.

14.2 INTERACTIONS WITH THE URBAN WATER CYCLE

Climate is the principal driver of the urban water cycle, through its influence on precipitation and evapotranspiration (Table 14.1). Urban water managers therefore require data on prevailing weather patterns, their degree of regularity and the nature of extreme events that could impact on urban water system components. This chapter focuses on the implications for monitoring the interactions between urban climate and water cycle management and the specific requirements and applications of urban meteorological monitoring. It can clearly be seen from Table 14.1 that *meteorological data are fundamental to interpreting and understanding all urban water cycle components*. Any urban water system monitoring campaign must therefore include integrated monitoring of climate variables.

Urban meteorology, while sharing most of the topics with traditional meteorology, focuses on weather features and mechanisms specific to urban settings. Urban centres are not only recipients of, but also active contributors to the meteorological variations that take place in the vicinity of these population centres. The anthropogenic alteration of land cover and the economic activities that take place within and around urban areas all exert considerable influence on local weather and climate (see Marshall et al., 2004). Among the most extensively discussed are the effects of the 'urban heat island' effect, and the impacts of air pollution (Crutzen, 2004). The urban heat island effect is a consequence of land-use change that converts vegetated surfaces into buildings or pavement,

Table 14.1 Potential interactions between climate and urban water cycle components

Urban water system component	Potential impact of climate
Water supply	Precipitation, temperature and evapotranspiration will influence demand, particularly for outdoor uses. Evapotranspiration will affect losses from water supply stores. Temperature will influence chemical treatment and water quality in water stores and supply networks. Extreme events such as severe freezing, or floods, may result in damage to urban water supply infrastructure.
Wastewater and combined sewer systems	Temperature will have significant effects on chemical and biological treatment processes. Temperature and rainfall will affect pathogen growth and transport. Infiltration into wastewater system from surrounding soils will depend on soil moisture, which in turn will depend on evapotranspiration and rainfall. Rainfall patterns will determine the frequency of combined sewer overflows.
Stormwater	Stormwater runoff (volume, rate, shear stress, etc.) will depend on rainfall intensities (measured at short duration). Pollutant concentrations will depend on rainfall intensity, as well as antecedent dry weather. Evapotranspiration and antecedent rainfall will influence soil moisture, thus influencing rainfall-runoff response. Rainfall intensities and spatial patterns will determine the risk and frequency of flooding.
Groundwater	Groundwater levels depend on recharge rates which in turn depend on soil moisture, which is governed by rainfall and evapotranspiration. Groundwater pollutant concentrations will depend partly on evapotranspiration and subsequent concentration of soluble ions, etc.
Aquatic ecosystems and urban streams	A wide range of indicators of ecosystem function and health are dependent on climate. For example, ambient temperature will influence dissolved oxygen concentrations, photosynthesis: respiration ratios, and the metabolism of organisms. Hydraulic stresses on lotic aquatic ecosystems will depend on rainfall (see items under stormwater above).
Human health	Temperature and rainfall will affect water quality in the inflows to water supplies, and in storages. Long dry periods may lead to toxic algal blooms in receiving waters.
Society and institutions	Societal attitudes and behaviour will determine community responses to climate-induced impacts, such as drought and floods, with consequences for required infrastructure, among other elements.

thereby reducing the overall heat retaining capacity of the land surface and albedo. In the summertime, the differential heating capacity between the urban land and adjacent areas translates into a temperature gradient that in turn induces local circulation conducive to the development of convective rainfall. In contrast, the presence of air pollution, often in the form of urban smog, tends to reduce incoming solar radiation penetrating to the earth surface, which tends to suppress convection.

Over the longer term, urbanization may also impact the regional climate by altering the hydrologic cycle. The increasing area of impervious surface, for example, usually accelerates runoff generation and thus may decrease the amount of water infiltrated during a storm event, decreasing soil water content and subsequently the moisture available for evapotranspiration. With increased urbanization the time series of stream-flow maxima is altered, producing an increased frequency of floods of greater magnitude.

Human activities overall are deeply affected by weather and climate and vice versa in urban areas. This mutual impact and the associated, multiple feedbacks pose considerable challenges to both climate forecasting and to engineers attempting to manage risks by incorporating weather and climate pattern considerations into the design of wet-weather flow infrastructure.

14.3 CLIMATE CYCLES AND VARIABILITY

Prevailing meteorological conditions are often collectively referred to as the regional climate. The term 'climate' implies that the mechanisms underlying the phenomena operate at sufficiently long time scales that they can be considered deterministic. As an example, thunderstorms that are induced by sea-breeze circulation typically develop in response to temperature differences between the land surface and ocean, and occur on an approximately daily basis in places such as Florida following the solar cycle (Byers and Rodebush, 1948). The solar cycle in turn stems from the rotation of the earth which, for all practical purposes, can be thought as time-invariant. This assumption of time-invariance allows researchers to derive mathematic relationships on the basis of past observation, and to use these relationships to predict the future weather phenomena. In the case of sea-breeze circulation, historical seasonal and diurnal temperature profiles can be used in the absence of other information as predictors of summertime rainfall.

On the other hand, extreme events are usually induced by disturbances within limited spatial and temporal scales. These disturbances translate into a particular weather phenomenon, though only when coupled with conditions that favour their propagation (see examples pertaining to thunderstorm development in Houze, 1994). The multitude of conditions that have to be satisfied prior to an occurrence of an extreme event suggests that a large set of information is needed simply to determine the underlying causes. Prediction of tornadoes, for instance, requires detailed, real-time information about the spatial distribution of air pressure, temperature, humidity and wind patterns, the latter usually identified using Doppler radar to detect small-scale rotations. A slight deviation of the predicted fields from the actual ones would result in an incorrect forecast (see related discussions in Thompson and Edwards, 2000). Historical records of tornadoes over a given area may help illuminate factors related to tornado development under specific weather and geographic conditions, but render at best limited utility in forecasting the time and location of tornadoes. Often, stochastic modelling approaches are used (where the occurrence of a particular event is modelled using a probability distribution) because there is not adequate understanding currently, to predict deterministic cause-effect relationships. This does not necessarily mean that the process is 'random', but simply that current measurement and prediction methods are inadequate to understand and represent these relationships.

For IUWM, there is a need to measure and understand both the long-term 'trend' behaviour, as well as the behaviour (frequency, extent and duration) of extreme events.

Different monitoring systems and approaches will be needed in each of these two contexts. In particular, the spatial and temporal scale requirements will be quite different, as discussed in Section 14.5.

14.4 METEOROLOGICAL VARIABLES AND DATA SOURCES

14.4.1 Key variables

The meteorological variables most pertinent to integrated urban water management include temperature (of water, air and ground), precipitation (both total rainfall over periods such as a day, week, or month, and also rainfall intensity, measured at short durations such as one-minute time steps) and wind speed. Humidity and solar radiation are very useful, as interpretative variables. Also important is potential evapotranspiration (PET), a quantity critical to the modelling of the hydrologic cycle and usually estimated indirectly (although it can be measured using lysimeters) from humidity and temperature measurements. Of all these variables, precipitation is the most extensively instrumented. Precipitation is also highly variable, through both time and space (see Chapter 5). For example, rainfall measurements taken using a typical rain gauge (having an area of $0.125\,m^2$) will be different from rainfall measurements obtained over a $1\,m^2$ area.

14.4.2 Meteorological data sources

Most countries have a meteorological agency from which climate data, such as temperature statistics (mean, minimum, maximum, at different intervals), precipitation, evaporation, humidity and the like can be obtained. In many cities rainfall and some streamflow data are collected by an urban water management agency, generally for transport purposes, stormwater and sewer system monitoring and for flood forecasting and warning. There are also various global datasets for the aforementioned variables around the world, although these will not provide the spatial resolution that may be required for a local study. These include time series data from tens of thousands of stations worldwide (e.g., Global Historical Climatology Network, www.ncdc.noaa.gov/cgi-bin/res40.pl?page=ghcn.html (Accessed 02 July 2007)) and the gridded data products (e.g., $2.5° \times 3.75°$ precipitation data from the University of East Anglia Climate Research Unit (www.cru.uea.ac.uk/cru/data (Accessed 02 July 2007)) and $1.0° \times 1.0°$ precipitation data from the Global Precipitation Climatology Centre (www.dwd.de/en/FundE/Klima/KLIS/int/GPCC/ (Accessed 02 July 2007)).

14.5 CONSIDERATIONS FOR THE COLLECTION AND USE OF METEOROLOGICAL DATA

14.5.1 Spatial scales

Most cities and urban settlements have gauging stations that record precipitation at daily, and sometimes sub-daily, resolution. Since the station records are point measurements, it is often necessary to interpolate these records in order to obtain a spatial representation of precipitation (see also Chapter 5 for guidance on temporal and spatial scale issues). Figure 14.1 shows an example of the interpolation error for a 1 km

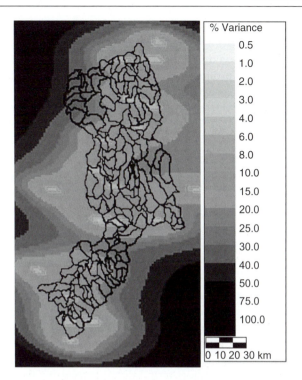

% Variance
0.5
1.0
2.0
3.0
4.0
6.0
8.0
10.0
15.0
20.0
25.0
30.0
40.0
50.0
75.0
100.0

0 10 20 30 km

Figure 14.1 Interpolation error expressed as a percentage of the rainfall variance for an interpolation of hourly rainfall from an operational flood-warning rain gauge network onto a 1 km grid for a catchment near Melbourne, Australia (See also colour Plate 14)

grid for a catchment near Melbourne, Australia. The accuracy of interpolated rainfall fields may vary with the nature of precipitation regimes, and caution should be exercised when utilizing the interpolated rainfall data in practical applications.

A relatively dense network of gauges is often required for accurate rainfall estimates over the spatial and temporal scales that are typical for urban forecasts. However, these networks are not common. As an alternative to expanding a gauging network to provide better spatial coverage, the use of remote sensing technologies to derive spatial-temporal rainfall fields has attracted recent attention. Among the remote-sensing instruments, radar provides rainfall estimates over relatively large spatial domains and at relatively high spatial and temporal resolutions (1 km to 4 km, 5-minute data for weather surveillance radar 1988 Doppler, or WSR-88D). In the United States, the network of WSR-88D radars is already playing a critical role in forecasting precipitation, tracking tornadoes and predicting flash floods. However, despite the strengths associated with radar products, the quantitative rainfall rates derived from radar signal reflectivity alone are subject to bias and high uncertainty. Research advances have been made on combining unbiased point rainfall intensity measurements from the rain gauge with the biased spatial details of storm patterns revealed by the radar (Todini, 2001; Zhang et al., 2001).

From the perspective of hydrological applications (Einfalt, 2005), precipitation data with a high spatial resolution may be required because of small-scale applications or because radar provides a spatial view on fast-response phenomena (Figure 14.2) and, thus, provides 'nowcasting' (short-term forecasting) information.

The spatial variability of quantities such as temperature, radiation and PET is, as previously discussed, not as pronounced as that of rainfall. One can typically employ the data from some distance away to represent the quantity for the area of interest. In addition, because the data do not vary as greatly from year to year, the use of long-term monthly averages may be possible for some applications. However, in this case, appropriate verification should be undertaken to see that the assumptions of 'stationarity' are met.

14.5.2 Temporal scales

The required temporal interval for monitoring depends on the phenomenon being observed, its variability and impacts. For example, average daily temperature (or maximum daily temperature) will provide the required temporal resolution for most urban water management applications. However, the time of concentration for urban catchments is generally very short, so short-duration rainfall intensity data is often needed for applications in stormwater management (Chapter 17) or flooding. For example, short-duration rainfall forecasts are critical for real-time control and flood management.

It is unfortunately beyond the current state of the art to use numerical weather prediction techniques to forecast rainfall (particularly convective rainfall which is largely responsible for urban flooding) at the space-time resolution which is needed for such applications. As a result, existing short-term rainfall forecasting techniques are largely based on spatially and temporally extrapolating the observed rainfall by weather radar (see, for example, Dixon and Wiener, 1993 for an algorithm designed for this purpose). Bertel et al. (2004) gives a good summary of the current use of radar data in hydrological forecasting and Ebert et al. (2004) provide details of forecast accuracy.

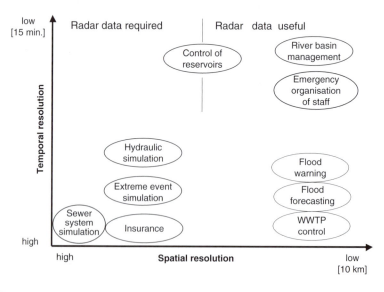

Figure 14.2 Utility of radar at different levels of spatial and temporal resolution for various hydrological applications

The difficulty lies with the fact that radar rainfall observations are uncertain, and fore-casts beyond 30 minutes are even more uncertain. The utility of radar data is especially reduced in urban settings by problems such as ground cluttering and beam blockage. Progress with the use of forecast rainfall data in hydrological applications depends on the *ability to represent the uncertainty* in the observation, forecast, and hydrological model to the user of the information in a way that is relevant to their decision making process (see also Chapter 6).

14.5.3 Climate change and its implications

In much of the world, information on weather and climate is already considered in the planning and implementation of urban infrastructure. As an example, in the United States, the capacity of stormwater conveyance and detention facilities is usually tied to the probable maximum precipitation (PMP), that is, the anticipated maximum rainfall that can possibly occur within a given time interval. Similarly, Australian design stan-dards for both stormwater quantity and quality management focus on the average recurrence interval (ARI) of a given rainfall event (based on the time of the concentra-tion in the catchment of interest). Water harvesting and storage systems are also sized primarily on a probabilistic basis, taking into account the frequency, duration and severity of drought conditions (and concurrent water demand levels, which are also affected by climate).

As the earth's climate is in a state of constant change, it is important to examine the assumptions underlying the derivation of statistical quantities such as PMP, and take into account the possible departure of the actual outcome from the prediction due to climate variability. Climate variability is defined with respect to time scales (inter-annual, 3-year to 7-year El Niño/Southern Oscillation (ENSO), inter-decadal and climate change). For example, the annual rainfall time series of Melbourne in Figure 14.3 highlights the considerable year-to-year variation and inter-decadal periods that are considerably drier or wetter than others.

The use of time series data with sufficient temporal coverage (two to three times the lifecycle of the design object) is recommended. When such data are unavailable, methods need to be established to address the uncertainty associated with the use of short-duration

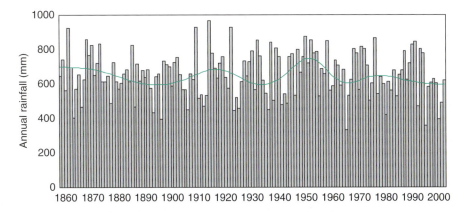

Figure 14.3 Annual rainfall recorded at Melbourne, Australia, 1860 – 2000

data. One way of quantifying the uncertainty is by incorporating synthetic climate data derived from stochastic climate model simulations. Examples of stochastic climate models can be found in www.toolkit.net.au/scl (Accessed 02 July 2007.), Wilks (1998) (multi-site daily rainfall model), Heneker et al. (2001) (single-site sub-daily rainfall model) and Seed et al. (1999) (space-time design rainfall model).

Over the longer term, there may be a need to consider the impacts of climate change on urban water systems. It is difficult to statistically confirm a trend in climate change given hydrologic time series data due to the short length of the record and high variability (Chiew and McMahon, 1993; Kundzewicz and Robson, 2000). However, there is strong evidence now that global warming is occurring and most of the warming over the last 50 years is attributable to human activities. Global warming is likely to lead to more intense rainfall, but the direction of change in total rainfall (seasonal and annual) will differ regionally across the globe (Intergovernmental Panel on Climate Change, 2001). Caution also needs to be exercised in using urban streamflow as an indicator of climate change since, with the effect of urbanization, the resulting non-stationarity of the time series may mask some of the hydrologic consequences of climate change.

Despite the uncertainty in climate change projections, it is useful to incorporate these projections because the impacts on urban water infrastructure can be significant (Intergovernmental Panel on Climate Change, 2001; World Meteorological Organization, 2004). A balanced approach would be to consider the range of plausible changes estimated by global climate models (that take into account the range of projections of greenhouse gas concentrations and global climate sensitivity to enhanced greenhouse gas concentrations). In the absence of more sophisticated models, one can assess the potential impacts of climate change by simply scaling the input data (historical or stochastically derived) by the range of projected changes to reflect a greenhouse-enhanced climate and compare the modelling results from the scaled input data to the original input data (Chiew and McMahon, 2002). Ongoing research is needed to develop methods which can more accurately determine the potential influence of climate change on the performance, durability and failure risk of urban water systems, including stormwater, wastewater, water supply and even aquatic ecosystems.

REFERENCES

Bertel, V., Kris, C., Cheze, J.-L., Annas, J., Moore, R., Jonas, O., Milan, S. and Jans, S. 2004. *Evaluation of Operational Flow Forecasting Systems that use Weather Radar.* Norrköping, Sweden, Swedish Meteorological and Hydrological Institute. (COST717–Working Group 1, Task 6.)

Byers, H.R. and Rodebush, H. 1948. Causes of thunderstorms of the Florida Peninsula. *Journal of Meteorology,* Vol. 5, pp. 275–280.

Chiew, F.H.S. and McMahon, T.A. 1993. Detection of trend or change in annual flow in Australian rivers. *International Journal of Climatology,* Vol. 13, pp. 643–53.

Chiew, F.H.S. and McMahon, T.A. 2002. Modelling the impacts of climate change on Australian streamflow. *Hydrological Processes,* Vol. 16, pp. 1235–45.

Crutzen, P.J. 2004. New directions: the growing urban heat and pollution 'island' effect - impact on chemistry and climate. *Atmospheric Environment,* Vol. 38, No. 21, pp. 3539–40.

Dixon, M. and Wiener, G. 1993. TITAN: Thunderstorm identification, tracking, analysis, and nowcasting – a radar-based methodology. *Journal of Atmospheric and Oceanic Technology,* Vol. 10, No. 6, pp. 785–97.

Ebert, E., Wilson, L., Brown, B., Brooks, H., Bally, J. and Janeke, M. 2004. Verification of now-casts from the WWRP Sydney 2000 forecast demonstration project. *Weather and Forecasting*, Vol. 19, pp. 73–96.

Einfalt, T. 2005. A hydrologists' guide to radar use in various applications. Paper presented at the Tenth International Conference on Urban Drainage, 21–26 August 2005, Copenhagen.

Heneker, T.M., Lambert, M.F. and Kuczera, G. 2001. A point rainfall model for risk-based design. *Journal of Hydrology*, Vol. 247, pp. 54–71.

Houze, R.A. 1994. *Cloud Dynamics*. New York, Academic Press.

Intergovernmental Panel on Climate Change. 2001. *Climate Change 2001: Impacts, Adaptation and Vulnerability – Contribution of Working Group II to the Third Assessment Report*. J. J. McCarthy, O. F. Canziani, N.A. Leary, D.J. Dokken and K.S. White (eds). Cambridge, Cambridge University Press.

Kundzewicz, Z.W. and Robson, A. (eds). 2000. *Detecting Trend and Other Changes in Hydrological Data*. Paris/Geneva, World Climate Programme – Water, UNESCO/WMO. (WCDMP-45, WMO/TD 1013.)

Marshall, C.H., Pielke, R.A., Steyaert, L.T. and Willard, D.A. 2004. The impact of anthropogenic land-cover change on the Florida Peninsula sea breezes and warm season sensible weather. *Monthly Weather Review*, Vol. 132, No. 1, pp. 28–52.

Seed, A.W., Srikanthan, R. and Menabde, M. 1999. A space and time model for design rainfall. *Journal of Geophysics Research*, Vol. 104, pp. 31623–31630.

Thompson, R.L. and Edwards, R. 2000. An overview of environmental conditions and forecast implications of the 3 May 1999 tornado outbreak. *Weather and Forecasting*, Vol. 15, No. 6, pp. 682–99.

Todini, E. 2001. Bayesian conditioning of radar to rain gauges. *Hydrology and Earth System Sciences*, Vol. 5, pp. 225–32.

Wilks, D.S. 1998. Multisite generalization of a daily stochastic precipitation generation model. *Journal of Hydrology*, Vol. 210, pp. 178–91.

World Meteorological Organisation. 2004. *Climate: Into the 21st Century*. W. Burroughs (ed.). Cambridge, Cambridge University Press.

Zhang, Y., Smith, J.A. and Baeck, M.L. 2001. The hydrology and hydrometeorology of extreme floods in the Great Plains of Eastern Nebraska. *Advances in Water Resources*, Vol. 24, No. 9–10, pp. 1037–50.

Chapter 15

Water supply

D. Prodanović

Institute for Hydraulic and Environmental Engineering, Faculty of Civil Engineering, University of Belgrade, Bulevar Kralja Aleksandra 73, 11000 Belgrade, Serbia

15.1 INTRODUCTION

From the very beginnings of urbanization, the water supply system (WSS) performed a central role. The ancient Egyptians, Greeks and Romans constructed systems to capture, store and distribute water using the same basic approach as is used today: capture, lifting, storage, treatment and distribution to consumers. Valves and pumps are used to manage the distribution, while storage reservoirs are used to balance temporal differences in supply and demand.

In the earliest WSSs, the only information collected was whether adequate water availability existed. Over time, monitoring levels increased, partly as water supply has gone from an entirely public-good function to a commonly commercial operation, where profit optimization drives the need for data (Obradović, 1999). In such an environment, real information about captured and delivered water, as well as accurate data regarding water quality issues enables both system-control and evaluation of economic viability.

In most cases measurements in contemporary WSSs are integrated within existing informatics support systems (Maksimović and Prodanović, 1995) (e.g. Figure 15.1). While these systems offer great capability, they require ongoing maintenance and upgrading, and their original design, if not carefully thought out, will limit performance.

Water supply systems are just one component of the complex urban water system, and the interactions between components are critical (see for example Chapters 1 and 13). Accordingly, the WSS needs to be able to exchange data with other external systems (see for example Chapters 9 and 10). The data exchange can be either continuous online as data are acquired (for example, the WSS can send to local authorities or can publish directly on the internet the content of chloride and turbidity of water for selected sites within the water distribution network) or off-line, as exchange of historical data, at specified (systematic or ad-hoc) times. Regardless of the online or off-line data exchange, it is important to follow the recommendations given in Part I of this book: the measuring site and equipment have to match the monitored variable requirements, the uncertainty has to be assessed, measured data validated and metadata used to store the sensor position, measuring conditions, calibration curves and validation results.

The water supply system is a complex system with a number of separate but connected components: water intake (withdrawal), conveyance of untreated water, water treatment, conveyance of clean water, water quality conditioning, reservoirs (storage), distribution network, and finally, the consumers (or water customers). Each subsystem has its

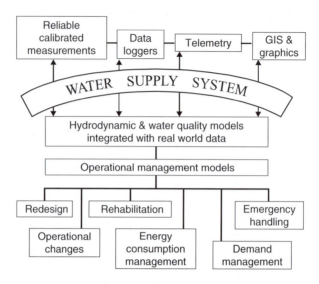

Figure 15.1 Components of an integrated informatics support system to a water supply system (WSS)

Source: Maksimović and Prodanović, 1995.

own required variables for status evaluation, control and optimal operation. In the next subsections a short description of the monitoring requirement for each of these components will be given.

Within water supply systems, water is often a 'product', and thus legally regulated metrology has to be used to quantify its movement and consumption. There are also other important roles of measurement in WSS, namely, *for process control, water balance check, modelling purposes* and for *diagnosis of system details.*

15.2 INTERACTION WITH OTHER URBAN WATER SYSTEM COMPONENTS

The water supply system makes up just one component within the complex urban water environment. It is closely coupled with other urban water systems (Table 15.1). The WSS monitoring programme must therefore take into account those interactions and acquire enough data to evaluate them. Collected data should be shared between systems and merged within one common database (Chapter 9). Storage of metadata (e.g. place and conditions of measurement, sensor type and manufacturer, calibration data, real time and date of measurement) is essential for effective use and sharing of such data.

15.3 SPECIFIC REQUIREMENTS WITHIN WATER SUPPLY SUBSYSTEMS

Each subsystem (intake, storage, treatment, etc) within the WSS has to be monitored and managed in order to optimize its performance. Data requirements for each subsystem are provided in the following sections. Of critical importance to all components is

Table 15.1 Potential interaction between water supply systems and other urban water system components

System component	Potential interactions with water supply system
Urban climate	Precipitation, temperature and evapotranspiration will influence demand, particularly for outdoor uses. Evapotranspiration will affect losses from water supply stores. Temperature will influence chemical treatment and water quality in water stores and supply networks. Extreme events such as severe freezing, or floods, may result in damage to urban water supply infrastructure.
Wastewater and combined sewer systems	Improper wastewater disposal may pollute drinking water supplies, resulting in reduced water availability and/or increased treatment cost. Increased consumption of drinking water means larger flow rates in the wastewater system, whilst a reduction in drinking water consumption will reduce dilution and affect treatment processes. In leaking water supply systems, wastewater can be sucked into the pipe and mixed with clean water if pressures fall below zero. Clean water from leaking water supply networks can infiltrate the wastewater system, changing the quantity and quality of wastewater and affecting the operation of treatment facility.
Stormwater	Stormwater may impact drinking water quality, especially in mixed land-use catchments. Stormwater may provide an alternative water supply, for potable or non-potable purposes (e.g. rainwater tanks, large-scale stormwater harvesting). Potential interaction of stormwater and water supply networks, where leakages in water supply pipes exists.
Groundwater	Leakage from the water supply network will increase the groundwater recharge and potentially increase the groundwater table. Urban or suburban groundwater is the main source of clean water for many systems, so monitoring and protection of groundwater is vital.
Aquatic ecosystems and urban streams	The discharge of polluted and cold water from mains or reservoirs can degrade aquatic ecosystems. If urban streams are in direct or indirect connection with clean water withdrawal, the quantity and quality parameters of streams will affect water supply system. Extraction of water from waterways will affect the flow regime.
Human health	Operation of the water supply system will determine the level of risk to human health (from toxic substances, or from bacteria and/or viruses). Cross-connections or leaks into the water supply distribution network could result in contamination, causing human disease or long-term health effects.
Society and institutions	Community preferences and attitudes will determine the required (a) security of supply and (b) water quality. For example, community attitudes will determine whether use of recycled water is acceptable, and will affect water conservation during times of drought. The roles and responsibilities of institutions will affect how water is supplied. For example, a government agency with responsibility for water, wastewater and waterway management may have a different approach than a private company with responsibility only for supply of drinking water.

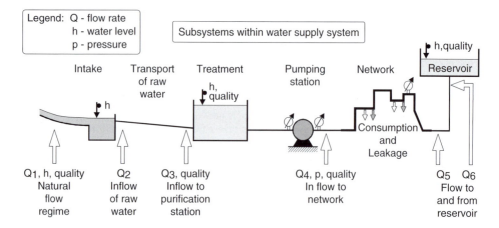

Figure 15.2 Diagram of subsystems within a water supply system and required measurements for water balance calculations

Source: Maksimović, 1994.

the necessity to have a mass balance. This means that the flow rate, pressure or level, and important quality parameters of input and output must be continuously measured (Figure 15.2) and recorded, to allow instantaneous and integrated values to be obtained.

15.3.1 Water intake

The water intake (withdrawal) is the input point to the WSS, and can be either from natural streams (as shown in Figure 15.2), from groundwater or from other connected WSSs (i.e. as part of a regional system). The intake is usually considered as a 'point of exchange', since the WSS typically has to pay for taken water from the point of withdrawal.

   The measurements at the intake should cover all important processes, including potential disturbances in water flow and quality in the system from which water is being extracted. Depending on the type of intake point, the following monitoring should be undertaken:

- *Natural water streams*: continuous measurement of upstream and downstream water levels, flow rates (if possible with direct measurement, otherwise calculated using rating curves) and upstream quality parameters (including point-source pollutant sources). Periodic measurements of bed profile near intake structure, to identify any siltation or erosion processes.
- *Lakes or dams*: continuous measurement of inflow and water level, as well as downstream flow rate (particularly if downstream environmental flows are stipulated). Influences on the groundwater table (see also Chapter 19) have to be monitored using boreholes and piezometric wells. If siltation of the lake is important, the bed profile has to be regularly checked. Changes in water quality parameters in

a lake occur with seasonal variations and are depth dependent, so it is important to continuously measure at least the temperature, turbidity and dissolved oxygen at several water depths.

- *Groundwater*: continuous measurement of level and flow rate in each well, and quality parameters for the whole wellhead. Monitoring of the groundwater table using boreholes and piezometric wells, near the wellhead and in the wider area subject to extraction.
- *Water intake from another (larger, regional) WSS*: water level in reservoir, quality parameters and extracted flow rate. Usually, two flow meters are used: one owned by the owner of the regional WSS and the other one by the WSS that takes the water.

The water intake subsystem is closely linked with the other water components. The availability of water depends in most cases on rainfall (and evaporation), surface water inflow or recharge of groundwater from other systems. Monitoring of those resources is mostly carried out within other systems and organizations, so efficient data sharing is needed. To minimize the effect of water pollution on WSS operation, a set of proactive measures are needed (catchment protection, back-up sources, etc.), which can be assisted by efficient collection, processing and exchange of monitored upstream water quality data.

Some common problems associated with the measurements of water intakes are:

- In most cases, measuring positions are far from urban areas, so vandalism, lightning damage, availability of electricity and data communication reliability all present challenges.
- The raw, captured water can contain dissolved gases (even explosive) which can be aggressive to certain type of sensors, can be corrosive, or sometimes can result in deposition. The selection of measuring sensors has to take into account the nature of the raw water, and the frequency of sensor recalibration will need to be increased.
- The exchange of data between different water related systems means that each system has an important role to play. Commonly, the water supply utility will be primarily interested in only current data for its own part of the water system. However, this precludes understanding of potential impacts on the water intake quality and quantity by other parts of the water cycle. In other words, this approach is often shown to be short-sighted.

15.3.2 Conveyance of untreated and clean water

The water transport system can be either for raw water (from the water intake to treatment plant) or for clean water (between the source of clean water and distribution network). Associated pumping stations are also considered as a part of transport subsystem.

Assuming that there is no consumption along the distribution line, water losses can be simply calculated by measuring inflow and outflow of the transport subsystem (Figure 15.3, Alegre et al., 2000). Even in more complex pipe systems, with several

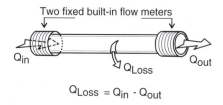

Two fixed built-in flow meters

Q_{in} Q_{Loss} Q_{out}

$$Q_{Loss} = Q_{in} - Q_{out}$$

Figure 15.3 Locating flow meters to monitor water loss in water transport pipes

Source: Maksimović and Prodanović, 1995.

main pipes, it is advisable to have a pair of flow meters for each pipe section. With such an arrangement, it is easy to monitor the pipe performance and its leakage rate. If a sudden pipe burst occurs (during earthquakes, for example, or during excavation works) it will be detected and shutting down the service valves thus prevents excess leakage (Gotoh et al., 1993).

The quantities to be continuously monitored in the water transport system are:

- *Flow rates*: apart from the instantaneous flow rate, the cumulative volume of passed water has to be recorded. Small and cheap sensors can also be used for direct monitoring of leaks (Halsey et al., 1999).
- *Pressure at selected points*: where the number of sensors and their positions mostly depend on the type of the flow regulation. Using pressure and flow data, a pipe's hydraulic conductivity deterioration can be assessed.
- *True position of regulating valves*: regardless of control type (manual or automatic).
- *Pump operation parameters*: pressure, upstream and downstream of the pump, voltage, current and the cumulative working period; especially, in systems with variable frequency control, the operating frequency must also be monitored.
- *Water quality*: the selection of water quality parameters to be monitored depends on, among other factors, the transport water. In general, the quality of water deteriorates in long and large transport pipes with small velocities. New developments in micro-sensor design for water quality monitoring (Stuetz, 2001) will reduce the cost of such equipment and will allow more widespread application in pipe networks.

Some important general considerations to be borne in mind for monitoring water transport systems are:

- Flow measuring device should be capable of measuring bidirectional flow, since that can occur in some systems.
- Flow meter calibration, critical in transport systems, can be a challenge, since in most cases the diameters of pipes are large.
- When selecting the number of measuring positions and types of sensors, it is good to have redundant measurements using different sensor types for quality control and to assess the overall measurement accuracy.

- Long and large pipes are prone to water hammer and oscillation (often induced by inadequate functioning of control elements). In such systems it is wise to have a transient pressure monitoring device (Prodanović et al., 2004). Standard pressure loggers are not suitable, however, due to long sampling intervals.

15.3.3 Water treatment and water quality conditioning

Treatment is required in practically all water supply systems. As the availability of clean water is decreasing and the demand by customers for high quality water is increasing, the need and the complexity of treatment plants is also increasing.

Water treatment plants are complex units with (in most cases) automatically controlled processes. A number of quantities are continuously measured and the operation of a treatment plant depends on the accuracy and availability of that data. The overall efficiency of a treatment plant is controlled at its outlet, where at least two separate monitoring systems are used: one continuous system based on built-in sensors, and the other one through regular sampling of water and manual (laboratory) water testing. Through regular data comparisons, the possibilities for errors in treatment plant operation are reduced and water quality is maintained within specified limits.

In spite of having a number of sensors, Supervision, Control, Data Acquisition and Data Analysis (SCADA) systems and manual tests of water in laboratory conditions, sharing of resulting data with other water cycle managers is often difficult, particularly because SCADA systems are often closed and the data are often regarded as 'commercial-in-confidence'.

Apart from water treatment plants, in large WSSs *water quality conditioning units* are also often used. They are situated along the clean water network and their role is to maintain the water quality, in most cases by controlling the amount of residual treatment, such as chlorine.

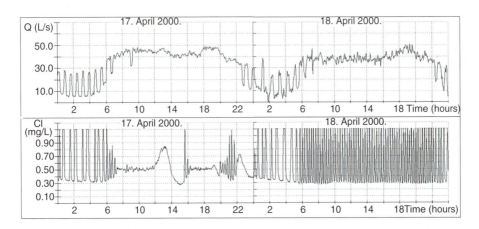

Figure 15.4 **An example of dynamic mismatch between pipe flow and automatic chlorine dosing station.** The sudden changes of flow (Q) through pipe (upper diagram) due to nearby pump causes the oscillation of automatic chlorine dosing system (measured residual chlorine Cl on lower diagram)

Source: Prodanović et al., 2001a.

Issues related to the monitoring of water treatment plants and water quality conditioning units include:

- Measurements within treatment plants are used for real-time control and most systems are not designed to store historical data.
- Measurements are vital for real time control. For example, inaccurate flow measurements (or an incorrectly positioned turbidity meter) could result in overdosing of chemicals.
- The dynamic response of the water treatment system to sudden changes of input parameters and dynamic response of measuring equipment should be matched. Fast reacting ('nervous') dosing controllers are dangerous (see Figure 15.4, where an automatic chlorine dosing station failed to keep the residual chlorine level at a constant 0.5 mg/L, due to sudden changes in flow rate – upper diagram, and starts to oscillate, raising the average residual chlorine much above the limited value – lower right diagram) (Prodanović et al., 2001a).
- There are commonly a number of chemical parameters measured off-line within laboratories. It is possible to establish a correlation among those parameters and parameters that are continuously monitored online, so that online monitoring can be used as a surrogate measure. The format of output data should include the metadata, to allow easy exchange with other data users.

15.3.4 Reservoirs

Reservoirs are used for temporary storage of water, allowing the operation of the water treatment plant to be independent of current water consumption, and balancing temporal variations in supply and demand. The storage volume of the reservoir and its position within the network are typically the subject of a detailed study using network simulation models. Different layouts are possible (DOH, 2006) and the optimal solution based on given criteria should be found (Kapelan et al., 2005).

For monitoring reservoir behaviour, the necessary parameters include inflow, outflow and water level. Depending on the type of reservoir, inflow and outflow pipes are either separated (two flow meters are needed) or the same pipe is used. A bidirectional flow meter is needed, with separate counters for direction toward the reservoir and direction from the reservoir to the network. Adding the continuous water level measurement, a redundant system is created, where accuracy checks of measured quantities can be performed.

In some WSSs the residual chlorine is checked within reservoirs and is increased, if needed. This is especially important within reservoirs where water may remain for days without recirculation, as is the case with large reservoirs constructed for potential future water needs, or during the low water consumption season, within systems with large seasonal variations.

15.3.5 Distribution networks

From the main transport pipe, water is distributed to the customers through the distribution network. The distribution network can be divided into several levels: primary (the largest pipes that are connected to the main transport pipe, with connections of important and large customers and mostly without house connections), secondary (with less important connections, fire connections) and tertiary (small diameters with hose connections or network within large houses). In order to limit the maximal pressure, the distribution network is always divided into pressure zones, where the number of zones depends on topography.

The orthodox approach to the distribution network is to design it with redundant connections (looped system) and not to monitor its behaviour. Repairs and maintenance will then often be based on customer complaints or observed leaks. However, if the network is to be operated in an optimal way, with the minimal number of bursts, reduced leakage and maximized output performance (increased reliability of clean water delivery to the customer), a proactive approach must be used.

The proactive approach means that the distribution network has to be continuously monitored. Measured quantities (pressure, flow, turbidity, residual chlorine, valve status) from numerous positions should be a part of complex online telemetry system. A sampling interval of 10 minutes to 15 minutes should be used in smaller systems, or even shorter for larger pipes. SCADA systems can be used to automatically check the water balance at each time step. If a mismatch occurs, the alarm signal should be sent to the field crew. In well monitored networks, a large number of false alarms can disavow a complete monitoring system. To reduce the number of false alarms due to communication problems, sensor malfunctioning or other reasons, the system should be equipped with some kind of self-learning mechanism, such as neural networks or expert systems (Kohonen, 2001).

The prerequisite for a good distribution network monitoring programme is its division into smaller parts, with known input(s), output(s) and with flow meters installed (Figure 15.5). Such areas are commonly known as district metering areas (DMA) (Thornton, 2002), and they should cover about 150 to 200 house connections, although this number may considerably vary from one WSS to another. Within each DMA the input flow (and output, if any) is continuously monitored and the difference between night (low flow) and day (high flow) readings are analysed on a daily basis.

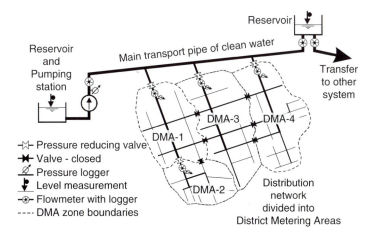

Figure 15.5 Water balance must lie measured and calculated within manageable units of the distribution system (e.g. District Metering Areas, DMAs)

Source: Obradović, 1999

Connecting the databases of house water meter readings and continuous measurement of inflow (and outflow) will enable evaluation of the water balance for each DMA. To reduce network leakage, the DMA can be also equipped with pressure regulating valves (PRV) which will maintain the pressure at the minimal acceptable value (Figure 15.5).

There are two possible approaches to network management after the introduction of DMAs. The first one is to keep each DMA separated from the rest of the system, and continuously monitor its parameters. This situation is presented in Figure 15.5, with black valves in closed status. Another possibility is to prepare the network for the DMA, equip the necessary pipes with flow and pressure meters, but allow certain redundancy by opening some interconnection valves. This will increase the reliability of the network, but make the detection of sudden bursts more difficult. In such an approach, the field crew should continuously test all DMAs by separating one at a time from the network for at least 24 hours and afterwards checking the water balance.

Figure 15.6 presents some possible outcomes of DMA monitoring. The left-hand graph is the result of continuous measurement of flow within one DMA (Obradović, 1999), where it is noticeable that a pipe burst during December (and was repaired the next year). The right-hand graph presents one 24-hour test of a DMA where measured flow was almost constant, without any night flow reduction (estimated water loss up to 90%; Prodanović et al., 2001b).

Recommendations regarding measurements and data analysis on the distribution network include:

- Calibration of flow measurement sensors is difficult within the water distribution network, due to irregular flow conditions. Other techniques (such as the velocity method or volume method) will be needed to calibrate flow sensors, in order to compute water balance within required accuracy.
- To optimize the value of data collected by the WSS, a 'data management plan' should be prepared at the commencement of monitoring (Babović et al., 2002).

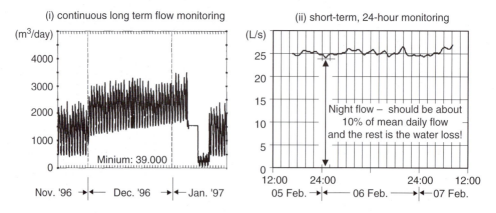

Figure 15.6 The results of monitoring two district metering areas: (i) continuous monitoring can reveal the development of a crack in a pipe (Obradović, 1999) and (ii) short-term (24-hour) diagnostic monitoring of system with high water losses (Prodanović et al., 2001b)

- Water balance computation has to be undertaken within each DMA, several DMAs, as well as an entire pressure zone. When doing water balance calculations, it is helpful to keep track of flow meters with lower accuracy, and to use redundant measurements to correct the overall result.
- Advanced data-mining techniques (Chapter 10) are needed to manage the large quantity of measured data. Using redundant measurements, the number of false alarms can be reduced, by pattern detection methods.
- Apart from the monitored hydraulic and quality data, geometric data about pipes (i.e. position, diameter, material, roughness, etc.) and other objects within the network (i.e. valves, pumps, etc.) should be as accurate as possible. Any changes in the network should be updated within databases and simulation models.

15.3.6 Consumers

Monitoring customer consumption is possibly the most important requirement of the water supply system. Using mechanical, mostly turbine water meters, the total volume of water is measured, often by manual meter reading. The interval between readings depends on the significance of the customer and varies between once per month and once per year. In such systems the rate of successful meter readings can be below 60% to 70%, due to a large number of blocked or damaged meters, or even meters with unknown location. Knowing that the average age of installed water meters is usually well above the regular service period (five years in most countries), it is reasonable to be suspicious regarding any water balance calculations that are derived from aggregation of customer meter readings.

The most common errors in customer readings are either propeller blockage or a reduction in the propeller rotation speed, both resulting in reduced reading and large negative errors in the measured volumes. Another problem with water meters is low sensitivity to small consumption (for example, continuous pipe leakage will not be measured). There are several solutions to this problem, such as using smaller water meters that are more sensitive to low flow, or using combined flow meters (one for small flows, and one for larger flows).

Most WSSs have some kind of database which maintains the water meter readings. The input of data is variously manual, semiautomatic (from handheld data input devices) or automatic (using remote meter reading systems). Apart from the production of invoices for water consumed, the data provides an opportunity for analysing system operation and performance. Different socio-economic information is easily extractable from this type of data. Extraction of demand patterns in large WSSs will be facilitated by use of the data-mining techniques.

Integration of the water meter readings database with the other information systems within the WSS is a difficult task, and is therefore rarely undertaken. For example, the time steps for water meter reading are variable (making matching to other data more difficult), as is the accuracy (decreasing with meter age). Matching the individual meter readings to computing nodes used in simulation model of the entire network can also be difficult.

15.4 THE ROLE OF MEASUREMENT

Measurements play a very important role in the operation of water supply systems. Measurements may have a simple objective, such as pressure measurement to test a pump's normal operation, or can be used for several tasks (the same pressure measurement at the pump's outlet can be used to evaluate its performance, to monitor the status of a downstream reservoir, to monitor the overall consumption, and be logged for later use as calibration data for a numerical simulation model).

In general, measurements in WSSs can be classified as:

- measurements for the purpose of charging for the water
- measurements for continuous checking of water balance and calculation of water losses
- measurements for process control and diagnostic purposes.

To achieve better data integration, it is important to identify all possible data users and their needs for additional metadata before a monitoring system is established.

15.4.1 Measurement for selling water

The WSS is usually the only component of a complex urban water system where water is regarded as a product. It is extracted from nature as a material, and the WSS adds new value to it through its treatment (quality improvement), pumping (raising its energy), storage and delivery to the customer.

In order to sell the product, it has to be legally measured, so legal metrology has to be applied at the customer's connection. In most countries water meters must comply with certain standards and regulations. For example, in Europe, in 2004 the EU Parliament adopted the Measuring Instruments Directive (MID) 2004/22/EC for different meter types, where general conditions for their application, installation, accuracy, calibration and ranges are given. The OIML (*Organisation Internationale de Métrologie Légale*) and all EU countries must follow the MID.

Legal metrology must also be applied in water withdrawal, since the WSS typically has to pay for water extraction. Depending on local policy, the price of taken water will consist of the maximal flow rate, the volume of water taken per period, the potential energy of water and the quality of water. In certain countries, the local authority can introduce an additional charge as a function of WSS leakage factor, to encourage the water supplier to reduce leakage and preserve the water resources.

15.4.2 Calculating the water balance

Measurements of input and output from the system, as needed for billing purposes, may produce only a total water balance of the whole system. The difference between input and output, as total water loss, cannot be allocated to any specific WSS subsystem without additional measurements within those subsystems. Figure 15.2 gives the minimum requirements for flow and level measurements so that the water balance can be calculated. There are also guides (e.g. Alegre et al., 2000) on how to assess the water balance and how to express it using performance indicators.

Since the data from different subsystems are used for water balance calculation, it is important to have enough metadata, mainly regarding the type of meter, its diameter, measuring range and typical accuracy within the range, low-flow threshold, calibration results and position within the network. The flow measurement accuracy on large pipes has to be better than 1% (typically 0.5%), level measurement accuracy better then 0.5% (typically 0.2%) and pressure 1% (if it will be used for pipe condition, the monitoring the accuracy has to be better than 0.2%, it must have atmospheric pressure correction, and its height must be known with an error of less then 0.05 m). For smaller pipes, the flow measuring accuracy can be in the range of 1% to 2%.

The period for data acquisition (or sampling rate) depends on the rate of flow and pressure changes at the measuring position. If the sampling rate is low (e.g. 5 minutes to 15 minutes) the error in flow integration will rise during periods of rapid changes in flow rate. As an example, a typical flow log is presented in Figure 15.7 (left), sampled with short time steps (continuous line, sampling every second) and with more infrequent, every 100 seconds (diamond dots). The difference is not so obvious, but if the volume of water is calculated (Figure 15.7, right) then significant differences will appear during periods of rapid flow rate change. However, very high sampling resolution will increase requirements for data storage (see also Chapter 5 for temporal scale considerations). The solution for this problem is to acquire two types of data from each flow measuring device: one is the continuous flow rate that can be used for monitoring and control purposes, and the other is the total volume passed through the meter and internally integrated, to be for use for water balance calculations.

Understanding (and subsequently managing) water balance is the critical requirement in a WSS, since it determines how efficiently the clean water, a scarce resource, is used. A number of organizations are working hard on the education of water companies and local authorities (WHO, 2001), creation of 'best management practice' for water conservation (Water Forum, 2006) or definition of water audit methodology and water loss control (AWWA, 2006; UKWI, 1994). Tables 15.2 outline the water balance scheme and definition of water balance components as defined by IWA/AWWA. Redundant data will help in the calculation of water balance. Also, pressure measurements, while not directly used in calculating the water balance, can

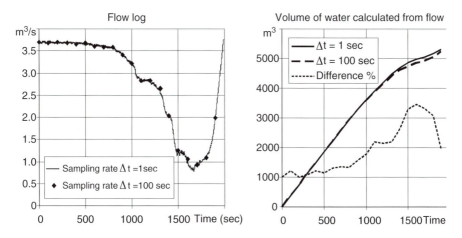

Figure 15.7 Illustration of the impact of the choice of sampling rate on the recorded flow and calculated water volume

Table 15.2 Tracking water use for calculation of a water balance[1]

System input volume[2] (corrected for known errors)			
Authorized consumption[3]	Billed authorized consumption	Billed metered consumption (including water exported) Billed unmetered consumption	Revenue water[7]
	Unbilled authorized consumption	Unbilled metered consumption Unbilled unmetered consumption	
Water losses[4]	Apparent losses[5]	Unauthorized consumption Customer metering inaccuracies Data handling errors	Non-revenue water[8] (NRW)
	Real losses[6]	Leakage from transmission lines and distribution mains Leakage and overflows at utility's storage tanks Leakage at service connections up to point of customer metering	

Notes:
[1] All data in volume for the period of reference, typically one year.
[2] Annual input volume to the water supply system.
[3] Annual volume of metered and/or unmetered water taken by registered customers, the water supplier and others who are authorized to do so.
[4] Difference between system input volume and authorized consumption (apparent losses plus real losses).
[5] Unauthorized consumption, all types of metering inaccuracies and data handling errors.
[6] Annual volumes lost through all types of leaks, breaks and overflows on mains, service reservoirs and service connections, up to the point of customer metering.
[7] Components of the system input volume which are billed and produce revenue.
[8] Difference between the system input volume and billed authorized consumption.
Source: Farley and Trow, 2003.

help to understand water balance results, since water losses are a function of pressure (Tabesh et al., 2005).

15.4.3 Process and water quality control

Each automatic control system is based on some kind of measurements. It may be as simple as control of inflow into a reservoir where level indicators (discontinuous level measurement) are used to close and open the input valve, or as complex as a chemical dosing system within water treatment plants. The general problem with such measurements is that data are usually not accessible by other users (the system is closed to increase its reliability).

Therefore, during the design stage of the automatic control system, it is important to decide what quantities are potentially needed by other users, as well as their required time scale and format. A secure system can then be constructed to transfer these data to other users, without impacting on security of the primary system.

15.4.4 Diagnostic measurements

Continuous measurements used for WSS management and water balance calculations are usually inadequate for analysis of the system details or system operation in irregular conditions. Additional measurements, targeted to the given problem, are commonly referred to as *diagnostic measurements*. Such measurements, when combined with the existing monitoring system, will provide valuable additional data about the behaviour of individual system elements, and help in solving the given problem (water hammer when the pump is switched off, for example). The measurements can also provide the additional data needed for calibration and verification of numerical simulation models.

Diagnosis of problems within the WSS must comply with precisely defined requirements and planned objectives which are to be achieved (Prodanović and Pavlović, 2003). Depending on the final objective, a suitable selection of measuring methods and equipment can be undertaken, as well as selection of data processing tools. Diagnostic measurements consist of several steps:

- *Measurement*: the choice of measurement equipment mostly depends on the problem to be addressed. In general, more accurate equipment is needed than the equipment typically used for continuous measurement. Logistical challenges (such as changes to normal operating procedures) will often make diagnostic measurements more difficult.
- *Analysis*: data preprocessing, validation and advance analysis of system behaviour in both static and dynamic conditions. The analysis gives information for a proposed solution.
- *Accuracy assessment*: each measured variable has a certain error. Using accuracy assessment, the overall error in the final result can be computed. To reduce the total error, the measured data has to be as accurate as possible, but also some redundant data is needed.
- *Cost–benefit analysis*: presents an important part of the diagnostic measurement. The water authority needs to know a technical solution to the given problem, as well as the costs and benefits of a proposed solution.

Diagnostic measurements cover a wide field of applications. In general, several types of diagnosis can be identified:

- *Standard flow and pressure measurement on distribution network*, for calibration of numerical simulation model or for network performance assessment. This is the most common type of measurement. The number of flow and pressure loggers depends on the available resources. If there is no previous knowledge about the network or the expected measured range and flow direction are unknown, at least two series of continuous measurements should be conducted. The first one will give general information about the system. Rough calibration of the numerical model should be performed based on that information. Using the model and a site visit, the possible critical parts of the WSS are detected. In the second series of measurements, some new measurement locations should be selected, focusing on the observed problems. Figure 15.8 presents an example of flow (velocity) measurement, in a case where there was no previous information about the system. Velocities were measured with an electromagnetic probe, which is generally able to measure velocities up to 10 m/s, but during the signal conditioning procedure they were limited to 3 m/s. Since obtained velocities were much above 3 m/s, some

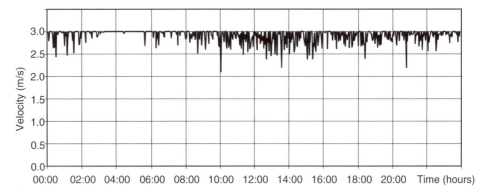

Figure 15.8 Example of velocity measurement on previously unmonitored system (with incorrectly–selected instrument range, limited to 3 m/s). The example shows that for systems where there is no prior-knowledge, two measurement campaigns will be needed with the first providing the information to refine the second

Source: Prodanović and Pavlović, 2003.

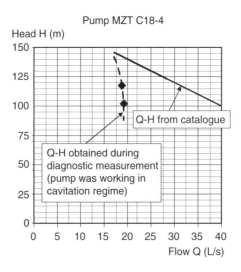

Figure 15.9 Diagnostic measurements could reveal the true pump specification. Diagram shows the result of measurements on the pump that works in cavitation regime, thus having the Flow-Head (Q-H) curve much different then expected (H = height, Q = discharge)

(Source: Prodanović and Pavlović, 2003).

checks were done with a mathematical model. It turned out that two level zones were directly connected through that pipe and water was flowing from high pressure zone to low pressure zone, producing velocities significantly beyond those which were expected.

- *System details diagnosis,* such as flow measurement in large pipes for recalibration of flow meters that are working in non-standard conditions. For example, Figure

15.9 shows a plot of true, or measured, pump stage-discharge characteristics, compared to the standard curve as from the pump's specification. Diagnosis may also be used for flow in irregular conditions, water hammer analysis, water mass oscillations in long pipes, filter performance in regular and overload conditions, or searches for partly closed valves within system or connections. Such measurements must be undertaken with more accurate equipment than would be used for standard operational monitoring.

• *Leakage loss diagnosis.* Continuous monitoring supported by occasional diagnostic measurements provides the best approach for this purpose, with change detection algorithms used to detect the occurrence of leaks. For small leaks, monitoring systems capable of accurately measuring low flows will be required.

• *Diagnostic measurements during abnormal circumstances*, such as pipe breaks, bursts, water hammer, and other events. To be most useful in this context, monitoring of the pre-incident network state is required (so that changes can be detected).

REFERENCES

Alegre, H., Hirner, W., Baptista, J. and Parena, R. 2000. *Performance Indicators for Water Supply Services: Manual of Best Practice*. London, International Water Association.

AWWA (American Water Works Association). 2006. *Water Loss Control, Conservation by Water Suppliers*. Available at: http://www.awwa.org/WaterWiser/waterloss/index.cfm (Accessed 02 July 2007.)

Babovic, V., Drecourt, J.P., Keijzer, M. and Hansen, P.F. 2002. A data mining approach to modelling of water supply assets. *Urban Water,* Vol. 4, No. 4, pp. 401–14.

DOH (Division of Environmental Health, Office of Drinking Water). 2006. *Water System Design Manual*. Available at: http://www.doh.wa.gov/ehp/dw/publications/design.htm (Accessed 02 July 2007.)

Farley, M. and Trow, S. 2003. *Losses in Water Distribution Networks – A Practitioner's Guide to Assessment, Monitoring and Control*. Cornwall, UK, IWA Publishing.

Gotoh, K., Jacobs, J.K., Hosoda, S. and Gerstberg, R.L. 1993. *Instrumentation and Computer Integration of Water Utility Operations*. Denver/Tokyo, American Water Works Association Research Foundation/Japan Water Works Association.

Halsey, G., Lewin, P.L., Chana, G. and White, P.R. 1999. Intelligent sensor for water leak detection and location. Paper present at Sensor and Transducer Conference, 17–18 February 1999, Birmingham, UK.

Kapelan, Z., Savic, D. and Walter, G.A. 2005. Multiobjective design of water distribution systems under uncertainty. *Water Resources Research*, Vol. 41, pp. W11407.

Kohonen, T. 2001. *Self-Organizing Maps*, 3rd edn, Vol. 30. Berlin, Springer.

Maksimović,. Č 1994. Measurements for diagnostics and rehabilitation of water supply systems. *ASI: New Technologies for Large Water Supply Projects*. Varna, Bulgaria.

Maksimović, Č. and Prodanović, D. 1995. Hydroinformatic support to poorly documented leaking water supply systems – an integrated approach. Paper presented at International Conference on Advanced Technologies for Saving Water, 6–7 February 1995, Athens, Greece.

Obradović, D. 1999. *Contemporary Water Works. Informatics and Operational Management*. Belgrade, Serbia,Yugoslav Association for Water Technology and Sanitation (in Serbian).

Prodanović, D. and Pavlović, D. 2003. The role of diagnostic measurements in poorly maintained and documented water supply systems. *Proceedings of XXX IAHR Congress, 24–29 August 2003*. Thessaloniki, Greece, Aristotle University of Thessaloniki and International Association of Hydraulic Research,.

Prodanović, D., Pavlović, D. and Ognjanović, M. 2001a. Automatic control within water-works – two examples. Paper presented at Twenty-first National Congress of Waterworks and Sanitation. 17–19 October 2001, Novi Sad, Serbia (in Serbian).

Prodanović, D., Ljubisavljević, D., Pavlović, D. and Babić, B. 2001b. *Diagnostic Measurements on Herceg Novi WSS for Numerical Simulation Model Calibration. Field Study.* Faculty of Civil Engineering, University of Belgrade, Serbia (in Serbian).

Prodanović D., Ivetić, M. and Savić, D. 2004. Dynamic monitoring and leakage detection by induced transients. *Proceedings of 6th International Conference on Hydroinformatics, 21–24 June 2004,* Singapore, Hydroinformatics Society of Singapore.

Stuetz, R. 2001. Using Sensor Arrays for online monitoring of water and wastewater quality. *American Laboratory Journal*, Vol. 33, No. 2, pp. 10–14.

Tabesh, M., Asadiani Yekta, A.H. and Burrows, R. 2005. Evaluation of unaccounted for water and real losses in water distribution networks by hydraulic analysis of the system considering pressure dependency of leakage. Paper presented at Conference Computing and Control in the Water Industry (CCWI) – Water Management for the Twenty-first Century, 5–7 September 2005, University of Exeter, UK.

Thornton, J. 2002. *Water Loss Control Manual.* New York, McGraw-Hill.

UKWI (UK Water Industry). 1994. *Managing Leakage Reports.* Swindon, UK, WRc plc/Water Service Association/Water Companies Association.

Water Forum. 2006. *Water Conservation Annual Report.* Available at: www.waterforum.org/WaterConservAnnualReport.pdf (Accessed 02 July 2007.)

WHO (World Health Organization). 2001. *Leakage Management and Control: A Best Practice Training Manual.* Available at: http://www.who.int\docstore\water_sanitation_health\leakage (Accessed 02 July 2007.)

Chapter 16

Wastewater

F. Clemens

Delft University of Technology, Faculty of Civil Engineering, Sanitary Engineering Section, 2600 GA, Delft, The Netherlands

16.1 INTRODUCTION

This chapter discusses the behaviour and effects of wastewater, its interactions with the urban water cycle, and its monitoring requirements. Given the close relationship between wastewater and stormwater, this chapter should be read in conjunction with the chapters on stormwater (Chapter 17) and combined sewers (Chapter 18).

While there are many differing definitions of wastewater, it is defined here as water[1] that either:

- is or has been used in an industrial or manufacturing process
- conveys, or has conveyed, sewage; or
- is directly related to manufacturing processing or raw materials storage areas at an industrial plant.

The term 'sewage' usually refers to household wastes, but this word is now more commonly replaced by the term 'wastewater'. Municipal wastewater is a general term applied to sanitary wastewater and urban runoff collected and transported by a sewage collection system to be treated in a municipal treatment plant. Industrial wastewater is normally referred to separately because, due to its composition, it may require special treatment processes, or can pose special demands on the transport system (in terms of resistance against corrosion, ability to carry high temperatures, etc.).

Monitoring of wastewater systems is an important factor in the successful operation and maintenance of these systems. Unfortunately, however, regular monitoring of these systems is often not undertaken, and even when implemented, is often not done in a rigorous manner. Furthermore, integrated monitoring of wastewater with other parts of the urban water cycle is rarely even considered.

This deficiency can be explained from an historical context, since most systems in use nowadays were designed and built decades ago. In those days there was little recognition of the need for monitoring, and perhaps more importantly, there were few if any means (e.g. sensors, data-acquisition and communication networks) to put monitoring into effect on a large scale. With increasing environmental pressure, more strict demands on the effectiveness of urban wastewater system, the need to verify and calibrate computer models, the introduction of risk management and direct operational

---

[1] see www.emgcorp.com/whatshot/dictionary.php (Accessed 03 July 2007.)

activities like real time control, there is now a widely recognized need to obtain more reliable data on the processes taking place in wastewater systems.

16.2 INTERACTIONS WITH OTHER URBAN WATER CYCLE COMPONENTS

Ideally there should be not interaction between wastewater and most other parts of the urban water cycle. In practice, however, many interactions exist, including many that are not intended (Table 16.1). For example, most wastewater treatment plants use the following techniques to purify the raw wastewater:

- primary settling
- activated sludge (subdivided in an anaerobic, anoxic and aerobic reactor)
- secondary settling.

These processes are more or less vulnerable either to changes in hydraulic load, which may come from water supply/demand variation, or infiltration of stormwater. Wastewater treatment is also affected by variations in the concentration and load of pollutants. The removal of nitrogen, in particular, is known for its vulnerability to variations in influent (Durchschlag et al., 1991). Another potential interactive effect is the dilution of wastewater during dry weather flow, due to infiltration of groundwater; in some cases up to 50% of the influent in treatment plants may be groundwater. This decreases the effectiveness of the treatment plant in terms of increased energy and chemical consumption and a decreased efficiency of the bio-chemical reactions (Langeveld, 2004).

16.2.1 Water supply

The most unwanted interaction is that between wastewater systems and systems for drinking water supply. A direct interaction is highly unlikely since normally there is a sound physical barrier between both systems (separate piping systems, normally constructed underground). In cases, however, where a wastewater system shows leakage (exfiltration) to the groundwater systems an indirect interaction may occur by two possible mechanisms:

- When the water supply system shows a temporary negative pressure, contaminated groundwater may enter and influence the quality of the drinking water. This may have substantial effects, since the contaminated water is directly supplied to the end-users who may become ill from ingesting bacteria or viruses.
- When the groundwater is used for the production of drinking water it is possible that even after treatment in the drinking water production plant, micro-pollutants, such as traces of pharmaceuticals, are present in the drinking water.

16.2.2 Combined sewer systems and storm sewer systems

Interactions between wastewater, stormwater and combined sewers systems are discussed in detail in Chapters 17 and 18, respectively, and so will be discussed only briefly

Table 16.1 Potential interaction between wastewater systems and other urban water system components

Urban water system component	Possible interactions with wastewater systems
Urban climate	Temperature will strongly affect chemical and biological treatment processes. Temperature and rainfall will affect pathogen growth and transport. Infiltration into wastewater from surrounding soils will depend on soil moisture, which in turn will depend on evapotranspiration and rainfall. Rainfall patterns will determine the frequency of combined sewer overflows.
Water supply	Although very unlikely, a water supply system may have temporary and local instances of low pressure, allowing contaminated groundwater to enter the water supply system. A more likely interaction between potable water and wastewater is the use of a wastewater-contaminated source used for the production of drinking water. In such cases the main risk is in micro-pollutants, such as medications, insecticides and herbicides.
Stormwater	In cases where separate systems are used, the occurrence of crossed or faulty connections is a prime concern. A small percentage of these cross connections (up to 2%) may be present directly after construction, and when nothing is done to remedy it, this percentage will grow with time due to construction activities etc. Stormwater inputs can adversely affect the efficiency of a wastewater treatment plant (Langeveld, 2004).
Combined sewers	Combined sewers interact with surface water bodies as well as with the groundwater system. Since it is (economically) not feasible to design a combined sewer system in such a manner that overload due to storm water never occurs, incidental spills via sewer overflows (termed Combined Sewer Overflows, CSOs) occur. These spills have adverse effects on receiving water bodies, through either groundwater entering the system (infiltration), or wastewater contaminating the groundwater (exfiltration) (Rutsch et al., 2006).
Groundwater	Due to leaking joints groundwater may infiltrate into the wastewater collection system, and this may have adverse effects on the wastewater system (including the efficiency of the treatment plant) as well as on the groundwater system. In the former case, the wastewater is diluted, resulting in an increased load on the wastewater treatment plant, while in the latter case the groundwater table may locally be lowered by a significant amount resulting in structural damage to buildings or urban infrastructure systems such as roads, pipelines, etc.
Aquatic ecosystems and urban streams	Wastewater systems influence aquatic ecosystems in several ways. In cases where untreated wastewater is diverted into surface waters, the influence is obvious: depletion of oxygen and toxic substances directly affecting biodiversity and functioning of aquatic ecosystems. Where sewer overflows occur, there may be both short-term effects (depletion of oxygen directly after a spill) and long-term effects (due to accumulation of conservative matter).
Human health	Cross connections may result in wastewater inputs to drinking water, posing a threat to human health. Wastewater discharge to waterways may pose a threat, through the spread of pathogens, or through bioaccumulation of micro-pollutants and toxicants in the food chain, ultimately ending in food (e.g. fish) consumed by humans.
Society and institutions	Trends and variability in water use will affect wastewater production. Community attitudes will also determine the required level of treatment of wastewater, and the manner of its disposal. Similarly, community attitudes will determine what frequency of combined sewer overflow is acceptable. Recycling of wastewater will depend on the ability of the wastewater agency to be involved in water supply, and on the constraints or facilitation provided by other institutional stakeholders.

here. The interactions between wastewater and stormwater are governed by the mismatch in hydrology and water quality of the two systems. While stormwater itself may often be highly polluted (Chapter 17), wastewater generally has higher pollutant concentrations. However, the highly variable flow rates in stormwater systems means that when the two systems are combined, overflows of highly polluted water will occur (Aalderink and Lijklema, 1985). Furthermore, leakage of wastewater from sewers may contaminate groundwater directly posing a possible threat to water sources used for the production of drinking water.

The combined sewer system is the most common type of urban drainage system in the developed world. During dry weather flow (DWF), pure wastewater is collected and transported to the wastewater treatment plant, where the wastewater is treated in such manner that discharge into (large) surface water bodies does not pose a direct severe threat to the water quality. Typical standards are 20 mg/L O_2 biological oxygen demand (BOD), 10 mg/L of nitrogen, 30 mg/L total suspended solids (TSS) (van Nieuwenhuijzen, 2002). During DWF it is possible that some of the wastewater exfiltrates into the soil and contaminates the soil and groundwater. It is extremely hard to identify such a process and quantification is even harder. Consequently, knowledge of this phenomenon is limited, although general opinion varies between a view that the overall impact is not very significant (Barrett et al., 1999), to those who believe that is poses a significant risk to environmental and human health (Eiswirth and Hötzl, 1997). Fenz (2003) estimated that about 1% to 5% of the DWF may be lost to exfiltration when the sewer system is above the groundwater table.

Whilst storm water sewer systems have, theoretically, less impact on groundwater and surface waters, there is the potential that, due to faulty connections, storm water systems may carry significant amounts of wastewater eventually resulting in similar effects to those induced by combined sewer systems.

16.3 MONITORING REQUIREMENTS

16.3.1 Parameters to monitor

The primary requirement for wastewater management is to understand its composition in terms of organic matter and nutrients. The biological oxygen demand (BOD) and chemical oxygen demand (COD) give an indication of the amount of oxygen needed to digest organic matter present in the wastewater. Knowing the level of BOD/COD is of importance when designing and operating a wastewater treatment plant because of the biological processors in the treatment process. In operational circumstances, an influent sample is taken at least once every week and is analysed for COD/BOD, phosphorus (typically total and dissolved), and nitrogen (including species of nitrogen such as nitrate and ammonia).

Measuring oxygen demand is also important for assessing the impacts of sewer spills on receiving waters, as the load of organic matter may cause a complete depletion of oxygen in the receiving surface waters, impacting on aquatic species. Taking samples during a combined sewer overflow (CSO) event is relatively complicated and is normally only done in the framework of research activities. In day-to-day practice only water level is monitored and an indication is given when CSO events occur. Detailed information on monitoring of CSOs is provided in Chapter 18.

16.3.2 Quantifying sewer leakage

Leakage from wastewater into groundwater can be very difficult to detect. Determining the location(s) of the leakage is even more difficult. Assessments of the magnitude and consequences of leakage vary. For example, Barrett et al. (1999) estimate that only a small percentage (1% to 5%) of the DWF is lost due to leakage, with only moderate impacts on soil and groundwater. On the other hand some authors speak of significant impacts, even quantifying leaking sewers as the main source of groundwater contamination (Wakida and Lerner, 2005).

In general, measuring infiltration (groundwater entering the wastewater system) is much easier than measuring exfiltration. Monitoring leakage can be done either directly or indirectly (Rutsch et al., 2006). Indirect methods include groundwater modelling, to identify the contribution of leakage using a calibrated groundwater model (see, for example, Lerner, 2002), and calculation of a water balance.

Both methods suffer from a high level of inaccuracy because of uncertainty in the data, limited accuracy of the models applied, sampling errors and the limitation of sampling density in time and space.

Direct methods to quantify leakage are:

- pressure testing, through experiments on a specific reach or a section of the system
- evaluating time series data on groundwater, precipitation and DWF discharge
- analysing 'tracers', such as oxygen isotopes, viruses or micro-pollutants.

In the pressure test method, a reach or a section of the wastewater system is closed (using inflatable seals) and water is pumped into the section to such a level that the section gets pressurized. During a period of time (typically 24 hours) the variation in water level is monitored. In this manner an accurate indication of the exfiltration is obtained (Dohmann et al., 1999). This method has disadvantages, for instance, it is necessary to shut down a part of the wastewater system, which cannot be done for too long a period. Furthermore, it is hard to obtain a clear picture of how leakage varies with time.

When time series data for wastewater discharge, precipitation and groundwater level[2] are available, and a strong correlation between precipitation, groundwater level and DWF discharge is present, it is possible to positively identify if a wastewater system is leaking. Quantification, however, remains difficult, and identifying the exact location of leakage can only be done using additional information, that is, closed circuit television (CCTV) recordings.

In the European Commission Project on Assessing the Performance of Urban Sewer Systems (APUSS), several methods to quantify leakage have been developed and tested. They include methods that potentially allow for a quantification of the leakage as well as for quantification of the inaccuracy in using tracers (Rieckermann et al., 2006). One of the most recently investigated tracer methods is the use of oxygen isotopes. When drinking water in an urban area has a different source than the groundwater in that urban area, there is a good possibility that the isotopic composition (in terms of the relative presence of O^{16}, O^{17} and O^{18} in the water molecules) differs

---

[2] Typical sampling frequencies are once per hour for DWF, once per day to once per week for groundwater level; precipitation is normally measured once per five minutes, once per hour, or on a daily basis.

enough to be able to quantify the mix of wasted drinking water (wastewater) and infil-trating groundwater (see for example, Kracht et al., 2003). Theoretically the method is, when applicable, reliable and allows for quantification of the accuracy of the result, a characteristic that classical methods lack (see e.g. Weiss et al., 2002). The isotope method has the advantage that no tracer has to be introduced in the system. However, the method has important requirements:

- There should be a stable isotopic composition of the drinking water, which can be a problem since many water supply systems obtain their water from different sources in changing proportions over time.
- The groundwater composition should be relatively homogeneous, which is defi-nitely not the case in some countries, e.g. in the Netherlands (see Schilperoort, 2004).
- The isotopic composition of the wastewater is assumed to be (nearly) identical to the composition of the drinking water, although in some cases, when significant proportions of the water are heated, or when natural gas is used for heating, this may not be the case (see Schilperoort et al., 2006).

16.3.3 Detecting cross-connections

One of the biggest problems with wastewater management is the occurrence of cross-connections between wastewater systems and stormwater systems. For example, due to mishaps during construction a certain percentage (undocumented estimates range from 0.5% to 2%) of connections is faulty. There are two possible instances of this: (i) the wastewater system is collecting and transporting stormwater, which, in extreme cases, may lead to hydraulic overload and result in decreasing the efficiency of the wastewater treatment plant; (ii) the reverse situation, in which wastewater is dis-charged into a stormwater system with the main effect being on the receiving surface waters. Methods to identify the effect of stormwater discharges on wastewater systems are relatively easy by evaluating the correlation between precipitation and the amount of water being discharged into the wastewater treatment plant. Techniques such as smoke testing can be used to determine the location of such cross-connections; smoke is introduced under pressure into the wastewater system, and when stormwater mani-folds are connected to this system, smoke will escape and a visual inspection of the direct surroundings will suffice.

Estimates of incorrect wastewater discharge into storm water systems are less easy to detect, however, especially when relatively low discharges are present. The most appropriate method is regular inspections in which the presence of wastewater is rela-tively easily detected (particularly during baseflow), while it is also possible to identify the manifolds from which the wastewater is discharged.

16.3.4 Measuring impacts on receiving waters

When wastewater is discharged into surfaces waters it affects the ecological health of the receiving water and it limits the potential uses of the water as a source for drink-ing water or for recreational purposes. Monitoring the interaction between waste-water and its receiving waters is a critical component of managing the overall urban water system. A brief summary of relevant aspects is provided here, and should be read in conjunction with the relevant chapters on monitoring potential impacts on aquatic ecosystems (Chapter 20) and human health impacts (Chapter 21).

A distinction should be made between short-term and long-term effects when developing a monitoring programme of the interaction between wastewater systems and surface waters. Typical short-term effects, that is, with time scales ranging from minutes to days (Aalderink and Lijklema, 1985) include:

- depletion of oxygen levels in the receiving water
- bacterial contamination
- unpleasant odours
- visual disturbance.

Such short-term effects can normally be monitored using the following parameters:

- oxygen level in the discharged wastewater and in the surface water
- COD and/or BOD level in the discharged wastewater and in the surface water
- discharge of wastewater being spilled
- direction and velocity of flow (for lentic systems)
- faecal contamination, assessed using an indicator species (see Chapter 21), normally through samples taken either for *E. coli* or for faecal streptococcus.

The sampling interval is ideally no more than half the time characteristic of the processes one is interested in. Depending on the variability of the discharge, either a more or less continuous flow or in the case of a CSO a more or less sudden, relatively short-lasting spill, sampling frequencies for oxygen and COD/BOD should preferably be in the order of 4 times to 6 times per hour. (For more specific guidance on measuring the pollutant load during a CSO, refer to Chapter 18). Such high sampling frequencies were until recently not practically and economically feasible. However, the development of sensors for oxygen, COD, TSS and nitrate based on optical principles provides promising potential for high temporal resolution monitoring (Gruber et al., 2005). When assessing, for instance, the quality of surface water in terms of bacterial or viral contamination, it is imperative to take a sample for subsequent laboratory analysis. Using this method results are obtained after 24 hours to 48 hours, depending on the analytical method applied. Online monitoring of these parameters is not yet feasible. It has been shown that the variability in the number of *E. coli* being discharged is very high. Even in urban stormwater discharges, a variation factor of 5 to 6 can occur in a time span of 10 minutes to 20 minutes, indicating that to estimate, for example, an event mean concentration, a sampling frequency of 6 samples per hour is necessary. Normally, water management agencies perform a routine sampling of surface waters used for recreational purposes during the summer, taking samples of the surface water regularly with the sampling frequency determined based on the estimated risk.

Long-term effects on surface water quality are caused by:

- nutrients (nitrogen, phosphorus) resulting in eutrophic conditions in the surface waters
- heavy metals, disrupting aquatic life

- herbicides, potentially poisonous for fauna as well as flora
- pharmaceuticals with largely unknown effects
- endocrine disrupters with effects on the health of aquatic lifeforms.

For these parameters it is more important to obtain estimates of loads over a longer period of time (months to a year) than to obtain information on the variation of the composition of the discharges. Sampling frequencies can thus be reduced. Ideally, discharge-proportional samples are taken, thus enabling, in conjunction with a correct discharge measurement, estimates for the load over a period of time. However, in practice it turns out to be difficult to obtain reliable estimates in this manner (Ort and Gujer, 2006). For practical purposes it may be advisable to take samples from the sediments in the surface water to find out if problems are to be expected, rather than trying to precisely quantify the loads from each and every discharge into the surface water. Better yet it may be possible or to use a combination of both techniques to establish a correlation between them.

REFERENCES

Aalderink, R.H. and Lijklema, L. 1985. Water quality effects in surface waters receiving stormwater discharges. *Water in Urban Areas*. TNO Committee on Hydrological Research, pp 143–159. (Proceedings and Information, No. 33)

Barrett, M.H., Hiscock, K.M., Pedley, S. Lerner, D.N., Tellam, J.H. and French, M.J. 1999. Marker species for identifying urban groundwater recharge sources: a review and case study in Nottingham, UK. *Water Resources*, Vol. 33, No. 14, pp. 3083–97.

Dohmann, M., Decker, J. and Menzenbach, B. 1999. Unterzuchungen zur quantitativen und qualitativen belastung von untergrund, grund- und oberflächenwasser durch undichte kanäle. M. Dohmann (ed.), *Wassergefahrdung durch undichte Kanäle*. Dordrecht, Springer Verlag.

Durchschlag, A., Härtel, L., Hartwig, P., Kaselow, M., Kollatsch, D., Otterpohl, R. and Schwentner, G. 1991. Total emissions from combined sewer overflow and wastewater treatment plants. *European Water Pollution Control*, Vol. 1, No. 6, pp. 13–23.

Eiswirth, M. and Hötzl, H. 1997. The impact of leaking sewers on urban groundwater. J. Chilton, *Groundwater in the Urban Environment*, Vol. 1: *Problems, Processes and Management*. Rotterdam, A.A. Balkema, pp. 399–404.

Fenz, R. 2003. Strategien der Kanalinstandhaltung [Strategies for channel maintenance]. *Wiener Mitteilungen [Vienna Reports]*, Vol. 183, pp. 91–118.

Gruber, G., Winkler, S. and Pressl, A. 2005. Continuous monitoring in sewer networks: an approach for quantification of pollution loads into surface water bodies. *Water Science and Technology*, Vol. 52, No. 12, pp. 215–23.

Kracht, O., Gresch, M., De Bénédittis, J., Prigiobbe, V. and Gujer, W. 2003. Stable isotopes of water as a natural tracer for infiltration into urban sewer systems. *Geophysical Research Abstracts*, Vol. 5, Abstract 07852.

Langeveld, J.G. 2004. Interaction within wastewater systems. Ph.D. thesis, Delft University of Technology, Delft, The Netherlands.

Lerner, D.N. 2002. Identifying and quantifying urban recharge: a review. *Hydrology Journal*, Vol. 10, pp. 143–52.

Ort, C. and Gujer, W. 2006. Sampling for representative micropollutant loads in sewer systems. *Water Science and Technology*, Vol. 54, No. 6–7, pp. 169–76.

Rieckermann, J., Borsuk, M.E. and Gujer, W. 2006. Using decision analysis to determine optimal experimental design for monitoring sewer exfiltration with tracers. *Water Science and Technology*, Vol. 54, No. 6–7, pp. 161–8.

Rutsch, J., Rieckermann, J. and Krebs, P. 2006. Quantification of sewer leakage: a review. *Water Science and Technology*, Vol. 54, No. 6–7, pp. 135–44.

Schilperoort, R.P.S. 2004. Natural water isotopes for the quantification of infiltration and inflow in sewer systems. M. Sc. thesis, Delft University of Technology, Delft, The Netherlands.

Schilperoort, R.P.S, Clemens, F.H.L.R., Meijer, H.A.J. and Flamink, C.M.L. 2006. Changes in isotope ratios during domestic wastewater production. *Proceedings of the Seventh International Conference on Urban Drainage Modelling*, 2–7 April 2006, Melbourne, Australia, Vol. 2, pp. 23–9.

van Nieuwenhuijzen, A.F. 2002. Scenario studies into advanced particle removal in the physical pre-treatment of wastewater. Ph. D. thesis, University of Technology, Delft, The Netherlands.

Wakida, F.T. and Lerner, D.N. 2005. Non-agricultural sources of groundwater nitrate: a review and case study. *Water Resources*, Vol. 39, pp. 3–16.

Weiss, G., Brombach, H. and Haller, B. 2002. Infiltration and inflow in combined sewer systems: long term analysis. *Water Science and Technology*, Vol. 45, No. 7, pp. 11–9.

Chapter 17

Stormwater

W. Shuster[1], T.D. Fletcher[2] and A. Deletić[2]

[1]National Risk Management Research Laboratory, Office of Research and Development, US
Environmental Protection Agency, Cincinnati, OH 45268, USA
[2]Department of Civil Engineering, Institute for Sustainable Water Resources, Building 60, Monash,
Monash University, Melbourne 3800, Australia

17.1 INTRODUCTION

The process of urbanization causes significant changes to the hydrologic regime of catchments by increasing the area of impervious surfaces (roads, roofs, etc.) and altering the natural drainage network. The urbanization process includes increasing surface area of road networks; increasing connectivity between impervious areas and rivers and streams via pipes (Walsh et al., 2005); fragmentation and drainage of wetlands; decreasing drainage capacity through floodplain development; and channelization and engineered water exchanges among major surface water components. Increased hydraulic efficiency in urban catchments can diminish capacity for infiltration of precipitation, resulting in shorter concentration times and increased volume of runoff, greater flood peaks, and in some instances due to decreased infiltration, reduced recharge of ground water (Terstriep et al., 1976).

Urbanization also results in increased generation and mobilization of pollutants within catchments (Taylor et al., 2005a). Changes in stormwater hydrology and water quality are also likely to have consequent impacts on other aspects of the urban water cycle. For example, groundwater resources may be affected (in terms of quantity, quality or both) by changes to catchment hydrology, while stormwater intrusions may also affect wastewater sewers (if a separate system is used). Consequently, data are required on the production, conveyance and management of urban stormwater runoff in order to be able to mitigate the effects of urbanization-induced stormwater on the aquatic environment and the urban water cycle.

Among many calls for improved information in this area, especially with regard to urban water systems, Bertrand-Krajewski et al. (2000) point out the many shortcomings of current understanding and practice associated with stormwater runoff management. The combination of alterations in land use and accordingly in the hydrologic cycle change the timing of runoff flows with significant and serious implications for flood prediction and risk management. Therefore, *data collected at short-intervals and from locations with significance in the drainage network is the key to understanding the full implications of storm events* with widely varying intensity and duration, falling on landscapes with equally variable antecedent conditions.

The general extrapolation of basin-specific parameters is not advisable (Hollis, 1975). An attempt is made in this chapter to present data needs for stormwater management by detailing the theory of stormwater production and how this drives the management of this predominantly urban runoff regime. For each section the questions of *what* needs to be monitored, *why*, *where* and *when* are answered. To some extent, this chapter will address the questions of *how* to monitor urban stormwater, although the reader is referred to further detail provided in Part I, as well as local guidelines on stormwater monitoring.

17.2 INTERACTIONS WITH OTHER URBAN WATER CYCLE COMPONENTS

Monitoring of urban stormwater needs to take into account not only the specific stormwater-related phenomena, but also the interactions with other urban water cycle components (Table 17.1). If properly integrated urban water management is to be achieved, then monitoring of both the stormwater behaviour and its interactions with other components needs to be undertaken.

17.3 MONITORING REQUIREMENTS

Determining what to monitor clearly depends on the objectives of the monitoring programme (refer to Chapter 3). However, management of urban stormwater is likely to require data on at least some of the following aspects:

- Catchment characteristics, including size, imperviousness, slope, land use cover
- Drainage and treatment network, namely, position, size, slope of pipes, pits and channels, as well as size and specification of engineered stormwater treatment systems
- Urban meteorology, including precipitation (rainfall intensity and duration), and evapotranspiration
- Stormwater runoff quantity, including flow rate and volume, duration, and frequency for outfalls from both pervious and impervious surface areas
- Stormwater runoff quality, including various water quality parameters, both chemical and biological (concentrations, loads, etc.) throughout the drainage system, and upstream, downstream and within treatment systems
- Aquatic ecosystem health or biocriteria status, including hydrology, water quality, and measures of biological condition and/or function
- Stormwater treatment design, including the quantity and quality of stormwater (e.g. Taylor et al., 2006)
- Land-use and construction regulations, including the effectiveness of enforcement of regulations and their impact on particular development activities on hydrologic and water quality impacts.

17.3.1 Catchment characteristics

At the start of any stormwater monitoring programme, the basic information needed, is size and slope of catchment, land-use, and nature and extent of impervious surfaces. *Landscape form*: These characteristics are usually measured from maps (manually), GIS models of the area, or Digital Terrain Models (DTM). Aside from manual delineations, there are some specific tools that can be used to delineate boundaries of catchments and sub-catchments, such as found within HYSTEM-EXTRAN (Prodanović, 1993) and a number of commercially available GIS-driven modules. Satellite pictures and aerial

Table 17.1 Potential interactions between urban stormwater and other urban water system components

Urban water system component	Possible interaction with stormwater
Urban climate	Stormwater runoff (volume, rate, shear stress, etc.) will depend on rainfall intensities (measured at short duration). In turn, pollutant concentrations will depend on rainfall intensity, as well as antecedent dry weather periods. Evapotranspiration and antecedent rainfall will influence soil moisture, thus influencing rainfall-runoff response. Rainfall intensities and spatial patterns determine flood risk and frequency.
Water supply	Stormwater may impact on drinking water quality, especially in a mixed-land use catchment. Stormwater may provide an alternative water supply, for potable or non-potable purposes (e.g. rainwater tanks, large-scale stormwater harvesting). Potential interaction of stormwater and water supply networks, where leakages in water-supply pipes exists.
Wastewater and combined sewer systems	In separate systems, stormwater may infiltrate into wastewater sewers (or vice-versa). Sewer system overflows into stormwater drainage network. Greywater discharges into stormwater drainage network. In combined systems (CSSs), stormwater quantity and quality will affect the dimension and design of treatment facilities.
Groundwater	Exfiltration from stormwater sewers recharges groundwater and raises its level and may cause geotechnical problems by decreasing soil stability. The effect on groundwater quality depends on the quality of stormwater.
Aquatic ecosystems and urban streams	Increased peak flows redefine channel size and shape and isolate floodplains. Water quality of urban streams will be greatly affected by aquatic life. Aquatic life in lentic and lotic receiving waters will be affected by hydrologic and hydraulic impacts, as well as water quality.
Human health	Polluted stormwater could lead to direct pollution of drinking water supplies, particularly where water is extracted from downstream surface waters. Indirect pollution will occur where groundwater is affected by sewer overflow. Poor quality stormwater poses a risk to humans where downstream surface waters are used for recreation or fishing. Risk is greatest where high levels of pathogens are present (usually due to contamination by wastewater).
Society and institutions	Values placed on receiving waters by communities will influence how stormwater is managed (i.e. the degree to which environmental values of receiving waters are taken into account). Similarly, community values regarding property, etc, will determine the impact of flooding, and thus the importance of flood management. The responsibilities for management of stormwater may be shared between multiple agencies with potential for gaps and inconsistencies.

photography are very often used as data sources for these models. In recent times, LIDAR (Light Detection And Ranging) datasets can satisfy data requirements for high-resolution contour, canopy coverage and stream identification. It is important to note, however, that in the urban context, the hydrologic catchment boundary may be significantly different from the topographical definition, due to the presence of underground pipe networks.

Land-use: Impervious coverage is a key predictor of stormwater runoff, and is thus a critical measure of urbanization impacts. The most general measure of imperviousness is total impervious surface area (TIA) expressed as a proportion or percentage of total catchment area. TIA represents all impervious areas within catchments, including roads, roofs, carparks, and footpaths. However, the hydrologic consequences of impervious areas vary, depending on the nature of their hydraulic connection to receiving waters. Effective impervious area is defined as the sum of impervious areas hydraulically connected (i.e. piped or similarly connected) to a drainage system. Effective impervious area includes streets with kerbs or gutters that are sewered to an outfall, or parking lots that produce runoff that is routed to a constructed drainage network which transport flows to surface waters or treatment plants. An ineffective or disconnected impervious surface (Alley and Veenhuis, 1983; Booth and Jackson, 1997) routes runoff to pervious surfaces where overland flows are partially or wholly infiltrated. Some examples of ineffective impervious areas include a roof with downspouts that drain onto either a grassed area or streetside bioretention systems.

It is the effective or connected impervious area that has the most pronounced effects on catchment hydrology (Lee and Heaney, 2003). Accurate data about effective impervious area is thus critical for stormwater management. Historically, total and effective impervious area has been derived from direct measurement (Lee and Heaney, 2003), expressed as correlations of each other (Alley and Veenhuis, 1983), or commonly determined as average proportions for a certain land-use type (Wibben, 1976; Dinicola, 1989). An alternative integrated measurement of typical sources of imperviousness and common urban stormwater infrastructure was developed by Sauer et al. (1983) who used a basin development factor (BDF) to scale the magnitude of urbanization on the basis of extent of sewerage, channel improvements, kerb-guttered roadways, drainage area, channel slope, rainfall intensity, detention facilities, and basin lag time. In a GIS-based impervious surface characterization approach informed by hydrologic principles, Walsh et al. (2003; 2004a) provide a structured approach to the determination of both connected and disconnected impervious coverage. Walsh's approach utilizes rigorous ground-truthing, and involves input from drainage engineers with extensive local experience.

Given the importance of effective imperviousness as an indicator of urban impacts, there will often be the need for in-depth surveys or evaluations of stormwater-sewer pipe networks with ground-truthing work to check the accuracy of map data. Direct measurements of the extent and location of impervious surfaces are best made through manual surveys. An example of this would be a door-to-door survey of a residential neighbourhood, wherein data are taken on connectivity of rooftop drainage, driveway dimensions and slope, among other attributes.

It is important to note that the approach of assuming a 'typical effective imperviousness' for a given land use is likely to produce significant errors. Hydrologic impacts of urbanization are not well predicted by land-use zoning (Sartor and Boyd, 1972, Bradford, 1977). In fact, the best assessment of effective imperviousness is given by direct analysis of rainfall and runoff data, as proposed by Chiew and McMahon (1999). In this approach, the proportion of effective imperviousness within a catchment is estimated from the slope of the relationship between rainfall and runoff, for the part of the slope which is derived from impervious area runoff alone (i.e. excluding the events where runoff comes from both impervious and pervious surfaces, Figure 17.1).

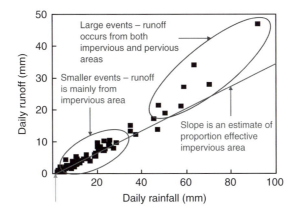

Figure 17.1 Estimating effective imperviousness from hydrologic data (See also colour Plate 15)

Source: Chiew and McMahon, 1999.

17.3.2 Drainage and treatment network

Stormwater drainage networks can contain household pipes, pits (grate or kerb-side inlets, catch-pits, gully pots), drainage pipes (separate for stormwater or combined for both sewage and stormwater), inspection openings, and other outlets. However, they can also have systems for flow attenuation-detention (e.g. detention basins and tanks) and treatment. The latter can be very complex systems with large areas of engineered wetlands, ponds, swales, bioretention systems, infiltration systems (soakaways, French drains or drainage tiles, infiltration wells, infiltration basins), among other elements. These systems are known by a range of terms, including best management practices (BMP), low impact design (LID), water sensitive urban design and sustainable urban drainage systems.

The mode of conveyance is an important aspect in stormwater management. From source to terminus of stormwater flows, structural integrity of infrastructure can vary significantly. It is well known that exfiltration and hydraulic efficiency of stormwater networks is highly dependent on their age and maintenance (US EPA, 2003). The age of an urban drainage system may also help to identify which flows and interactions might predominate. For example, older systems may exhibit evidence of sanitary sewer exfiltration along a greater extent of the sewer network, itself perhaps a combined sewer system. To monitor the condition of the pipes, cameras may be deployed (usually attached to a floating object introduced into the system), but simple data collection on the ground (via inspection of the system or monitoring of greywater, *E. coli*, or other factors) may be sufficient.

Layout, size and slope of collection and conveyance structures have to be determined right at the start of monitoring. Usually detailed drainage and topographic maps are combined with design specifications to arrive at a specific plan. It is essential to obtain the cooperation of the authorities (i.e. local councils, sewer districts, water authorities or water companies) that maintain the data and in the case of new systems, the developers, construction companies and investors. If extensive infrastructure exists already, plans and blueprints are typically available, attached to licences or permits. For new or old projects alike, detailed drawings of BMP systems will illustrate their purpose in the built environment and at the same time help establish the scope of the monitoring programme. Specific data on maintenance regimes should be obtained. These data might cover frequency of maintenance, costs, as well measured runoff attenuation, water quality and pollutants removed.

17.3.3 Urban meteorology

Meteorology is a critical determinant of stormwater behaviour. While Chapter 14 provided a detailed consideration of data needs for urban meteorology, a short summary of the considerations specific to stormwater is provided here.

As the main driver of runoff production (Church et al., 1999), characterization of *both spatial and temporal aspects of precipitation* (Berne et al., 2004) is critical, especially knowledge about the spatial distribution of precipitation monitoring (i.e. the location of rain gauges); and the temporal distribution of precipitation monitoring (i.e. minimum time interval of rainfall recording). The resolution of a rainfall monitoring system is also important, since it can influence the calculated overall rainfall precipitation (see Chapters 5 and 14).

17.3.4 Monitoring the spatial and temporal distribution of precipitation

Given the advancement in radar technology for rainfall measurement, local guidelines and services should be consulted. A comprehensive overview of recommendations and requirements for collecting, documenting, and reporting precipitation measures is offered by Church et al. (1999). Berne et al. (2004) concluded that integrated radar and field rain gauge measurements have sufficient resolution at 5 minutes and 3 km range for 1000 ha catchments, and 3 minutes and 2 km for 100 ha areas. However, field rain-gauge networks should be denser within smaller catchments, both to make up for the shortfall in radar effectiveness below 2 km to 3 km, and to account for the greater relative spatial heterogeneity in rainfall at small scales. The location of rain gauges will also be important, particularly in the case of a very small catchment, where at least one rain gauge should be located at the centroid of the catchment. If not, significant discrepancies between measured rainfall and runoff will inevitably occur.

17.3.5 Temporal distribution of precipitation monitoring

The purposes for which monitoring data are to be used guide the selection of the time resolution of data. For example, if the purpose of monitoring is to model stormwater quality (and consequently the performance of stormwater treatment), then rainfall should be recorded at an interval of no coarser than 6 minutes (e.g. CRCCH, 2005). Rainfall data should be recorded at the lowest time intervals appropriate to the accuracy and logging capacity of the equipment used. For example, if a 0.2 mm-per-tip tipping bucket is used, the time when each tip takes place may be recorded. In this way no information is lost during the recording process. Consistent interpolation methods would be required to truncate continuous raw rainfall accumulation data into even-interval time units (e.g. 5 minutes).

Additional climate characteristics that may be needed for full understanding of urban stormwater behaviour are listed below:

- *Evapotranspiration (ET):* actual evapotranspiration (AET) is preferred to potential evapotranspiration (PET), though each has its own utility, depending on assessment objectives. At least daily average ET estimates are crucial for monitoring the performance of some BMPs (e.g. ponds, swales, bioretention systems) and for understanding the changes to water balances brought about by urbanization. Recent advances in post-processing of remotely sensed data (e.g., SEBAL) offers good performance compared with traditional estimates, which tend to be more data-intensive and oriented to the more specific and perhaps less useful scale of the microclimate.
- *Snow accumulation and snowmelt patterns:* This is crucial in cold climate regions where snow is a major, if not seasonal, contributor to runoff flows.
- *Temperature (air and ground):* In some regions the biggest floods are produced during heavy rainstorms falling onto frozen ground, where the whole drainage area then behaves like a directly connected impervious area, having a runoff coefficient almost equal to one.
- *Rainfall quality:* This may be important for monitoring stormwater quality, and quantifying pollutant sources and sinks.
- *Wind direction:* This factor helps to predict hotspots of pollutant deposition and accumulation on hard surfaces.
- *Antecedent dry weather period:* The time since the last rainfall, or the last rainfall of a given magnitude, is often used as a predictor of pollutant buildup in stormwater quality models.

17.4 STORMWATER QUANTITY

For each catchment it is important to understand the process of rainfall to runoff generation: where runoff is produced, how it is routed, when is, or the timing of, the onset of runoff production and what is the means of conveyance. To gain such understanding, a comprehensive set of measurements might include:

- *measurements within the drainage system,* including inflow into drainage pits, flow in pipes and channels, discharge at stormwater outlets, frequency and location of sewer surcharges, combined sewer overflows, and exfiltration rates from the system;
- *measurements within the catchment,* including soil water content, hydraulic conductivity and initial runoff losses;
- *measurements within stormwater treatment systems,* including inflow and outflow into the systems, flow at various points within the system, water level in the system, exfiltration rates, and groundwater depths;
- *measurements of flow and storage,* including within both landscape locations (e.g. abstraction of rainfall by interception or concavity in topography) and receiving surface water bodies.

In determining which variables to measure for stormwater quantity, the previously discussed interactions with other parts of the urban water system will need to be considered.

17.4.1 Measurements within the drainage system

Where possible, both water head and velocity (hence flow rate) should be measured at each outlet to assess discharged volumes, water quality fluxes and gradients indicating water exchanges amongst different components of the hydrologic cycle. These measures should be made in the water column, downstream of any biological sampling transect (since they will disturb organisms), between water column and interstitial water in substrates and/or stored in banks (hyporheic zone). Ideally, this data should be complemented by correspondent information on biological and water quality status. Interactions such as inputs of wastewater or groundwater, or extractions for water supply, need also be considered in the location of flow measurement within the drainage system. For example, to determine a mass balance of pollutant sources within a catchment, flow (and water quality) measurement upstream and downstream of a combined sewer overflow (CSO) point may be required. Ideally, the monitoring system for stormwater, wastewater, groundwater and water supply will be designed as one integrated package. Failing this, negotiation with designers and operators of the other systems should be an absolute minimum requirement.

Flow in pipes is commonly measured using an ultrasound Doppler or transit-time flow meter. Simple weirs (rectangular or V-notch) are used to backwater the flow or take advantage of any available hydraulic head. Standard literature may be used to design the weirs (e.g. French, 1985) while pressure sensors or ultrasound level sensors are used to monitor level either upstream or at the weir (i.e. the head over the weir). Critical-depth flumes, a very good solution for continuous flow measurement, should be used where their installation is feasible.

Continuous recording of flow rates is now very common. This allows monitoring of both base flow (during dry periods) and wet-weather flow (during storm events) in the system. An appropriate flow range of the flow meter will need to be considered, to ensure that it has accuracy both for small flow rates during dry weather periods and for large flow rates during storm events. In some cases, two separate flow meters may be required, one for low flows, and one for high flows. Similarly, the sampling interval of the flow sensor should be proportional to the rate of change in flows. One approach is to use a rain gauge to trigger increased higher sampling rates at the commencement of a storm.

Knowing the flow rate allows an easy integration of flow passing through the system and calculation of total volumes and pollutants loads. Specification of monitoring equipment has been discussed in Chapter 7. However, we note that indirect gauging of stormwater discharge in conveyances via depth measurement across a suitable flow control is a cost-effective way of measuring discharge in well-characterized cross sections, reducing the need for expensive monitoring infrastructure for discharge. Of course, the constant checking of rating curves (Q/H) must be done both during dry weather flow and during storm events.

Given the importance of flow data in explaining other processes, particular attention should be given to calibration and validation of flow monitoring. Redundant sensors should be considered to militate against impacts of instrument failure (either non-operation or production of erroneous readings) upon data contiguity and quality (refer to Chapter 7 for further information on selection of monitoring equipment). Readers are also referred to Chapter 8 for considerations regarding data validation.

17.4.2 Measurements within the catchment

Soil moisture content and saturated hydraulic conductivity of pervious areas need to be monitored in certain cases, given that runoff can be generated from pervious as well as impervious surfaces. The drying of soil during periods of low or no rainfall will affect the potential for runoff production during the next rain event. This implies a certain duration of effectiveness for BMPs that rely on infiltration as their basis for performance. Measurements of initial losses on pervious surfaces are also important in the modelling of stormwater flows, and hence predictions of flooding.

Soil moisture content may be measured continuously using time domain reflectrometry (TDR) or calibrated tensiometers. Hydraulic conductivity is usually measured by standard techniques for in situ double-ring methods or other well-documented approaches (e.g. Guelph permeameter, rainfall infiltrometer, etc.). Given the variability of soil properties within a catchment, a pilot study may be necessary to determine the appropriate spatial resolution of soil hydraulic conductivity measurements. Soil survey map units might serve well as the spatial scale of resolution for these measures. However, variability in soil properties in urban catchments is likely to be much greater than that of a 'natural catchment', due to the intensity of soil disturbance (e.g. importation of soils, excavation and re-shaping of soil profiles, etc.), and so pilot-testing will almost certainly be necessary to determine the required density of soil samples.

Wear and damage (i.e. age) of impervious surfaces will increase infiltration and water storage in the catchment and may increase the possibility of cracks, which may provide additional pathways for preferential flow, with the combined effect of reducing or delaying production of surface runoff. Streets and other transportation surfaces are assumed to have a finite water storage or capacity of 0.8 mm to 2.3 mm water for cool and hot weather respectively (Albrecht, 1974), with 1.0 mm assumed in many urban rainfall-runoff models (e.g. Chiew and McMahon, 1999; CRCCH, 2005).

17.4.3 Measurements within stormwater treatment systems

Most guidance tends to discount hydrologic monitoring, usually due to the perceived expense of monitoring to the required level. However, if a BMP is ineffective, then the cost to the surrounding environment will quickly mount as damages due to BMP failure increase. To monitor performance of stormwater treatment systems it is generally necessary to monitor inflow and outflow rates (e.g. exfiltration rates and water levels) over time. Monitoring of flow rates and average storage at intermediate points within the system may also be necessary. Despite detailed guidelines on monitoring BMPs in many countries (e.g. ASCE/US EPA, 2002), we cannot overemphasize the importance of proper monitoring of inflows and outflows so that system effectiveness may be evaluated. Without accurate flow data, all other data (e.g. water quality) will be of little value. Conversely, the significant differences often apparent between the upstream and downstream hydrographs of a stormwater treatment system makes it relatively 'easy' to come to the mistaken conclusion that the system is not effectively reducing pollutant loads. For example, in a 'flashy' catchment (i.e., a catchment prone to flash floods), the inflow hydrograph may have receded relatively quickly, increasing the chance of any sampling approach actually 'missing' the inflow load (Figure 17.2 shows hypothetical inflow and outflow hydrographs, subjected to a uniform-time sampling regime).

Downstream, where the response is highly attenuated, detection is far more likely to occur at a temporal resolution sufficient to characterize the response. Sampling downstream of a BMP that may offer significant flow attenuation will require data logging for a much longer period than when sampling upstream. Overall, this discrepancy may result in an underestimate of inflow loads, leading to a relative underestimate in treatment performance. In most cases therefore, continuous recording of flow data is a must because of potentially long detention times in the BMPs, and highly variable flow dynamics (depending on the outlet design), which may complicate determination of what is inflow and outflow within a single event. However, the data-logging can utilize a variable time-step according to the rate of change in flows.

To monitor inflow into a large retention (e.g. pond or wetland) or detention BMP, the measurements will usually be made within the inflow pipes. Doppler flowmeters could be used as well as flumes. If an objective is to measure treatment performance (i.e. pollutant trapping in the BMP), it is usually not good practice to use weirs at the inflow into a large BMP (i.e. due to sediment trapping). Flow gauging equipment can normally be connected so that it will trigger other monitoring equipment (e.g. water quality auto-samplers), using either wired or wireless connection. However, it is important to consider the flow regime (shape and duration of the hydrograph) at inflow and outflow and to adjust data logging periods accordingly.

In the case of infiltration systems, assessment of exfiltration can be done on the basis of monitoring inflow rates and water levels in the system through a simple mass balance approach. Monitoring clogging of infiltration systems can be achieved by measuring the exfiltration over lengthy time periods (or over a number of short spells of time spaced over a long period). Clogging tests could be done by filling the system artificially

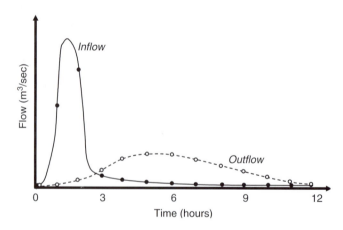

Figure 17.2 Hypothetical hydrographs upstream (solid line) and downstream (dashed line) of a stormwater treatment system (e.g. wetland), showing the different periods of flow and water quality sampling necessary to accurately quantify the upstream and downstream pollutant loads. Circles show uniform time-interval sampling upstream (closed) and downstream (open), which captures a representative sample of the outflow, but not the inflow (missing its peak)

every six months to the same point and then monitoring the exfiltration, or by observing exfiltration after storms (Le Coustumer et al., 2006).

Monitoring of flows through 'treatment-trains' or parallel arrangements of several treatment systems (e.g. wetlands, ponds, swales, bioretention systems, French drains, infiltration trenches) requires particular attention. Because inflow may be distributed along the length of the treatment network, it will need to be measured at a range of points within the network, in order to be able to track the mass balance. It may be possible to measure inflow at a small number of representative inlets, but this should only occur after a pilot study demonstrates that the monitored inlets adequately represent the behaviour of the entire network.

17.5 STORMWATER QUALITY

We will not recast here the full range of important topics that underpin a well-structured water quality monitoring programme; rather readers are referred to local monitoring guidelines. The emphasis here is to identify key considerations for monitoring stormwater quality within the overall framework of integrated urban water management.

Choosing which water quality variables to measure will depend on the objectives of the monitoring programme. Careful consideration should be given to the potential interactions of the various system components. For example, urban stormwater runoff may result in accumulation of heavy metals in downstream estuaries, threatening fisheries. Similarly, leaking wastewater systems may contribute pathogens to stormwater. Before choosing target water quality variables, therefore, a map of potential system interactions and impacts should be developed.

Many of the parameters listed in Chapter 4, such as temperature, pH, redox potential, and turbidity, can be measured using continuous sensors with appropriately deployed instruments that automatically log data. A wide variety of water quality sensors are available, and these are typically deployed as an integrated suite (e.g. temperature, turbidity, redox potential, dissolved oxygen, etc.) in what is known as a 'data sonde'. A sonde can be deployed in a stream channel, within or downstream of a BMP, within a conveyance pipe, among other settings where questionable safety, access issues, or data collection objectives dictate that it is important to have automatic, unattended measurement and logging of water quality parameters.

The importance of rigorous quality control and calibration protocols for water quality sondes (and flow measurement equipment) cannot be overstated. Provided that sampling periods are long enough, uncertainties in estimated loads of pollutants will be minor if errors in measurement are *random*. Where there are *systematic errors* (e.g. due to drift of a sensor), however, the level of uncertainty may become very large (Fletcher and Deletić, 2006) (see Chapter 6).

New sensor technologies are being developed, allowing more and more water quality parameters to be continuously measured using sensors. However, the majority of the water quality parameters require collection of water samples and subsequent laboratory analyses. Locally approved standard methods should be followed for sample collection and analysis (e.g. Greenberg et al., 1999). For example, proper poly-aromatic **h**ydrocarbon (PAH) analysis requires the use of glass bottles for sample collection. In particular, special care should be taken in the sub-sampling of field samples ('sample splitting'), usually done when water collected in one bottle has to be used for wet analyses of more than one variable. For example, sub-sampling for measurements of particle-size distribution or particle concentration is a very difficult job (particularly where there are small numbers of large particles). There are various methods of splitting samples (cone or churn splitter), and the consensus among the authors is to adopt and adhere to one type of splitting procedure applying this uniformly throughout a monitoring campaign. However, it is advisable also to check the chosen method against the alternatives as part of the data validation and quality control procedure (Chapter 8).

In some situations, it may be desirable to use a water quality parameter which can be used as a surrogate for other parameters which cannot be measured continuously or automatically with sensors. For example, turbidity can provide a useful surrogate measure of total suspended solids (TSS), *provided* that careful calibration is undertaken using a site-specific water sample and empirical equation (Bertrand-Krajewski, 2004), and the sampling frequency and duration are adequate as determined by pilot studies (Fletcher and Deletić, 2006).

17.5.1 Monitoring water quality within the drainage network

As discussed in Chapter 7, samples for water quality analyses may be taken by manual grab sampling, by an automatic sampler, or by using continuous sensors. After defining the objectives of the stormwater monitoring programme, the following needs to be clearly specified: (i) where to sample within the system; (ii) how often to sample, e.g., continuous and discrete sampling, wet and dry weather monitoring, monitoring event mean concentrations (composite sampling) and pollutographs; (iii) how many events to sample, that is, the number of events that is representative for the site and point of measurement; and (iv) what will be the trigger event for data logging (i.e. will it be rainfall, flow volume, level, and how much).

Samples can be taken at any point within the system depending on the aims of the monitoring programme. It is usually done on the outflows from catchments, overflows from the system, inflows into pits, and in the pipes and channels. If water is sampled using an auto-sampler, due consideration should be given to the position of the intake pipe to ensure that a representative sample has been taken (Deletić et al., 2000). Gravity-fed samplers are well-adapted for collecting samples from inflow into pits, as no deposition can accumulate prior to the sampling. If measurements (e.g. turbidity, electrical conductivity, pH, temperature, etc.) are taken continuously, it is critical that the probes are placed in the running stream of water, within a box that has a very small volume, or that provisions are made for a recirculating pump so that the probe is always immersed. Self-cleaning probes (mechanically or using ultrasound) should be used wherever possible.

Water samples for further laboratory analyses can be taken on a flow-weighted basis (e.g. after each approximately 100 L, depending on the size of targeted storms) or at regular time intervals (i.e. between 5 minutes and 30 minutes). This is usually done when the aim is to construct and monitor pollutographs. However, if the aim is to monitor event mean concentrations (EMC), the same samples can be combined (in the case of samples taken at regular time intervals, flow measurements should be used to guide the composition using the flow-weighted approach). Composite samples may be taken directly, during the event (i.e. by using an automatic device capable of taking a sampling volume proportional to the flow volume). If the aim is to determine the site mean concentration (SMC), it is important to note that the simple mean of the collected EMCs will not give the representative value, because it does not take into account the *relative weight* (i.e. volume) of each storm event. The SMC should thus be calculated as the *flow-weighted average* of the EMCs.

Determining how many events (N) are needed to yield a representative sample at one site is a very difficult question and can only be answered by conducting a pilot study of the site and water quality parameters in question (or another site which is taken to be representative of the target site). Recent research in Australia (Francey et al., 2004) showed that to record a SMC with a reasonable confidence (such that the 95% confidence interval is no greater than 50% of the mean) in a single catchment, sample sizes of 50, 30, and 25 are required for TSS, TP (total phosphorus), and TN (total nitrogen) respectively. These numbers should be used only as a general guide, however, as they have been derived from experiments in a number of catchments in Melbourne and Brisbane only. Similar work has been undertaken in France and elsewhere (Francey et al., 2004).

17.5.2 Monitoring water quality within stormwater treatment systems

Beyond attenuation of flow, stormwater treatment systems often have the function and capacity of improving water quality. This capacity is variously realized as providing a sink for dissolved pollutants (e.g. nutrients and metals) and settling sediment particles. Thus, it is important to plan for water quality monitoring of inputs and of the corresponding outputs of stormwater treatment systems so as to calculate relative effectiveness and efficiency. Continuous water quality measurements can provide valuable data that might explain transients in water quality during certain periods of climate or demand. Although many pages could be filled with details on how one might go about doing this, there exists specific summaries of BMP monitoring matters that serve the needs of monitoring practitioners quite well.

As in the case of monitoring of water quality of stormwater in drainage networks, it is important to collect water samples during both dry and wet weather. Dry weather samples are usually collected as grab samples, while storm events are mainly monitored using auto-samplers. Monitoring at the inlet and the outlet are essential to assess removal performance (or calculate performance indicators), but monitoring at the intermediate points will enable a better understanding of the treatment processes occurring within the system.

Traditionally, monitoring of stormwater treatment systems has focused primarily on wet weather performance, distinguishing 'storm event performance' from 'base flow performance'. Under this regime, sampling normally stops soon after the inflow hydrograph has receded to baseflow (Figure 17.3). This approach will risk missing secondary (post-storm-event) peaks in concentration, which may occur as a result of interflow processes (Figure 17.3). To avoid this, a pilot study should always be undertaken, using regular *sampling during both wet and dry weather*. Findings from this should guide the timing and frequency of sampling.

Build-up of sediment, its quality and impacts on benthic processes can be of high importance in large BMP systems (e.g. ponds, wetlands, detention basins). Sediment flows and deposition can be characterized over time (e.g. surveyed a few times a year), as well as sediment cores taken for further lab analyses (e.g. particle size distribution,

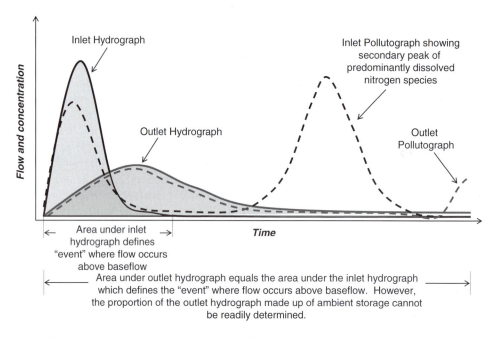

Figure 17.3 Post-storm event secondary peak in pollutant concentration

Source: Derived from Taylor et al., 2006.

TP, etc.). This will help in providing a picture of the long-term behaviour of both the catchment and the treatment system.

In order to understand treatment processes for nutrients (nitrogen and phosphorus), monitoring should extend beyond the total concentration [i.e. total nitrogen (TN) and total phosphorus (TP)], including data on individual species and compounds (e.g. ammonia, nitrate, total dissolved nitrogen). Such data are critical to interpreting the observed behaviour in concentrations (Taylor et al., 2005a) of nutrient species. For example, data on organic nitrogen (particulate and dissolved), ammonia and nitrate, will allow the relative rates of mineralization, nitrification and denitrification process to be assessed. The composition of nutrient species will also have significant implications for receiving waters, which will be most sensitive to bioavailable forms of nitrogen, such as ammonia. Similarly, data on heavy metals in both dissolved and particulate form will provide information on rates of sedimentation, adsorption and desorption, which are prerequisite to improving the design of treatment systems (Taylor et al., 2005b).

17.6 STORMWATER IMPACTS ON AQUATIC ECOSYSTEM HEALTH

Urban stormwater has major impacts on the functioning and health of urban streams and aquatic ecosystems. Chapter 20 provides specific detail on monitoring aquatic ecosystems. Provided here, therefore, are some key points on the monitoring of these systems, in relation to potential interactions with urban stormwater.

Stream channel status has been typically evaluated with ecosystem endpoints as indicators, some examples of which are benthic macroinvertebrate metrics such as taxa found to be tolerant or intolerant to pollution. Other measures are based on ecosystem functioning, such as measures of nutrient retention and energy cycling (e.g. Walsh et al., 2005). On the other hand, measures of physical parameters such as bank full width and depth, and bed particle size characteristics, can help determine the extent and quality of habitat available for stream biological communities. When these factors are measured over time, monitoring results can signal changes in the morphology of the stream channel and in habitat quality.

The USEPA has developed highly detailed guidance on the assessment and evaluation of stressors in stream networks [http://www.epa.gov/bioindicators/ (Accessed 02 July 2007.)]. Other useful guides include the Environmental Monitoring and Assessment Program materials [http://www.epa.gov/emap/ (Accessed 02 July 2007.)], principally those offered by Kaufmann et al. (1998, 1999). A useful guide on urban stormwater and the ecology of streams is also provided by Walsh et al. (2004b).

One critical consideration in monitoring the relationship between urban stormwater and aquatic ecosystems is to *ensure that the correct scale is used*. For example, application of 'water sensitive urban design' or 'low impact design' approaches in one small subcatchment, are unlikely to have significant water quality or ecological benefits for the broader catchment simply because the impacts of urbanization in the remainder of the catchment will overwhelm the small scale of beneficial works. Monitoring to demonstrate the benefits of 'best management practices' should therefore occur at a scale where all, or at least most, of the monitored subcatchment, has been treated using these practices. Monitoring of stormwater impacts on waterways and aquatic ecosystems will also need to consider linkages on the larger water system (e.g. wastewater, groundwater intrusions, extractions for drinking water), again reinforcing the need for an integrated approach to monitoring design.

REFERENCES

Albrecht, J.C. 1974. Alterations in the hydrologic cycle induced by urbanization in Northern New Castle County, Delaware: magnitudes and projections. [*US Environmental Protection Agency Library Report Number*: DI-14-31-0001-3508; DI-14-31-0001-3808; OWRR-A-017-DEL; W74-07729; OWRR-A-017-DEL(2)].

Alley, W.M. and Veenhuis, J.E. 1983. Effective impervious area in urban runoff modeling, *Journal of Hydraulic Engineering*, Vol. 109, No. 2, pp. 313–19.

ASCE and US EPA – American Society of Civil Engineers and US Environment Protection Agency. 2002. *Urban Stormwater BMP Performance Monitoring: A Guidance Manual for meeting the National Stormwater BMP Database Requirements*. US Environment Protection Agency (EPA-821-B-02-001).

Berne, A., Delrieu, G., Creutin, J.-D. and Obled, C. 2004. Temporal and spatial resolution of rainfall measurements required for urban hydrology. *Journal of Hydrology*, Vol. 299, No. 3–4, pp. 166–79.

Bertrand-Krajewski, J.-L. 2004. TSS concentration in sewers estimated from turbidity measurements by means of linear regression accounting for uncertainties in both variables. *Water Science and Technology*, Vol. 50, No. 11, pp. 81–8.

Bertrand-Krajewski, J.-L., Barraud, S. and Chocat, B. 2000. Need for improved methodologies and measurements for sustainable management of urban water systems. *Environmental Impact Assessment Review,* Vol. 20, pp. 323–31.

Booth, D.B. and Jackson, C.R. 1997. Urbanization of aquatics – degradation thresholds, stormwater detention, and limits of mitigation. *Journal of the American Water Resources Association,* Vol. 33, No. 1, pp. 1077–90.

Bradford, W.L. 1977. Urban stormwater pollutant loadings: a statistical summary through 1972. *Journal of the Water Pollution Control Federation,* Vol. 49, No. 4, pp. 613–22.

Chiew, F.H.S. and McMahon, T.A. 1999. Modelling runoff and diffuse pollution loads in urban areas. *Water Science and Technology,* Vol. 39, No. 12, pp. 241–48.

Church, P.E., Granato, G.E. and Owens, D.W. 1999. Basic requirements for collecting, documenting, and reporting precipitation and stormwater-flow measurements. Washington DC, US Geological Survey. (Open-file report 99-255, *USGS National Highway Runoff Data and Methodology Synthesis*).

CRCCH (Cooperative Research Centre for Catchment Hydrology). 2005. *MUSIC (Model for Urban Stormwater Improvement Conceptualisation* – Version 3.0.1). Melbourne, Cooperative Research Centre for Catchment Hydrology. Available at: www.toolkit.net.au/music (Accessed 02 July 2007.).

Deletić, A., Ashley, R. and Rest, D. 2000. Modelling input of fine granular sediment into drainage systems via gully-pots. *Water Research,* Vol. 34, No. 15, pp. 38363–44.

Dinicola, R.S. 1989. Characterization and simulation of rainfall-runoff relations for headwater basins in Western King and Snohomish Counties, Washington State. Washington DC, US Geological Survey. (USGS Water Resources Investigation Report 89-4052).

Fletcher, T.D. and Deletić, A. 2006. *A Review of Melbourne Water's Pollutant Loads Monitoring Programme for Port Phillip and Western Port.* Melbourne, Melbourne Water Corporation.

Francey, M., Deletić, A., Duncan, H.P. and Fletcher, T.D. 2004. An advance in modelling pollutant loads in urban runoff. Paper presented at the UDM (Urban Drainage Modelling) Conference, 15–17 September 2004, Dresden, Germany.

French, R.H. 1985. *Open Channel Hydraulics.* New York, McGraw-Hill.

Greenberg, A.E., Clesceri, L.S. and Eaton, A.D. 1999. *Standard Methods for the Examination of Water and Wastewater,* 20th edn. New York, American Public Health Association, Water Environment Foundation and American Water and Wastewater Association.

Hollis, G.E. 1975. The effect of urbanization on floods of different recurrence interval. *Water Resources Research,* Vol. 11, No. 3, pp. 431–35.

Kaufmann, P.R. and Robison, E.G. 1998. Physical habitat characterization. J.M. Lazorchak, D.J. Klemm and D.V. Peck (eds), *Environmental Monitoring and Assessment Program – Surface Waters: Field Operations and Methods Manual for Measuring the Ecological Condition of Wadeable Streams.* Washington DC, US Environmental Protection Agency, pp. 77–188. (EPA/620-R94-004F).

Kaufmann, P.R., Levine, P., Robison, E.G., Seeliger, C. and Peck, D. 1999. *Quantifying Physical Habitat in Wadeable Streams.* Washington DC, US Environmental Protection Agency (EPA/620/R-99/003).

Le Coustumer, S., Barraud S. and Béranger, Y. 2006. Étude préliminaire du colmatage des systèmes d'infiltrations des eaux pluviales en milieu urbain. *Revue Européenne de Génie Civil,* Vol. 10, No. 4, pp. 263–78.

Lee, G.L. and Heaney, J.P. 2003. Estimation of urban imperviousness and its impacts on storm water systems. *Journal of Water Resources Planning and Management.* Vol. 129, No. 5, pp. 419–26.

Prodanović, D. 1993. Subcatchment delineation. L. Fuchs (ed.), *GIS for Water Engineers—Manual.* Hanover, Germany, COMETT Project, Institut für technisch-wissenschaftliche Hydrologie.

Sartor, J.D. and Boyd, G.B. 1972. *Water Pollution Aspects of Street Surface Contaminants* (No. EPA-R2-72-081). Washington DC, US Environmental Protection Agency.

Sauer, V.B., Thomas, Jr., W.O., Stricker, V.A. and Wilson, K.V. 1983. *Flood Characteristics of Urban Watersheds in the United States*. Washington DC, US Geological Survey (US Geological Survey Water-Supply Paper 2207)

Taylor, G.D., Fletcher, T.D., Wong, T.H.F. and Breen, P. F. 2005a. Design of constructed stormwater wetlands: influences of nitrogen composition in urban runoff, and dissolved nitrogen treatment behaviour. Paper presented at the Hydrology and Water Resources Symposium, 20–23 February 2005, Canberra, Australia.

Taylor, G.D., Fletcher, T.D., Wong, T.H.F. and Breen, P.F. 2005b. Nitrogen composition in urban runoff – implications for stormwater management. *Water Research*, Vol. 39, No. 10, pp. 1982–89.

Taylor, G.D., Fletcher, T.D., Wong, T.H.F. and Duncan, H. P. 2006. Baseflow water quality behaviour: implications for wetland performance monitoring. *Australian Journal of Water Resources*, Vol.10, No. 3, pp. 293–302.

Terstriep, M.L., Voorhees, M.L. and Bender, G.M. 1976. *Conventional Urbanization and its Effect on Storm Runoff*. Carbondale, IL, USA, Illinois State Water Survey.

US EPA. 2003. *Exfiltration in Sewer Systems*. Washington DC, US Environmental Protection Agency (US-EPA Report 600/R-01/034).Available at: http://www.epa.gov/ORD/NRMRL/pubs/600r01034/600r01034.htm (Accessed 02 July 2007.)

Walsh, C.J., Sim, P.T. and Yoo, J. 2003. *Methods for the Determination of Catchment Imperviousness and Drainage Connection*. Melbourne, Cooperative Research Centre for Freshwater Ecology, Monash University, Australia. (Project on Urbanization and the Ecological Function of Streams Report no. D210).

Walsh, C.J., Papas, P.J., Crowther, D., Sim, P.T. and Yoo, J. 2004a. Stormwater drainage pipes as a threat to a stream-dwelling Amphipod of conservation significance, *Austrogammarus australis*, in South-Eastern Australia. *Biodiversity and Conservation*, Vol. 13, pp. 781–93.

Walsh, C.J., Leonard, A.W., Ladson, A.R. and Fletcher, T.D. 2004b. *Urban Stormwater and the Ecology of Streams*. Melbourne, Cooperative Research Centre for Freshwater Ecology, Monash University and Institute for Sustainable Water Resources, Department of Civil Engineering.

Walsh, C.J., Fletcher, T.D. and Ladson, A R. 2005. Stream restoration in urban catchments through redesigning stormwater systems: looking to the catchment to save the stream. *Journal of the North American Benthological Society*, Vol. 24, No. 3, pp. 690–705.

Wibben, H.C. 1976. *Effects of Urbanization on Flood Characteristics in Nashville-Davidson County, Tennessee*. Washington, DC, US Geological Survey (USGS Water-Resources Investigation 76–121).

Chapter 18

Combined sewers

J.-L. Bertrand-Krajewski

Laboratoire LGCIE, INSA-Lyon, 34 avenue des Arts, F-69621 Villeurbanne CEDEX, France

18.1 INTRODUCTION

Combined sewer systems (CSS) are composed of collectors and pipes, simultaneously transporting wastewater and stormwater in a single network. As a consequence, most of the guidance given in Chapters 16 and 17 is also applicable in CSSs. The reader should therefore first read these chapters.

There are a number of special characteristics of combined sewer systems which have implications for data requirements and monitoring. These are the subject of this chapter:

- The high variability of water and pollutant loads reaching downstream wastewater treatment plants (WWTPs). These WWTPs are designed to treat wastewater and a limited fraction of the stormwater generated (e.g. up to 2 or 3 times the mean dry weather flow), with consequences for discharges of untreated stormwater during large storm events;
- The combined sewer overflow (CSO) structures designed to discharge excess combined sewage into surface water bodies. This occurs when flows exceed the maximum treatment capacity of the downstream WWTP and/or the hydraulic transport capacity of the CSS. As this combined sewage has a different composition to typical stormwater, typically more organic matter, i.e. higher chemical oxygen demand (COD), (biological oxygen demand (BOD), nitrogen (N) and phosphorus (P) loads, and bacteriological contamination), specific impacts on receiving surface water bodies are to be expected and must be considered in management and therefore in planning data collection activities.

18.2 INTERACTIONS WITH OTHER URBAN WATER CYCLE COMPONENTS

Monitoring of combined sewers needs to be coordinated and consistent with monitoring of other components of the urban water cycle that may interact with the CSS, especially waste water treatment plants (WWTPs) and receiving surface-water bodies. However, due to the separate historical, academic and technical development of these fields, it is unfortunately rare to find integrated monitoring of CSSs, WWTPs and water bodies. As a result unnecessary overlap in measurement of variables occurs often using different methods. Worse still, some important variables, which can be used to assess interactions between the components, remain unmeasured. For example, in sewer systems, it is most frequent to measure total, dissolved and particulate COD, while BOD_5 fractions, and more recently COD fractions, linked to their relative biodegradability are measured in WWTPs, and total and dissolved organic carbon fractions are measured in rivers. In addition, time scales vary often considerably, ranging from one minute to tens of hours in CSSs, from hours to days in WWTPs and from hours to weeks and months in rivers.

Coordination of monitoring objectives, variables, units and time scales in the various components of the urban water system from CSSs to receiving water bodies is still not common and remains a challenge. This challenge is not only present in monitoring, but also in many related activities such as operation, management and modelling (Rauch et al., 2002; Vanrolleghem et al., 1999). Additional difficulties arise due to the frequent administrative and institutional separation of the entities responsible for these components of the urban water cycle: one operator for the CSS, another for the WWTP, and a third entity in charge of managing water bodies. A summary of the most important interactions is provided in Table 18.1. These interactions should be taken into account when developing any monitoring programme for CSSs and can also be used to assist in developing a conceptual model of the overall system, which is necessary to determine what, where, when and how to monitor.

18.3 MONITORING REQUIREMENTS

As most monitoring requirements are identical to those for wastewater and stormwater, the reader should refer to the corresponding Chapters 16 and 17 for detailed information about catchment and network characteristics, urbanization, land use, wastewater and pollutant generation, meteorology, discharges and pollutant loads. Among the specific aspects of CSSs, the most important is probably the operation of CSOs. As CSOs represent a very significant contribution to impacts on and contamination of receiving water bodies, their monitoring is a key issue to understand how CSSs interact with the other components of the urban water system. As previously discussed in this document, monitoring shall be undertaken according to clearly defined objectives (see Chapters 2 and 3).

The French legal requirements for CSO monitoring are used throughout this chapter as an example. A national regulation published in 1994 defines the minimum monitoring requirements for CSSs, including CSOs and WWTPs (JO, 1994). This CSO monitoring protocol aims to evaluate volumes and some pollutant loads discharged into receiving water bodies. But it may also contribute to better knowledge of the functioning and modelling of the CSSs and to possible improvements of its structure (retrofitting) and operation to increase efficiency and to decrease negative impacts of discharges on aquatic ecosystems. These minimum legal requirements can of course be extended by any sewer operator, the ultimate state being a fully developed continuous measurement network including many monitoring locations within the CSS, named 'permanent diagnostic', possibly coupled to real-time control and regulation (see the concept and definition of permanent diagnostic in Joannis et al., 2006; other information including experience of CSS monitoring can be found e.g. in Bertrand-Krajewski and Ahyerre, 2006a, 2006b).

Combined sewer overflow monitoring requirements depend on the dry weather daily BOD_5 load (DBL, in kg O_2/d) transported in the CSS just upstream of the CSO structure. According to Appendix II of the national order, the regulation requires the following:

- $0 < DBL < 120$ (i.e. 2000 people equivalent), then no monitoring is required
- $120 < DBL < 600$, then estimation of overflow periods and discharge is required

Table 18.1 Potential interaction of combined sewer systems and other urban water system components

Urban water system component	Possible interactions with combined sewer systems
Urban climate	Temperature will significantly affects chemical and biological treatment processes. Temperature and rainfall will affect pathogen growth and transport. Infiltration into wastewater from surrounding soils will depend on soil moisture, which in turn will depend on evapotranspiration and rainfall. Rainfall patterns will determine frequency of combined sewer overflows.
Water supply	Combined sewer overflows (CSOs), exfiltration from sewer pipes and other untreated discharges from combined sewer systems (CSSs) may affect the water resources quality and then the drinking water supply.
Wastewater	Ideally all wastewater should be transported to the wastewater treatment plants (WWTP). Missing connections, leakage of house connections, and combined sewer exfiltration may lead to discharges of polluted wastewater into the soil and the environment. WWTP operation (of both water and sludge treatment components) is directly affected by water and pollutant loads coming from the upstream CSS, and especially by their fluctuations and variability. In particular, the different time scales, ranging from one minute for hydraulic aspects to weeks for treatment processes depending for example on the age of activated sludge, need to be considered.
Stormwater	Mixed with wastewater, a relatively small fraction of storm water is treated in WWTPs after having been transported by CSSs. The remaining portion is discharged (sometimes with some degree of treatment such as temporary retention and settling) through CSOs into receiving water bodies.
Groundwater	Interactions of CSSs with groundwater are bi-directional: (i) CSS exfiltration of wastewater towards the non-saturated and saturated zones may occur, leading to soil and groundwater contamination, with possible geotechnical consequences for buildings and constructions when exfiltration affects soil stability; (ii) infiltration of groundwater into CSSs may also occur, with various consequences including reduction of transport capacity of CSSs, premature and more frequent CSOs, degradation of WWTP performance and higher energy costs (due to pumping needs).
Aquatic ecosystems and urban streams	CSSs affect urban streams: CSOs can increase flow velocity, thus contributing to bed and bank erosion; CSOs may also lead to downstream sedimentation when high TSS loads are discharged by CSOs. Reciprocally, urban streams affect CSSs, e.g. during high flows or floods when water from rivers (or high tides) may enter the CSS. CSOs, WWTP outflow and all other discharges from CSSs affect the chemical, biological and ecological quality of receiving water bodies, sometimes hindering natural and anthropogenic uses of water resources.
Human health	CSOs can lead to direct pollution of drinking water supplies, particularly where water is extracted from downstream surface waters. Indirect pollution occurs where groundwater is affected by sewer overflows. CSOs pose a risk to humans where downstream surface waters are used for recreational or fishing purposes.
Society and institutions	Community attitudes will determine the frequency of CSOs acceptable to the public, and the level of treatment required for overflows. The degree of cooperation between the agency responsible for managing the combined sewers, and the agency involved in managing waterway health, will influence management of the interactions between these aspects.

- DBL > 600, then continuous measurement of discharge and estimation of TSS and COD loads, plus nitrogen loads (if the receiving water body is sensitive to eutrophication) is required.

If the CSS possesses numerous CSOs (e.g. in Lyon, France, a city with 1.2 million inhabitants and 2700 km of sewers, there are in total 370 overflow structures), all CSOs with a DBL > 120 kg/d cannot be monitored, mainly due to financial limitations. In this case, based on hydrologic-hydraulic modelling results, only the CSOs representing 70% of the total volume discharged during a given period of time (typically one year, or the most sensitive period for the receiving water body) are monitored. 'Measurement of discharge' means using water level sensors and, most frequently, flow velocity sensors to measure the spilled volumes by means of a locally calibrated and site-specific relationship. Due to the fact that many CSOs have complex structures and a highly variable hydrodynamic behaviour during storm events, simple relationships between water level and discharge, as given in textbooks, are not applicable in real sewers (Garcia-Salas, 2003; Wastewater Planning Users Group, 2006; Water Research Centre, 1994). A special effort must be devoted to a site specific investigation, with the use of multiple sensors sometimes coupled to computational fluid dynamics (CFD) modelling, aiming to understand the various possible flow regimes that may appear in the CSO during storm events. Detailed information about this type of investigation can be found, for example, in Vazquez et al. (2006).

For CSOs not fully monitored (DBL between 120 and 600 kg/d), the duration of overflows is measured and the discharge of spilled water is estimated. 'Estimation of discharge' means measuring the number of discharges, their duration, and a rough estimation (e.g. based on a local simple relationship between the water level in the CSO and the outflow) of the volumes spilled into the receiving water body.

'Estimation of TSS and COD loads' means that concentrations are not necessarily measured for all storm events: in this case, a site mean concentration (SMC) can be applied (see also Chapter 17 on the use of site mean concentrations), based on sampling campaigns carried out on site for some events. However, the SMC may be very badly estimated if only a few events are monitored (Mourad, 2005; Bertrand-Krajewski, 2006b). This is why traditional sampling and laboratory analyses can be advantageously replaced by continuous monitoring of surrogate parameters for the estimation of TSS and COD concentrations, such as online measurements of turbidity for TSS (Bertrand-Krajewski, 2004; Ruban et al., 2005) and UV-visible spectrometry for COD (Gruber et al., 2005).

Figure 18.1 illustrates the type of results obtainable by means of turbidity measurements. A turbidimeter was installed at the outlet of a 245 ha residential catchment in Ecully, France (in the Greater Lyon area) equipped with a combined sewer system. The data were stored with a two-minute time step. During the complete year 2004, 71 rainfall events occurred and 20 events among them generated overflows discharged into the small creek at the outlet of the catchment. All events were monitored continuously. For the 20 events generating overflows, the mean TSS concentration (mg/L) estimated from turbidity during the overflow periods and the corresponding volumes (m^3) are shown in Figure 18.1. Mean TSS concentrations in overflows show a very high variability, ranging from less than 50 to more than 850 mg/L. Discharged volumes are also very variable, from a few to more than 5500 m^3. Traditional sampling

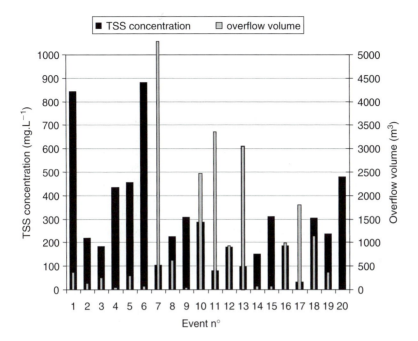

Figure 18.1 Mean concentration of total suspended solids (TSS) estimated from turbidity for twenty overflows, Ecully catchment outlet, 2004

Source: Bertrand-Krajewski, 2006b.

(manually or even with autosamplers) would not provide such a high resolution, and thus the greater variability of the observed volumes and pollutant loads would not be detected with these traditional approaches. Figure 18.2 gives an example of TSS and COD online concentrations estimated by means of a UV-visible spectrometer in another catchment during one storm event.

One of the most significant challenges in monitoring CSOs lies in the intermittent character of storm events. It is not always easy to ensure correct and reliable measurements under such conditions. Another difficulty is that CSOs have rarely been designed and built to facilitate measurements. In order to guarantee reliable measurements, it may be necessary to modify the structure itself after detailed analysis of the local measurement conditions and objectives. If at all possible (for construction of new systems, for example), infrastructure design should take into account potential installation of monitoring equipment, to avoid the need for later, expensive modifications.

Apart from CSOs, another important specific monitoring need in CSS is related to sediments and deposits. Many CSSs, even those designed to be self-cleansing, are affected by sediment deposition. The deposited sediments reduce the hydraulic capacity, generate nuisances (odours, hydrogen sulphide, etc.) and constitute a source of pollutants, which can be partly or totally scoured during storm events increasing the amount of pollutants spilled through CSOs into the receiving water bodies (Ashley et al., 2004). Monitoring sediment deposition is therefore an important issue (Figure 18.3)

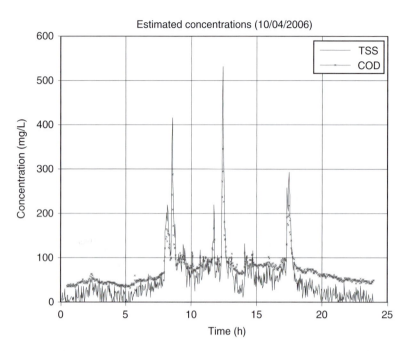

Figure 18.2 Total suspended solids (TSS) and chemical oxygen demand (COD) equivalent concentrations estimated from UV-visible spectrometry (see also colour Plate 16)

Source: Bertrand-Krajewski, 2006a.

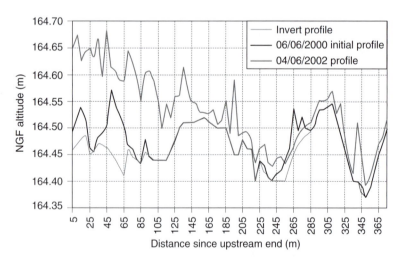

Figure 18.3 Example of sediment deposition monitoring in a 400 m long reach of the combined sewer system in Lyon, France (see also colour Plate 17)

Source: Bertrand-Krajewski et al., 2005.

for operational purposes. A sediment monitoring programme thus is a useful way to evaluate the efficiency of cleansing actions and to determine where, when, and how frequently sewer cleaning should be undertaken

In many cases, monitoring the inflow at the WWTP also yields a lot of information and knowledge about the upstream CSS. Unfortunately, as sewer and WWTP operators are frequently different institutions, the data and the knowledge are not shared. Reciprocally, data from the upstream CSS could be useful for the WWTP operator, especially to anticipate when real time control and regulation are applied. Ideally then, operators of CSSs should develop a 'data sharing plan' with operators of other parts of the urban water cycle, where the potential exists for interactions between those components and the CSS.

REFERENCES

Ashley, R.M., Bertrand-Krajewski, J.-L., Hvitved-Jacobsen, T. and Verbanck, M. (eds). 2004. *Solids in Sewers*. Cornwall, UK, IWA Publishing. (IWA Scientific and Technical Report No. 14).

Bertrand-Krajewski, J.-L. 2004. TSS concentration in sewers estimated from turbidity measurements by means of linear regression accounting for uncertainties in both variables. *Water Science and Technology*, Vol. 50, No. 11, pp. 81–88.

———. 2006a. Field data requirements for monitoring and modelling of urban drainage systems. Paper presented at the Conference on 'Cities of the future: Creating Blue Water in Green Cities', 12–14 July 2006, Racine, WI, USA.

———. 2006b. Influence of field data sets on calibration and verification of stormwater pollutant models. *Proceedings of the 7th International Conference on Urban Drainage Modelling, 2–7 April, Melbourne, Australia*. Melbourne, International Water Association, pp. 3–20.

Bertrand-Krajewski, J.-L., and Ahyerre, M. (eds). 2006a. Dossier – Autosurveillance, diagnostic permanent et modélisation des flux polluants en réseaux d'assainissement urbains. *La Houille Blanche*, Vol. 4, pp. 97–142.

———. 2006b. Dossier – Autosurveillance, diagnostic permanent et modélisation des flux polluants en réseaux d'assainissement urbains. *Techniques–Sciences–Méthodes*, Vol. 6, pp. 17–91.

Bertrand-Krajewski, J.-L., Bardin, J.-P. and Gibello, C. 2005. Long-term monitoring of sewer sediment accumulation and flushing experiments in a man-entry sewer. Paper presented at Tenth International Conference on Urban Drainage, 22–26 August 2005, Copenhagen, Denmark.

Garcia-Salas, J.-C. 2003. Evaluation des performances, sources d'erreur et incertitudes dans les modeles de deversoir d'orage. Ph.D. thesis, Institut National des Sciences Appliquées, Lyon, France.

Gruber, G., Bertrand-Krajewski, J.-L., de Bénédittis, J., Hochedlinger, M. and Lettl, W. 2005. Practical aspects, experiences and strategies by using UV/VIS-sensors for long-term sewer monitoring. Paper presented at the Tenth International Conference on Urban Drainage, 22–26 August 2005, Copenhagen, Denmark.

JO (1994). Arrêté du 22 décembre 1994 relatif à la surveillance des ouvrages de collecte et de traitement des eaux usées mentionées aux articles L.-372-1-1 et L. 372-3 du Code des communes. Paris, *Journal Officiel de la République Française*, 10 February 1995.

Joannis, C., Aumond, M., Rufflé, S. and Cohen-Solal, F. 2006. Détection et diagnostic d'anomalies affectant des résultats de mesures en réseaux d'assainissement. *Actes du Séminaire GEMCEA Validation de Résultats de Mesure en Continu Issus de Réseaux de Surveillance*, Paris, France, 24 March.

Mourad, M. 2005. Modélisation de la qualite des rejets urbains de temps de pluie: sensibilite aux donnees experimentales et adequation aux besoins operationnels. Ph.D. thesis, Institut National des Sciences Appliquées, Lyon, France.

Rauch, W., Bertrand-Krajewski, J.-L., Krebs, P., Mark, O., Schilling, W., Schütze, M. and Vanrolleghem, P. 2002. Deterministic modelling of integrated urban drainage systems. *Water Science and Technology*, Vol. 45, No. 3, pp. 81–94.

Ruban, G., Bertrand-Krajewski, J.-L., Chebbo, G., Gromaire, M.-C. and Joannis, C. 2005. Précision et reproductibilité du mesurage de la turbidité des eaux résiduaires urbaines. Paper presented at Conference on Autosurveillance, Diagnostic Permanent et Modelisation des Flux Polluants en Reseaux d'Assainissement Urbains, Societe Hydrotechnique de France (SHF) et Group de Recherche Rhône-Alpes sur les Infrastructures et l'Eau (GRAIE), Association Scientifique pour l'Eau et l'Environnement (ASTEE), 28–29 June 2005, Marne-la-Vallée, France, pp. 191–200.

Vanrolleghem, P., Schilling, W., Rauch, W., Krebs, P. and Aalderink, H. 1999. Setting up measuring campaigns for integrated wastewater modelling. *Water Science and Technology*, Vol. 39, No. 4, pp. 257–68.

Vazquez, J., Zug, M., Phan, L. and Zobrist, C. (eds). 2006. *Guide technique sur le Fonctionnement des Déversoirs d'Orage*. Strasbourg, ENGEES, Available at: http://www-engees.u-strasbg.fr/index.php?id=714 (Accessed 02 July 2007.).

Wastewater Planning Users Group. 2006. *Guide to the Quality Modelling of Sewer Systems*. Version 1.0. London, UK. Available at: http://www.wapug.org.uk/ (Accessed 02 July 2007.)

Water Research Centre. 1994. *Sewage Rehabilitation Manual*, 4th edn. Wiltshire, UK, Water Research Centre.

Chapter 19

Groundwater

D. Pokrajac

Department of Engineering, Kings College, University of Aberdeen, Aberdeen, Scotland, AB24, 3UE, UK

19.1 INTRODUCTION

Groundwater is an important component of the urban water cycle, providing water supply, as well as impacting aquatic ecosystems (for example, by determining baseflow levels). At the same time, many other water cycle components and activities impact groundwater, such as the migration of pollution or the over-extraction of water. Despite these interactions and their potential consequences, groundwater tends often to be 'forgotten' in many monitoring programmes. This chapter discusses the interactions between groundwater and the urban water system and their implications for monitoring. It then outlines specific considerations for monitoring groundwater and finally presents an example of an integrated groundwater monitoring programme.

19.2 INTERACTION OF URBAN GROUNDWATER WITH OTHER URBAN WATER SYSTEMS

Depending on the characteristics of a particular hydrogeological setting, typically urban subsurface infrastructure lies either within or above an urban aquifer and can thus interact with the aquifer (Table 19.1). The infrastructure may be part of:

- urban water system, such as sewers, water mains, abstraction wells, recharging wells and ponds, building drainage and land-fill leachate drainage, all of which have strong interactions with urban groundwater aquifers serving as a sink for or source of groundwater and a source of pollution;
- other buried infrastructure transporting or storing liquid (e.g. petroleum fuel), possibly a point or linear source of pollution;
- urban infrastructure systems with significant void volumes, such as transportation tunnels, galleries for electrical and telecommunication cables, deep building cellars, underground stations and garages, which can act as sinks for groundwater; and
- 'passive' buried elements, such as deep foundations, properly sealed landfills, in which although not involved in mass exchange or as a pollution source, can affect the local hydrodynamics.

Traditionally these systems have not been regularly monitored to identify and quantify possible interactions with groundwater. Thus the aim of this chapter is to raise awareness of the significant effects these systems can have on groundwater and vice versa

Table 19.1 Potential interaction of groundwater and other urban water cycle components

Urban water system component	Possible interactions with groundwater
Urban climate	Groundwater levels depend on recharge rates, which in turn depend on soil moisture. Soil moisture is governed by rainfall and evapotranspiration. Groundwater pollutant concentrations will depend partly on evapotranspiration and subsequent concentration of soluble ions, etc.
Water supply	During the regular operation of the water supply system, water losses act as a source of recharge for urban groundwater and hence raise its level, which may have geotechnical consequences through decreased soil stability. Generally, such interactions will not adversely affect groundwater quality. During network maintenance, groundwater may accidentally infiltrate into some pipes and pollute drinking water.
Wastewater and combined sewers	Faulty connections, leakage of house connections, and wastewater sewer exfiltration may lead to discharges of polluted wastewater into the unsaturated soil or directly into the groundwater, leading to its contamination. In cases where wastewater sewers are below groundwater level, groundwater will infiltrate into them, reduce their capacity, increasing the WWTP load and hence operational costs. Groundwater infiltration into the sewers increases quantity of water that is treated and hence increases the cost of energy and other operational costs. Infiltration of groundwater into the CSS may also occur, with various consequences including reduction of transport capacity of CSS, premature and more frequent overflows, degradation of WWTP performance, and higher energy costs (due to pumping requirements).
Stormwater	Exfiltration from stormwater sewers recharges groundwater and raises its level and may cause geotechnical problems through decreased soil stability. The effect on groundwater quality depends on the quality of stormwater. In the case of infiltration into stormwater sewers, the transport capacity of the sewers is reduced. This may lead to more frequent local flooding.
Aquatic ecosystems and urban streams	Groundwater may discharge into urban streams affecting their quantity and quality, or recharge from them, in which case the stream affects the quality of groundwater.
Human health	As described for water supply, leakage from wastewater and stormwater (including in combined sewers) poses potential risks to human health. Long-term contamination of groundwater (e.g. heavy metals) will result in or cause chronic health problems, where groundwater is used for water supply.
Society and institutions	Community understanding of interactions between surface water and groundwater may limit attempts to achieve more integrated management. Responsibility for groundwater management may not rest with the institutions who manage other aspects of the water system that affect groundwater quantity and quality. Institutional constraints may limit collaboration between the agencies.

and to identify requirements for monitoring these interactions as part of an integrated urban water management framework.

Hypothetically a drop of water from an urban aquifer may 'travel' through an abstraction well into a water supply pipe to be used in a household, disposed to a wastewater sewer, and discharged into an urban stream, which may recharge the same urban aquifer and eventually return to the same cycle again. Alternatively, the drop of water could return to the aquifer from a leaking water main or a leaking sewer. In either case, on its way the water drop may undergo dramatic changes in quality. This illustrates the variety of urban water systems which exist in a contained sub-surface space below a city, and the complexity and magnitude of associated problems. Other urban water systems interact with groundwater by either receiving water from, or discharging it into, urban aquifers. This interaction may be the result of a deliberate action, such as groundwater abstraction for water supply, or the unintended consequence of aging urban infrastructure (e.g. water losses from the water supply mains).

As a result of these interactions, the management of urban aquifers is (or should be) closely related to the management of other elements of the urban water systems. Their requirements are often conflicting, so an integrated approach is needed, where all these systems are considered simultaneously and their mutual interactions taken into account. Standard urban groundwater problems – inadequately controlled ground-water abstraction, excessive urban infiltration and excessive subsurface contaminant load (Morris et al., 1997) – are often initiated by the requirements or poor management of another water system such as water supply or wastewater. The proposed management requirements and targets always include actions within the other system. *Therefore, the relationships between all these systems need to be quantified in order to define the methodology for an integrated approach.*

Numerous studies have been conducted in order to quantify the risks involved in the failure to recognize the close interrelation of urban systems (see for example Chadha et al., 1997; Misstear and Bishop, 1997, Eiswirth and Hotzl, 1997). The pollution of groundwater below rapidly growing cities is a widely reported phenomenon (Bruce and McMahon, 1996; Trauth and Xanthopoulos, 1997). The problem is especially pronounced in developing countries where the development of essential services, including water supply and sanitation cannot keep pace with rapidly growing populations and the accompanying urban and industrial development (Rahman, 1996; Massone, 1998). On health grounds, the provision of sewerage systems must remain a priority for cities in the developing world (Lerner and Tellam, 1992). However, although these hazards are genuine, good management, including effective management of water quality, can reduce the risks (Lerner, 1996). High costs, low investment capacity and a low general public awareness are the main reasons why an improvement in the situation is difficult to achieve (Mull et al., 1992). In other words, the most critical challenges are not technical; they relate more to social attitudes, a lack of institutional capacity and willingness (see also Chapters 11 and 22)

These problems are not confined to developing countries only. In several cases, after introduction of a new water resource (for example, water obtained by desalination) into a drinking water system, higher pressure would be sustained for longer time, leakage would increase, and the underlying aquifer would get a new 'source term'. In these cases, the groundwater level has increased significantly. Many cellars which were dry became flooded, and water started to emerge on the surface, creating ponds, lakes, new streams, etc. All of these problems are essentially avoidable, by:

- developing a sound conceptual model of the interactions between groundwater and other urban water cycle components

- monitoring these components based on the conceptual model
- using the monitoring data to refine the conceptual model and to inform integrated decision-making.

In summary, the characteristics of urban groundwater, including its quality, strongly depend on a range of activities in the city itself. A thorough review of these characteristics and a number of real world examples of urban pollution problems is presented by Lerner (2004). The essential prerequisite for defining and understanding such problems is the existence of data on groundwater quantity and quality. The following section outlines the principles of monitoring urban groundwater quantity and quality.

19.3 MONITORING URBAN GROUNDWATER QUANTITY AND QUALITY

The physical system to be monitored typically consists of the vadose zone and the underlying aquifer or a system of aquifers (saturated zones which allow groundwater flow) and aquitards (low permeability zones where the flow of groundwater between aquifers is restricted). In order to design groundwater monitoring one must first develop a conceptual model of the subsurface. The conceptual model consists of the main geological and hydrogeological units. While geological units are defined on the basis of the rock or soil type and their geological history, hydrogeological units are distinguished by their role in groundwater transport as either aquifers or aquitards.

The conceptual model of a particular location is based on all available borehole logs and geological profiles, combined, where necessary, with additional site investigations. All these data are compiled and interpreted to produce a clear picture of the hydrogeological units. An example of a geological profile is shown in Figure 19.1. It consists of four geological and three hydrogeological units (layers 2 and 3 can be interpreted as a single aquifer).

The groundwater monitoring system consists of a sufficient number of observation wells installed at appropriate locations and depths, sensors and data acquisition-storage, and/or data transmission units. This means that, regarding groundwater quantity, the number and the spatial distribution of the wells has to be sufficient for the reconstruction of the groundwater table or piezometric surface based on the measured groundwater

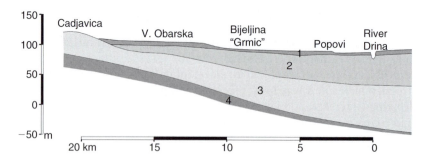

Figure 19.1 Geological cross-section below the city of Bijeljina, Bosnia. (1) swamp clays; (2) sand and gravel alluvial deposits; (3) Paludin sand and sandy gravel with layers of clay; (4) Pontian marly clays and marl (see also colour Plate 18)

Source: Pokrajac, 1999.

levels. It is important to remember that the movement of the groundwater is controlled by the hydrological conditions, not only across the domain of interest, but across the whole sub-surface catchment area. The monitoring wells network should therefore generally cover a wider area than the area of specific interest, ideally the whole catchment until its natural hydrogeological boundaries. If these boundaries are not known, then either (a) a pilot investigation will need to be conducted, or (b) there will need to be a search of relevant data held by other agencies, which may help delineate the catchment.

The positioning of the observation wells for monitoring groundwater quality depends on the general type, location and spatial extent of pollution sources (point or diffuse), and the direction of groundwater movement. In the case of surface point-sources such as a hazardous waste landfill sites, fuel tanks, inadequate sanitation facilities, or industrial discharge, a sufficient number of wells must be provided *up-gradient* from the site in order to quantify the background groundwater quality. The *down-gradient* wells will then be used for evaluating the influence of pollutants released from the site on the groundwater quality. In the case of diffuse pollution sources this requirement, a sufficiently dense network of observation wells distributed along the main directions of groundwater movement is required (see also Chapter 5 for a discussion of spatial scale considerations).

The interaction of urban groundwater with other urban water systems is very complex. For instance, depending on the groundwater level, as shown schematically in Figure 19.2, a stormwater or wastewater sewer can be either a source of groundwater recharge and pollution, or act as a sink. Variations in behaviour such as this were found, for example, in a test area of the city of Rastatt, Germany (Pokrajac et al., 2006). A rather dense network of observation wells is required in order to detect such interactions. While it may not be feasible to install such a dense network over the entire city area, it may be beneficial to establish some selected test areas for a detailed study. Any compromises in the required density should be considered against the objectives of monitoring and the consequences of not collecting the required data. For example, if the monitoring network is insufficient to detect sewer leakage into groundwater, causing contamination of groundwater which is used elsewhere as a water resource, the costs and consequences are likely to be much greater than the costs of having monitored at the required density in the first place (see also Chapter 12).

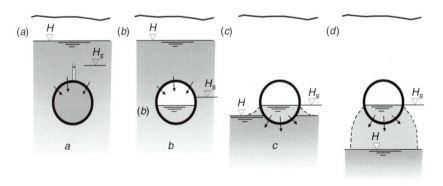

Figure 19.2 A sewer may drain (*a, b*) or recharge (*c, d*) urban groundwater, depending on the groundwater level (H) (see also colour Plate 19)

Source: Pokrajac et al., 2006.

Once the network of observation wells is established, groundwater monitoring may commence. It consists of measuring:

- *Groundwater levels.* These can be measured manually, using a manual level meter, or automatically, using pressure transducers connected to a data logger or telemetry system. The required frequency of measurement depends on the particular application and is typically much lower than for other water systems, e.g. streams, because of the much longer time scales of groundwater movement. This demands the use of batteries for power supply in areas where there is no access to the power grid (sensors and loggers will be able to be 'dormant' much of the time). The analysis and interpretation of observed groundwater levels are usually combined with numerical modelling to provide information about groundwater movement and evaluate parameters governing it (for further information, see Anderson and Woessner, 1991).
- *Groundwater quality.* Groundwater quality can be monitored (measured) online or off-line, as a part of a long-term monitoring programme or as an ad hoc survey, for example after accidental pollution. A limited number of variables can be measured online using conventional individual sensors or multiple parameter probes. Off-line measurement consists of taking samples from the observation wells and their analysis in the laboratory. The sampling frequency again depends on the application. Systematic sampling at regular intervals can be combined with campaign measurements during periods of special interest (such as after a major storm or after a combined sewer overflow). For further information on pollutant migration it is beneficial to use tracer tests. They provide useful data for evaluating parameters governing pollution transport.

For more detailed prescriptions on groundwater monitoring techniques, the reader is referred to local protocols. In addition Nielsen (1991) provides a very comprehensive summary of many aspects of groundwater monitoring, including the principles of the monitoring system design, details of monitoring well design and construction, and interpretation of groundwater level data. Nielsen also discusses principles and details of groundwater sampling and water quality data analysis, while Kass (1998) contains a detailed review of tracer test techniques.

In summary monitoring urban groundwater follows general principles of monitoring groundwater, however, additional but important complexity exists due to the interaction with other water systems. These interactions should not be ignored and should be the focus of urban groundwater monitoring programmes.

19.4 EXAMPLE OF AN INTEGRATED GROUNDWATER MONITORING PROGRAMME

An example of combined permanent and ad-hoc measurement of groundwater is described by Pokrajac et al. (2006). This example is based on an experimental survey of groundwater pollution carried out by the University of Karlsruhe, Germany, within the EU funded project 'Assessing and Improving Sustainability of Urban Water Resources and Systems' (AISUWRS). In this project, continuous monitoring of groundwater levels, undertaken in selected boreholes attached to the wastewater network (Figure 19.3) has been combined with ad hoc measurements of water quality aimed at quantifying the effects of leakage from sewers on the groundwater quality. Modelling was performed using the UGROW model.

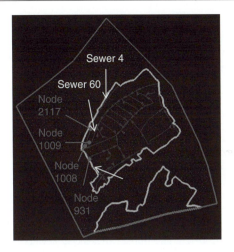

Figure 19.3 Selected area of the sewer network in Rastatt with positions of nodes selected for groundwater level inspection (see also colour Plate 20)

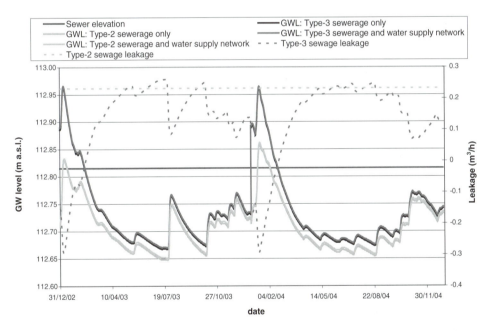

Figure 19.4 Groundwater (GW) levels and sewage leakage at node 931. The pipe invert level is shown by the red line. Leakage from the sewer is shown by the dotted line. Positive leakage means that the sewer is recharging groundwater (see also colour Plate 21)

The results of the modelling exercise showed that, in some parts of the sewer network, the seasonal variation in groundwater levels in the dry season caused sewer exfiltration, while in the wet season there was infiltration. An example of such a condition is shown in Figure 19.4. In the periods when groundwater levels are below the level of the bottom of the inside of the pipe, or pipe invert (red line), the 'leakage' from the sewer (dotted line) is positive, indicating that the pipes were recharging groundwater.

Conversely, in the periods when groundwater was below the invert, the pipes were act-ing as sinks for groundwater (negative leakage).

This example illustrates the complexity of the interaction of groundwater with other water systems and the need for integrated monitoring of the urban water system.

REFERENCES

Anderson, M.P. and Woessner, W. 1991. *Applied Groundwater Modeling, Simulation of Flow and Advective Transport*. New York, Academic Press.

Bruce, B.W. and McMahon, P.B. 1996. Shallow ground-water quality beneath a major urban center: Denver, Colorado, USA. *Journal of Hydrology*, Vol. 186, No. 1–4, pp. 29–151.

Chadha, D.S., Kirk, S. and Watkins, J. 1997. Groundwater pollution threat to public water sup-plies from urbanisation. P.J. Chilton et al. (eds), *Groundwater in the Urban Environment: Problems, Processes and Management*. Rotterdam, A. A. Balkema Publishers.

Eiswirth, M. and Hotzl, H. 1997. The impact of leaking sewers on urban groundwater. P.J. Chilton et al. (eds), *Groundwater in the Urban Environment: Problems, Processes and Management*. Rotterdam, A.A. Balkema Publishers.

Kass, W. 1998. *Tracing Technique in Geohydrology*. Rotterdam, A.A. Balkema Publishers.

Lerner, D.N. 1996. Urban groundwater – An asset for the sustainable city? *European Water Pollution Control*, Vol. 65, pp. 43–51.

Lerner, D.N. and Tellam, J.H. 1992. The protection of urban groundwater from pollution, *Journal of the Institution of Water and Environmental Management*, Vol. 61, pp. 28–37.

Lerner, D.N. (ed). 2004. *Urban Groundwater Pollution*. Paris/Rotterdam, UNESCO/A.A. Balkema Publishers.

Massone, H.E., Martinez, D.E., Cionchi, J.L. and Bocanegra, E. 1998. Suburban areas in devel-oping countries and their relationship to groundwater pollution: a case study of Mar del Plata, Argentina. *Environmental Management*, Vol. 222, pp. 245–54.

Misstear, B.D. and Bishop, P.K. 1997. Groundwater contamination from sewers: experience from Britain and Ireland. P.J.Chilton et al. (eds), *Groundwater in the Urban Environment: Problems, Processes and Management*. Rotterdam, A.A. Balkema Publishers.

Morris, B.L., Lawrence, A.R. and Foster, S.D. 1997. Sustainable groundwater management for fast-growing cities: mission achievable or mission impossible? P.J. Chilton et al. (eds), *Groundwater in the Urban Environment: Problems, Processes and Management*. Rotterdam, A.A. Balkema Publishers

Mull, R., Harig, F. and Pielke, M. 1992. Groundwater management in the urban area of Hanover, Germany. *Journal of the Institution of Water and Environment Management*, Vol. 62, pp. 199–206.

Nielsen, D.M. (ed.). 1991. *Practical Handbook of Ground-Water Monitoring*. Chelsea, Michigan, US, Lewis Publishers Inc.

Pokrajac, D., Stanić, M., Lazić, R., Jaiprasart, P. and Maksimović, Č. 2006. UGROW – a soft-ware tool for integrated urban water management. Proceedings of the 7th Conference on Urban Drainage Modelling, 2–7 April 2006, Melbourne, Australia.

Pokrajac, D. 1999. Interrelation of wastewater and groundwater management in the City of Bijeljina in Bosnia. *Urban Water*, Vol. 1, No. 3, pp. 243–55.

Rahman, A. 1996. Groundwater as source of contamination for water supply in rapidly grow-ing megacities of Asia: Case of Karachi, Pakistan. *Water Science and Technology*, Vol. 34, No. 7–8, pp. 285–92.

Trauth, R. and Xanthopoulos, C. 1997. Non-point pollution of groundwater in urban areas. *Water Research*, Vol. 31, No. 11, pp. 2711–18.

Chapter 20

Aquatic ecosystems

P. Breil[1], M. Lafont[2], T.D. Fletcher[3] and A. Roy[4]

[1]Unité Hydrologie-Hydraulique, Cemagref, 3 bis Quai Chauveau, Lyon, 69336 CEDEX09, France
[2]Unité Biologie des Écosystèmes Aquatiques, Cemagref, 3 bis Quai Chauveau, Lyon, 69336 CEDEX09, France
[3]Department of Civil Engineering (Institute for Sustainable Water Resources), Building 60, Monash, Monash University, Melbourne 3800, Australia
[4]National Risk Management Research Laboratory, Office of Research and Development, US Environmental Protection Agency, Cincinnati, OH 45268, USA

20.1 INTRODUCTION

Aquatic ecosystems are a vital part of the urban water cycle (and of urban areas more broadly), and, if healthy, provide a range of goods and services valued by humans (Meyer, 1997). For example, aquatic ecosystems (e.g. rivers, lakes, wetlands) provide potable water, food resources and recreational opportunities. Aquatic ecosystems also act to filter pollutants, decompose organic material and cycle nutrients. If functioning properly, aquatic ecosystems can mitigate the impacts of stormwater runoff and wastewater, including thermal disturbance, nutrient loading and sediment loading. However, these ecosystems have often been considered as receiving waters for discharging and diluting pollutants, ignoring the essential attributes of waterways necessary for providing ecosystem services.

With urbanization set to continue in many areas around the world, the impacts on aquatic ecosystems have the potential to grow. As a result, the extent to which these ecosystems can provide the goods and services on which humans depend will diminish. The condition of aquatic ecosystems in urban areas depends on the flow regime and quality of water coming from surrounding upstream areas. Of particular importance is the type and pattern of development in peri-urban areas (areas on the outskirts of existing urban development, which are often undergoing urbanization), which constitute the catchments for the urban aquatic ecosystems.

To protect, enhance, and restore aquatic ecosystems, we need to understand how they operate in the urban context. We need to monitor their condition, their processes, and the factors affecting them. To do this, we need to collect data – not only about the aquatic ecosystems – but about all the possible components of the urban water cycle which interact with the ecosystem (and vice versa). In brief, data are needed for:

- quantifying the condition of the ecosystem
- identifying and quantifying stressors to the system
- monitoring the effects of management strategies aimed at the protection or rehabilitation of aquatic ecosystems.

This chapter provides information on why, what, when, where and how to collect data about aquatic ecosystems, their condition and functioning, and interactions with other urban water components.

20.2 INTERACTIONS WITH THE URBAN WATER CYCLE COMPONENTS

Aquatic ecosystems are, by definition, the recipients of many impacts from other components of the urban water cycle, such as extractions of water for drinking water and discharges of stormwater or wastewater. Thus, aquatic ecosystems can be degraded by a range of processes such as flow regime changes (for both high and low flows), as well as changes in the chemical and physical characteristics of water (e.g. increased pollutant loads, perturbations in temperature, etc.). In addition, the physical habitat of aquatic ecosystems can be directly modified by works aimed at increasing the hydraulic capacity of waterways (e.g. channel straightening works). Table 20.1 summarizes some

Table 20.1 Potential interaction of aquatic ecosystems and other urban water system components

Urban water system component	Possible interactions with aquatic ecosystems
Urban climate	A wide range of indicators of ecosystem function and health are dependent on climate.
	Hydraulic stresses on lotic aquatic ecosystems will depend on rainfall.
Water supply	Discharge of polluted and thermally-regulated water from mains or reservoirs can degrade aquatic ecosystems.
	Surface water withdrawals for water supply will be affected by the quality and quantity of water in aquatic ecosystems.
	Extraction of water from waterways will affect the quantity and timing of water in aquatic ecosystems.
Wastewater	Wastewater treatment plant (WWTP) discharges often make up a significant portion of flows in streams and rivers, thus affecting surface water quality and quantity.
Stormwater	Increased peak flows will redefine channel size and shape, and isolate floodplains from aquatic ecosystems.
	Pollutant loading from stormwater runoff will reduce water quality of urban ecosystems.
	Aquatic life in aquatic ecosystems will be affected by stormwater quality and quantity.
Groundwater	Groundwater may discharge into aquatic ecosystems, and, conversely, surface water may recharge groundwater, thus affecting the quantity and quality of each.
Combined sewer systems	Combined sewer overflows (CSOs) can increase flow velocity, thus contributing to streambed and bank erosion.
	CSOs may discharge high amounts of total suspended solids, leading to sedimentation in aquatic ecosystems.
	CSOs affect the chemical, biological and ecological quality of receiving water bodies.
Human health	Eutrophic ecosystems may lead to toxic algal blooms, threatening human health either via (a) direct contact or (b) consumption of fish, etc. from affected waters.
	Bioaccumulation of pollutants in aquatic ecosystems may contaminate food resources.
Society and Institutions	The values placed on aquatic ecosystems will determine constraints on operation of other parts of the urban water system, in terms of acceptable impacts to aquatic health.
	Roles and responsibilities of agencies will affect the degree of integration of ecological management objectives with, for example, water supply objectives.

of the main interactions between aquatic ecosystems and other urban water cycle components. A table like this should be used to build a conceptual model of interactions to guide the required monitoring.

20.3 IMPERATIVES FOR AQUATIC ECOSYSTEM MONITORING AND MANAGEMENT

Throughout much of the world, there is increasing pressure from society to protect, enhance and restore aquatic ecosystems. Recent legislative and regulatory developments have reflected this pressure. For example, the European Framework Directive (EFD) promotes the objective of a good ecological status wherever economically sustainable (with assessment due in 2015). Based on this assessment, rehabilitation plans will need to be developed. The overriding assumption is that basic biotic functions must be preserved or rehabilitated where possible, to ensure a sustainable water resource for future generations. The ecological status assessment is based on three components: morphological, physico-chemical and biological characteristics (http://europa.eu.int/comm/environment/water/water-framework/index_en.html-Accessed 02 July 2007).

Similarly, the United States has instituted several comprehensive aquatic ecosystem monitoring programmes (http://www.epa.gov/ebtpages/wateaquaticecosystems.html Accessed 02 July 2007) through state and regional EPA offices, which are driven by the requirements of the Clean Water Act (1972), and more recent developments that strive to control urban non-point source pollution (namely NPDES Phase II; (http://cfpub.epa.gov/npdes/index.cfm-Accessed 02 July 2007). Qualitative and quantitative indicators of aquatic ecosystem condition are assessed comprehensively under EPA's Environmental Monitoring and Assessment Program (http://www.epa.gov/emap/-Accessed 02 July 2007). The US Geological Survey also has a National Water Quality Assessment (NAWQA) programme which samples aquatic biota, pesticides, nutrients, volatile organic compounds and trace metals (http://water.usgs.gov/nawqa/-Accessed 02 July 2007). Comparable assessments of aquatic ecosystems are occurring in much of the world.

20.4 APPLICATIONS FOR DATA ON AQUATIC ECOSYSTEMS

To understand the condition and functioning of an aquatic ecosystem, it is necessary to understand the stressors on that ecosystem. Since other components of the urban water cycle can act as stressors to aquatic ecosystems, previous chapters of this book, on water supply (Chapter 15), wastewater (Chapter 16), stormwater (Chapter 17), combined sewers (Chapter 18) and groundwater (Chapter 19) should be referred to for complete details on how to monitor each of these components. This chapter provides some specific considerations for monitoring the primary stressors to urban aquatic ecosystems: hydrology, geomorphology, and water quality. For detailed reviews of the impacts of urbanization on aquatic ecosystems, the reader is referred to Paul and Meyer (2001) or Walsh et al. (2005b).

20.4.1 Hydrologic and hydraulic stressors

Flow and its variability is a driving force in many aquatic ecosystems. It provides and renews nutrients, affects water quality and temperature, and determines the type and amount of habitat which is available at any given time. However, to understand these relationships (often involving complex interactions of biotic processes and abiotic factors), substantial amounts of data will often be required, including data on both surface and groundwater dynamics.

In the urban landscape, loss of natural vegetation, soil compaction, artificial drainage networks and extensive areas of impervious surfaces generate increased runoff and reduced infiltration during storm events, thus altering the magnitude, volume, frequency and duration of peak discharges in receiving streams. In addition urban aquatic ecosystems can experience water table depletion, or conversely increased volume due to inter-basin water transfers. Hydrologic alteration can impact channel morphology, erosion dynamics, pollutant loadings and, in turn, aquatic communities (Moscrip and Montgomery, 1997; Booth, 1990; Booth, 1991; Bledsoe and Watson, 2001; Schmitt et al., 2001; Fletcher et al., 2007). The nature and cause of flow disturbance depends on the nature of the catchment (e.g. channel shape, geology, and soils), and, importantly, the type and arrangement of the sewer network (e.g. combined or separate sewers), which are addressed separately.

In cities with combined sewers (see Chapter 18), sewer overflows are the main source of hydrologic disturbance: increases in the rate of intra-annual peak floods can be ten- to twenty-fold greater than otherwise and are very sensitive to catchment imperviousness (Hollis, 1975; Wong and Tan, 1995; Paul and Meyer, 2001; Beighley and Moglen, 2002). Combined sewer systems are the norm in Europe. Monitoring direct urban runoff or overflows to the receiving waters is therefore important to assess potential impacts. Monitoring in the stream and using a time-series filter (Figure 20.1) can then be used to separate the natural flow from elevated peaks caused by impervious area or combined sewer overflow (CSO) contributions.

In areas with separate stormwater (see Chapter 17) and sanitary sewers, runoff from impervious areas is routed directly to aquatic ecosystems, such that even very low levels of imperviousness can significantly change flow regimes (Walsh et al., 2005a). Here, the catchments created by topography may be less important than those created by storm sewer networks for describing hydrologic conditions in receiving water bodies (Graf, 1977). Unlike combined sewer outfalls, stormwater outfalls are numerous and

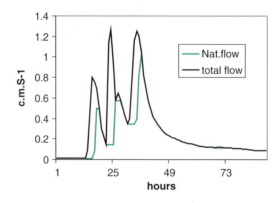

Figure 20.1 Using a moving minimum (here on a four time-step basis) to filter combined sewer overflow and natural components from a total-flow time series

dispersed throughout the landscape, so monitoring these discharges to instream flows is not feasible. Instead, monitoring should include comparison of urban flow regimes to forested reference catchments of similar size, and/or use of effective (connected) impervious area as an indicator of hydrologic alteration (see 'catchment stressor indicators' below).

20.4.2 Water quality stressors

Urban landscapes generate pollutant loadings to aquatic ecosystems from surface runoff (in separate systems), and sewage overflows (in combined systems). The wet weather contamination can be noxious for all aquatic living organisms, including hyporheic organisms and even human beings (due to pathogens, viruses and bacteria). Urban pollution accumulates in fine sediments (Rochfort et al., 2000) and the hyporheic layer when downwellings predominate (Lafont et al., 2006).

Common substances requiring monitoring include organic matter and humic compounds, ammonium salts, nutrients (nitrogen, phosphorus and their various species), PAHs (poly-aromatic hydrocarbons), pesticides, PCBs (poly-chlorinated biphenyls), hormones and xenobiotics, as well as heavy metals (mainly lead, zinc, copper and cadmium) and pathogens [for further detail, refer to local water quality monitoring guidelines, or the review by Paul and Meyer (2001)]. Other contaminants may be of local interest, and should be included in the monitoring programme. Selection of variables to monitor (see also Chapter 4) will also depend strongly on their potential interaction with other urban water cycle components.

Monitoring the concentration of many substances, such as organic matter, will need to take into account possible processes associated with food-web integration. For toxic materials, such as heavy metals, xenobiotics, PAHs and ammonium salts under certain conditions, monitoring of bioaccumulation may also be necessary. For example, assuming that regulations exist and are enforced to limit as far as possible toxic inputs from the urban landscape, toxic effects will not occur from sudden excess concentrations, but from accumulation in storage areas (e.g. fine sediment, dominant downwelling in hyporheic systems), affecting sensitive species over longer time-frames. Toxic effects can also occur along the trophic web when pollution-tolerant species are consumed by other species higher in the web; aquatic biota should be monitored to quantify this effect.

Monitoring of water quality effects needs to also take into consideration the temporal scale and persistence of any effect (see also Chapter 5). For example, some 'spikes' of elevated pollutant concentration (e.g. of nitrogen), may disappear quite quickly, either by simple dilution and flow, or by in-stream processing such as denitrification. On the other hand, some materials, such as heavy metals, may become attached to the benthos and have a long-lasting 'legacy' effect. Any monitoring programme will need to be based on a sound conceptual model of the behaviour of pollutants of concern, to ensure that the temporal resolution and duration of monitoring is suited to the behaviour of these pollutants.

20.4.3 Geomorphic stressors

The degradation of aquatic habitat, either due to altered hydrology (e.g. increased or decreased flows) or direct modification (e.g. channel straightening, concrete lining), has major implications for aquatic ecosystems (Walsh et al., 2001; Paul and Meyer, 2001;

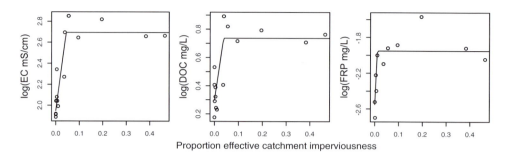

Figure 20.2 Use of a catchment-scale indicator to monitor water quality in streams; EC = electrical conductivity; DOC = dissolved organic carbon; FRP = filterable reactive phosphorus

Source: Walsh et al., 2005a.

Fatta et al., 2002). Interstitial benthic organisms, including hyporheic organics, such as crustaceans and oligocheates, are affected. Influxes of fine sediments from urban areas also alter physical habitat and, in turn, aquatic biota (Wood and Armitage, 1997).

Methods for measurement of geomorphic conditions as a source of stress for aquatic ecosystems have been developed in many countries throughout the world. Visual stream habitat indices are commonly used to rapidly assess conditions (e.g. US EPA Rapid Bioassessment Protocols, Barbour et al., 1999; Australian Index of Stream Condition, Ladson et al., 1999). Although these visual measures seem to adequately represent stream condition, quantitative measurements along transects within stream reaches provide more detailed and less subjective information regarding stream conditions. Typical geomorphic measurements include: channel morphology, stream gradient, bed texture, flow and sediment transport, channel and bank dimensions, and basin morphology. These measurements are described in detail in widely used stream manuals in the United States (Harrelson et al., 1994; Fitzpatrick et al., 1998; Lazorchak et al., 1998).

20.4.4 Catchment-scale stressor indicators

Often the stressors acting on urban streams can be highly confounded, such that monitoring of an individual stressor (e.g. peak flows, sediment or heavy metal concentration, etc.) cannot explain nor predict changes in aquatic biota or ecosystem function (Walsh et al., 2005b). Catchment-scale indicators have been successfully used to get around this problem. For example, Walsh et al. (2005a) demonstrated the proportion of impervious areas within a catchment that is directly connected to the drainage network (termed 'effective imperviousness') can be used to predict differences in pollutant concentrations (Figure 20.2), ecosystem composition and function (Figure 20.3). Walsh et al. (2005a) argue that in the catchments they were studying (with separate stormwater systems), these indicators were useful not only as explanatory variables, but as performance indicators, since appropriate guidelines and regulations could be used to control the amount of effective imperviousness in a catchment, for example, by ensuring that most impervious areas were 'disconnected' from the drainage network, through systems such as infiltration systems, biofiltration systems or even stormwater harvesting and re-use systems (Ladson et al., 2006; Fletcher et al., 2007). The selection of appropriate

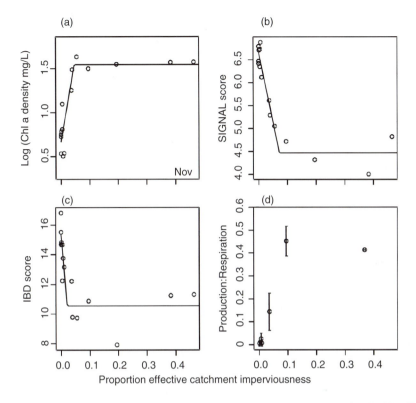

Figure 20.3 Use of a catchment-scale indicator to monitor aquatic ecosystem health. Chl a = chlorophyll *a*; SIGNAL = benthic macroinvertebrate index; IBD = *indice biologique diatomée* (diatom index)

Source: Walsh et al. 2005a.

catchment-scale indicators will depend on the nature of the water system and land-use within the catchment. For example, in a catchment with CSOs, the density of CSOs, or the mean annual discharge volume, may be an appropriate indicator.

20.4.5 Relative importance of stressors for different types of aquatic ecosystem

For lentic waters (lakes, ponds, pools, lentic river reaches, etc.), indicators such as temperature, carbon and oxygen gradients will often be critical as will exchanges with the benthic substrate. In lotic waters (streams, lotic river reaches, etc), the benthic substrate (including both bed and banks) remain very important and so indicators of substrate fluxes are often required. The hyporheic layer (below the benthos), when it exists in lotic systems, is key to understanding the dynamics of the biodegradation process, especially where downwellings and water flow in the substrates can result from riffles, meanders and water table depletion as a consequence of water supply extractions. Wetlands will tend to have intermediate characteristics, and thus indicators suitable for both lentic and lotic systems may be required.

Riparian zones (e.g. natural river banks and lake shores) are important buffers for nutrient retention and transformation, although this effect is substantially reduced within the urban context where the piped drainage and wastewater systems will often bypass the riparian zone with discharges directly from the urban catchment into the receiving waters. In this case, the natural processes in the riparian zone such as denitrification may be greatly reduced due to changes in moisture and chemistry (Groffman et al., 2002).

Considering the primary stressors of hydrology, water quality and geomorphology, it is generally understood that:

- still (lentic) waters will be more sensitive to long-term pollutant loads and thermal fluctuations than flowing waters, and
- flowing (lotic) waters will be more sensitive to geomorphic alterations and the destruction of habitat diversity, as well as to short-term fluctuations in water quality.

20.4.6 Additional data considerations for defining protection and rehabilitation strategies

Protection, restoration and rehabilitation strategies involve, to a greater or lesser degree, attempts to address the stressors affecting an aquatic ecosystem, and to make modifications within the ecosystem itself to make it more resilient to the stressors being experienced. To assist these strategies, data are thus needed on changes in stressors over time – before and after a specific intervention – and changes in indicators of ecosystem composition and function.

Data needs to be collected in a timely fashion to make a diagnosis of hydrological, chemical, and physical stressors, as well as biological characteristics. Rehabilitation and restoration of aquatic ecosystems is an inherently adaptive process, and therefore a dedicated environmental database should be developed so that the system's behaviour, fluxes and processes, and changes in these over time can be detected. Such a database must be organized to facilitate data sharing with those responsible for managing water supply, wastewater, stormwater, combined sewers and groundwater. In addition to identifying trends, the data can be used to identify potential thresholds where ecological responses to a given stressor accelerate (either positively or negatively), which can be used to determine targets for protection or rehabilitation.

Reasonable targets for rehabilitation must be set based on the existing condition and potential for improvement. For example, there may be a range of rehabilitation targets, depending on the density of urbanization surrounding a waterway (Figure 20.4). The choice of target will depend on a range of factors, not only environmental, but also social and economic. However, the *indicator* used to assess achievement against the target will generally remain the same. For example, different water quality targets may be determined for 'pristine' and 'significantly degraded' waterways. The same indicators (for example, concentrations of total suspended solids, dissolved oxygen, total nitrogen, total phosphorus, lead, zinc and copper) will be collected in each case.

In other cases, it may be that the indicators themselves need to change to reflect the type of target for given waterways. For example, in Figure 20.4, the target for the 'un-sustainable' waterway may simply be about water quality (and thus using a relatively simple set of indicators), while the target for the 'near natural' waterway may include

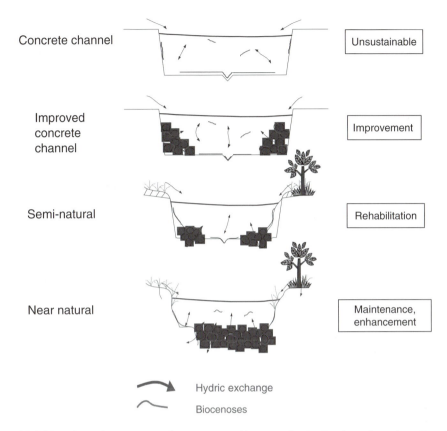

Concrete channel		Unsustainable
Improved concrete channel		Improvement
Semi-natural		Rehabilitation
Near natural		Maintenance, enhancement

Hydric exchange

Biocenoses

Figure 20.4 Hypothetical target states for a stream subject to urbanization (see also colour Plate 22)

a wide range of metrics, such as water quality, flow variability (relative to natural), macroinvertebrate composition and diversity, habitat diversity and bioavailability of nutrients. Based on a similar rationale, Walsh et al. (2004) propose a hierarchical series of targets for a network of waterways, starting with upland streams (typically in peri-urban areas), and ending with estuaries and bays. For the upland streams, they set a target of maintaining the effective imperviousness to less than 5% (of the total catchment area), thus minimizing disturbances to flow and water quality. Where such a target is not achievable, for example, in an already densely developed urban area with existing effective imperviousness of more than 50%, they suggest that an annual pollutant load target should be used instead.

20.5 MONITORING REQUIREMENTS

Based on the stressors to aquatic ecosystems outlined in the previous section, the following information is intended to provide guidance on 'why', 'what', 'where' and 'when' to monitor indicators for assessing aquatic ecosystems. We do not attempt here to provide great detail on *how* to measure these indicators, because these methods should follow local guidelines which have been written to suit local characteristics, be they physical, chemical or biological.

20.5.1 The 'why'

Indicators should be selected based on the question to be addressed, and thus on the objectives of the monitoring (i.e. the 'why'). For example, we may wish to know 'what is the existing condition of the ecosystem?' or, alternatively, 'what are the current impacts on the ecosystem?' These two questions will require quite different indicators, with the second requiring not only measurements of the ecosystem itself, but also of the interactions between it and other components of the urban water cycle. Only once we really understand the answer to the *why* question properly will we be able to consider the other questions. Therefore, careful consideration should be given to determining the primary objective, and consulting with those responsible for the management of potentially impacting water cycle components, such as stormwater, wastewater (and combined sewers), drinking water and groundwater.

20.5.2 The 'what'

There are three broad classes of indicators for use in urban aquatic ecosystems: physical, chemical, and biological. These classes were selected because they reflect the primary stressors to aquatic ecosystems (see Section 20.4), and must retain some natural characteristics to maintain essential ecosystem functions. Because the physical, chemical and biological components are tightly integrated, it is not sufficient to use just one class as an indicator of whole ecosystem function. Specific indicators within these classes should be determined based on the objectives of the monitoring and the resources available.

20.5.3 The 'where'

Aquatic ecosystems reflect their spatial location within landscapes, and their functions are closely tied with the physical conditions of the lotic or lentic environment. Within lotic (flowing water) environments, ecosystem characteristics are typically measured at the reach scale (e.g. 100 m to 1 km) and replicate habitat units (e.g. pools, riffles) within each reach. Thus, it is important to select a reach that is representative of the entire stream or river system and located near the mouth of the catchment of interest. In lentic environments (e.g. ponds, lakes and wetlands), ecosystem characteristics vary with the depth of the water and the proximity to banks, inlets and outlets. Thus, the sampling location within the lentic environment, as well as the physical conditions of the selected lake or wetland (e.g. size, depth, hydrologic permanence) must be considered.

20.5.4 The 'when'

Broadly speaking, indicators can be used to monitor ecological status, or internal or external processes. The temporal resolution of monitoring will depend on the questions being addressed, including whether the focus is on status or process. Because ecosystems are dynamic, *status monitoring* is likely to need a continuous monitoring approach to determine changes over time. This is particularly the case where a reference catchment is difficult to define (and so the degree of degradation of the observed system is not known with a high degree of certainty). In this case, a continuous monitoring approach should be undertaken to measure dynamic variability in ecosystem status. Monitoring should take into account such natural cycles. Similarly, in lentic waters, there will typically be quite significant seasonal cycles in temperature (and thus in respiration rates, oxygen gradients, etc.) which should be considered.

Where *process monitoring* is required, the appropriate temporal scale will need to consider the rate of the process. For example, it may be necessary to understand the movement of pollutants at a number of points in the catchment, if the resulting ecosystem is a function of multiple non-point sources of pollutants. Monitoring ecosystem processes can be undertaken using indicators, such as the water exchange factor which measures the potential for water exchange in the benthic and hyporheic layers depending on the bed texture (ranging from coarse material such as boulders, pebbles, gravel, to medium sands fine sediments and silts). For example, if there is only one-way transfer (downwellings), the result can be accumulated pollutant storage in the hyporheic layer.

Table 20.2 summarizes *why* to measure each of the various indicators recommended, *what* to measure, as well as *when* and *where* (the '4W approach'). The choice

Table 20.2 Aquatic ecosystems indicators: description and application

Indicator	Description
Physical indicators	
What	Hydrologic and hydraulic parameters (water depth, flow rates, shear stress, etc.). Geomorphic parameters (channel cross sections, bed texture, benthos particle size distribution). Habitat availability parameters (area, temporal permanence, etc.).
Why	Physical indicators express (directly or indirectly) the link between catchment hydrology and the resulting physical and habitat conditions of the ecosystem.
Where	Lotic systems: reach scale (100 m–1 km). Lentic systems: hydrology at inlet and outlet (if present) or consistent location.
When	Hydrological parameters: continuous with measurement interval dependent on catchment size and type of water body (e.g. 2 min to 30 min intervals for lotic systems; hourly or daily for lentic systems). Geomorphic parameters: approximately once per 12 months to 24 months or according to the rate of change in the catchment. Habitat availability parameters.
Chemical indicators	
What	Specific parameters dependent on catchment and land use, typically: sediment, nutrients (nitrogen and phosphorus), heavy metals, pathogenic indicators, pesticides, PCBs, xenobiotics, ions and ammonium salts, organic matter, oxygen demand, etc.; *local guidance should be sought.*
Why	Water quality integrates catchment hydrology and pollutant loading, and affects biotic condition and ecosystem functioning.
Where	1. Lotic systems: integrated, depth grab samples. 2. Lentic systems: inlet and outlet (and possibly one or more points at intermediate locations throughout the system).

(Continued)

Table 20.2 (Continued)

Indicator	Description
When	Dependent on hydrology and seasonal pollutant loadings. Typically baseflow (or non-storm conditions) monitoring undertaken at fortnightly-monthly intervals, and storm events sampled separately (Fletcher and Deletić, 2007). Intensive storm-event sampling (of at least 12 events per storm) may also be required to capture the variability in concentrations (see Chapters 17 and 18).
Biological indicators	
What	Composition of populations of macro-invertebrates, fish, algae, etc.
Why	Biological indicators provide an integrated measure of ecosystem quality, synthesizing the effects of a range of stressors.
Where	1. Lotic systems: macro-invertebrate and algal replicates within habitat units (e.g. riffles, pools); fishes at reach scale of 100 m to 1 km. 2. Lentic systems: range of locations, relative to water depth, temperature and oxygen gradients.
When	Annually or bi-annually based on hydrology, shading (leaf-on and leaf-off for algae), and life histories of animals (macro-invertebrates prior to mass emergence, fishes after spawning). Typically repeated yearly or every second or third year, depending on rate of change in catchment.

of indicators includes not only those that describe the ecosystem (its condition and processes), but its interactions with other components of the urban water cycle.

REFERENCES

Barbour, M.T., Gerritsen, J., Snyder, B.D. and Stribling, J.B. 1999. *Bioassessment Protocols for Use in Streams and Wadeable Rivers: Periphyton, Benthic Macroinvertebrates and Fish*, 2nd edn. Washington DC, US Environmental Protection Agency, Office of Water. (EPA 841-B-99-002)

Beighley, R.E. and Moglen, G.E. 2002. Trend assessment in rainfall-runoff behavior in urbanizing watersheds. *Journal of Hydrologic Engineering*, Vol. 7, No. 1, pp. 27–34.

Bledsoe, B.P. and Watson, C.C. 2001. Effects of urbanization on channel instability. *Journal of the American Water Resources Association*, Vol. 37, No. 2, pp. 255–70.

Booth, D.B. 1990. Stream-channel incision following drainage-basin urbanization. *Water Resources Bulletin*, Vol. 26, No. 3, pp. 407–17.

———. 1991. Urbanization and the natural drainage system – impacts, solutions, and prognoses. *Northwest Environmental Journal*, Vol. 7, pp. 93–118.

Clean Water Act. 1972. Washington DC, US Environmental Protection Agency. (Also referred to as *Federal Water Pollution Control Act of 1972*).

Fatta, D., Naoum, D. and Loizidou, M. 2002. Integrated environmental monitoring and simulation system for use as a management decision support tool in urban areas. *Journal of Environmental. Management*, Vol. 64, pp. 333–43.

Fitzpatrick, F.A., Waite, I.R., Arconte, P.J.D., Meador, M.R., Maupin, M.A. and Gurtz, M.E. 1998. Revised methods for characterizing stream habitat in the National Water Quality Assessment Program. Washington DC, US Government Printing Office. (US Geological Survey Water Resources Investigations Report 98-4052).

Fletcher, T.D. and Deletić, A. 2007. Observations statistiques d'un programme de surveillance des eaux de ruissellement; leçons pour l'estimation de la masse de pollutants [Statistical observations of a stormwater monitoring programme; lessons for the estimation of pollutant loads]. Paper presented at the NOVATECH 2007: Conference on Sustainable Techniques and Strategies in Urban Water Management, 24–27 June 2007, Lyon, France.

Fletcher, T.D., Mitchell, G., Deletić, A., Ladson, A. and Séven, A. 2007. Is stormwater harvesting beneficial to urban waterway environmental flows? *Water Science and Technology*, Vol. 55, No. 5, pp. 265–272.

Graf, W.L. 1977. Network characteristics in suburbanizing streams. *Water Resources Research*, Vol. 13, pp. 459–63.

Grapentine, L., Rochfort, Q. and Marsalek, J. 2004. Benthic responses to wet-weather discharges in urban streams in Southern Ontario. *Water Quality Research Journal of Canada*, Vol. 39, No. 4, pp. 374–91.

Groffman, P.M., Boulware, N.J., Zipperer, W.C., Pouyat, R.V., Band, L.E. and Colosimo, M.F. 2002. Soil nitrogen cycle processes in urban riparian zones. *Environmental Science and Technology*, Vol. 36, No. 21, pp. 4547–52.

Harrelson, C.C., Rawlins, C.L. and Potyondy, J.P. 1994. *Stream Channel Reference Sites, An Illustrated Guide to Field Technique*. Fort Collins, CO, USA, US Department of Agriculture Forest Service Rocky Mountain Forest and Range Experiment Station. (General Technical Report RM 245)

Hollis, G.E. 1975. The effect of urbanization on floods of different recurrence interval. *Water Resources Research*, Vol. 11, No. 3, pp. 431–35.

Ladson, A.R., Walsh, C.J. and Fletcher, T.D. 2006. Improving stream health in urban areas by reducing runoff frequency from impervious surfaces. *Australian Journal of Water Resources*, Vol. 10, No. 1, pp. 23–34.

Ladson, A.R., White, L.J., Doolan, J.A., Finlayson, B.L., Hart, B.T., Lake, P.S. and Tilleard, J.W. (1999). Development and testing of an index of stream condition for waterway management. *Freshwater Biology*, Vol. 41, No. 2, pp. 453–68.

Lafont, M., Vivier, A., Nogueira, S., Namour, P. and Breil, P. 2006. Surface and hyporheic oligochaete assemblages in a French suburban stream. *Hydrobiologia*, Vol. 564, pp. 183–93.

Lazorchak, J.M., Klemm, D.L. and Peck, D.V. 1998. *Environmental Monitoring and Assessment Program – Surface Waters: Field Operations and Methods for Measuring the Ecological Condition of Wadeable Streams*. Washington DC, US Environmental Protection Agency, Office of Research and Development. (EPA/620/R-94/00F)

Meyer, J.L. 1997. Stream Health: Incorporating the human dimension to advance stream ecology. *Journal of the North American Benthological Society*, Vol. 16, pp. 439–47.

Moscrip, A.L. and Montgomery D.R. 1997. Urbanization, flood frequency, and salmon abundance in Puget lowland streams. *Journal of the American Water Resources Association*, Vol. 33, No. 6, pp. 1289–95.

Paul, M.J. and Meyer, J.L. 2001. Streams in the urban landscape. *Annual Review of Ecology and Systematics*, Vol. 32, pp. 333–65.

Rochfort, Q., Grapentine, L., Marsalek, J., Brownlee, B., Reynoldson, T., Thompson, S., Milani, D. and Logan, C. 2000. Using benthic assessment techniques to determine combined sewer overflow and stormwater impacts in the aquatic ecosystem. *Water Quality Research Journal of Canada*, Vol. 35, pp. 365–97.

Schmitt, L., Maire, G. and Humbert, J. 2001. La puissance fluviale : définition, intérêt et limites pour une typologie hydro-géomorphologique de rivières. *Zeitschrift für Geomorphologie*, Vol. 45, pp. 201–24.

Walsh, C.J., Sharpe, A.K., Breen, P.F. and Sonneman, J.A. 2001. Effects of urbanization on streams of the Melbourne region, Victoria, Australia. I. Benthic macroinvertebrate communities. *Freshwater Biology*, Vol. 46, pp. 535–51.

Walsh, C.J., Leonard, A.W., Ladson, A.R. and Fletcher, T.D. 2004. *Urban Stormwater and the Ecology of Streams*. Melbourne, Australia, Monash University (CRC for Freshwater Ecology, Water Studies Centre, CRC for Catchment Hydrology and Institute for Sustainable Water Resources, Department of Civil Engineering).

Walsh, C.J., Fletcher, T.D. and Ladson, A.R. 2005a. Stream restoration in urban catchments through redesigning stormwater systems: looking to the catchment to save the stream. *Journal of the North American Benthological Society*, Vol. 24, No. 3, pp. 690–705.

Walsh, C.J., Roy, A.H., Feminella, J.W., Cottingham, P.D., Groffman, P.M. and Morgan, R.P. 2005b. The urban stream syndrome: current knowledge and the search for a cure. *Journal of the North American Benthological Society*, Vol. 24, No. 3, pp. 706–23.

Wong, T.S.W. and Tan, S.-K. 1995. Flood peak increase in an urbanizing basin subject to temporal varying and uniform storms. *Proceedings of Second Symposium on Urban Storm Water Management*. Melbourne. pp. 167–70.

Wood, P.J. and Armitage, P.D. 1997. Biological effects of fine sediment in the lotic environment. *Environmental Management*, Vol. 21, pp. 203–17.

Chapter 21

Human health

S. Haydon[1] and F. Clemens[2]

[1]*Melbourne Water, 100 Wellington Parade, Melbourne 3001, Victoria, Australia*
[2]*Delft University of Technology, Faculty of Civil Engineering, Sanitary Engineering Section, 2600 GA, Delft, The Netherlands*

21.1 INTRODUCTION

Sanitation is one of the most important human-related considerations in the management of the urban water cycle. Stringent attention must be paid to reducing the risk of contamination with infectious agents: bacteria, viruses or otherwise health-adverse substances which may be carried by water. As was acknowledged in the mid-nineteenth century, faecally-contaminated water used for consumption was a major carrier of several disease-causing agents (e.g. amoeba, giardia, cryptosporidium, etc.). The risk of actually getting ill is determined by the following factors:

- the chance of coming in contact with the agent
- the vulnerability of the person in question (particularly vulnerable groups are referred to as YOPIs (the young, old, pregnant or immuno-deficient)
- the span of time during which exposure occurs.

There are a large number of pathogens that typically infect people through the hydrologic cycle in some way. The obvious route of infection is via the consumption of contaminated water. Typically, water is contaminated with faecal material containing pathogenic organisms. Other modes of infection are:

- washing and preparation of food with contaminated water
- inhalation of droplets
- bathing
- recreational exposure
- opportunistic infection of wounds.

In classic sanitation, a more or less strict separation between wastewater and water for consumption is maintained, causing a situation in the developed world in which the average inhabitant has become ignorant of the risks involved with water-related diseases. The concept of sanitation is based not so much on a thorough risk-assessment, but on a very rigorous separation between wastewater on one hand, and water for consumption, municipal use, or recreation on the other hand. With the advent of Integrated Urban Water Management, this strict separation begins to be broken down, as re-use and recycling of water becomes part of the water supply solution. This change has substantial implications for monitoring human-health related aspects of the urban water cycle.

21.2 HUMAN HEALTH INTERACTIONS WITHIN THE URBAN WATER CYCLE

From a human health perspective, the ideal water system would keep the various components of the urban water cycle completely separate. Drinking water would be collected from catchments that had no human activities in them and all sewerage discharges would be made into isolated receiving waters remote from human activity, ideally after appropriate treatment. Stormwater from rural and urban areas would also be kept isolated from drinking water supplies to minimize contamination. Typically this is not the case and many water supplies are from multi-use catchments, often with untreated sewage discharging into the same waters that are ultimately used for drinking. This leads to the situation where 1.8 million people per year, mainly children, die from water-borne diarrhoeal disease (WHO, 2004a). Overwhelmingly, microbial diseases outweigh the disease burden from chemical or radiological contaminants (WHO, 2004b).

Human health is potentially impacted by all aspects of the urban water cycle (Table 21.1). For example, inappropriate sanitation directly contaminates drinking water. Poorly designed reservoirs and wetlands may promote mosquito breeding, which may lead to increases in malaria and other mosquito-borne diseases. Poor drainage practices may increase flooding and associated health impacts.

Recent developments in urban water management have resulted in more sophisticated systems in which, under the influence of several drivers like sustainability, re-use of water, water harvesting from roofs, and the growing demand on water in arid areas (e.g. Australia, parts of the USA), the possibility of contaminated water supplies is increasing.

An example in the Netherlands was described by Schaart (2003) and discussed by Ashley et al. (2004) and Clemens (2006), in which a comparison was made between a 'traditional' solution for sanitation and a 'sustainable' solution. The comparison focussed on the risk for users of these systems of being infected by pathogens related to water. The traditional system comprises the delivery of just one quality of drinking water and a combined sewer system to collect and transport wastewater. With its almost perfect separation of water streams, this system is tailored almost entirely for the objective of 'public hygiene', although clearly with little regard for the pursuit of sustainability.

On the other hand, is a more 'sustainable' system, which includes the re-use of stormwater (or rainwater collected locally at each house) and recycled wastewater for toilet flushing, garden watering, clothes washing and other non-potable uses. The use of such technologies may reduce demand on traditional potable resources use by up to 80%, also reducing the need for water collection, transport, treatment and disposal. In the Netherlands example, each house (or small number of houses) has a small tank for the storage of collected (and filtered) rainwater. The tank is topped up with potable water when storage volume falls below a defined minimum. In designing this system the objective 'public health' was given no thought at all. However, while the overall concept was good, this lack of attention to public health had significant consequences. There was at least one case of contamination due to cross-connection, with toilet water being mistakenly discharged into the storage tank. This cross-connection went undetected for six months (because high water temperature in the washing machine, and the effectiveness

Table 21.1 Potential interaction of human health and urban water system components

Urban water system component	Possible human health interactions
Urban climate	Weather will affect water quality and pathogen levels in drinking water and waterways.
Water supply	Water supply of poor quality water is a primary source of human disease.
Wastewater	Wastewater leaks or cross-connections could contaminate (a) drinking water and (b) recreational waters.
Combined sewer systems	Combined sewer overflows (CSOs) may indirectly enter into the water supply, either through infiltration into water pipes or into groundwater. Similarly, CSOs may spill into recreational waters (such as lakes, wetlands or creeks).
Stormwater	Stormwater could potentially leak into the water supply system. Stormwater can carry high pollutant and pathogen loads, thus affecting receiving waters, which may be used for recreational purposes.
Groundwater	Groundwater may be contaminated (by wastewater, stormwater, or combined sewer overflows), and in turn may infiltrate into water supply pipes.
	Groundwater may be the primary drinking water supply in which case its quality will have direct health implications.
Aquatic ecosystems and urban streams	The quality of water and its pathogen load will determine risk to public health through contact in recreation, or through food supply (e.g. fish). For example, accumulation of heavy metals in sediments could result in bioaccumulation through the food chain leading to human health risks.
Society and institutions	Community standards relating to hygiene and disease will affect (a) the risk of infection, as well as (b) community attitudes to use of recycled wastewater.
	The degree of transparency will affect the way in which the potential for health risk from waterways and/or drinking water is reported.

of clothes washing detergent reduced the pathogen load significantly). Unfortunately, the final rinse (with contaminated water), using lower temperatures, and without detergent, resulted in sickness of the residents. This incident drew significant media attention.

The lesson of this case is *not* that integrated sustainable approaches should be abandoned, but that such systems impose stricter monitoring requirements, because of their increased level of risk. As these so-called 'alternative' systems develop further, some level of risk awareness must be obtained by the designer, builder and by the user. However, the principal problem in achieving this is the fact that hardly any data are available on:

- the possible contamination routes for different types of agents (bacteria, viruses, poisons, etc.)
- the chance of being exposed to a minimum dose of the agent
- the failure risk of complex sanitation systems, e.g. data on failure rates of return valves, pumps, occurrence of cross-connections.

21.3 MONITORING REQUIREMENTS

21.3.1 Monitoring rationale

Most monitoring programmes related to water and human health are focussed on the quality of drinking water as it is distributed in classic systems only. Monitoring can be focused on two sides of the problem. Firstly, health surveillance programmes noting levels of disease in the community are usually run by health departments independent of water suppliers. Epidemiological studies are often used to determine waterborne outbreaks. However, these are often only useful in identifying a cause after or during an outbreak. Notably the first and most famous epidemiological study by Snow (1849) which identified water as the agent for cholera did not prevent the outbreak. These studies do add significant value in identifying failure modes in water supply systems, many of which are 'eminently preventable' (Hrudey and Hrudey, 2004).

Monitoring water directly rather than via health surveillance programmes is thus an obvious approach. For such monitoring, there are the questions of what to monitor, how often, where and when. The intensity, location and duration of monitoring will be dependent on the risk. For example, with regard to direct water quality in an alternative system (where recycling or multiple water sources of varying quality are used), there may be increased monitoring due to the risk of cross-connections.

The basic aim of monitoring should be to understand the extra risks of human illness when introducing alternative sanitation systems when compared to the classic systems. The first problem encountered in this objective is that the actual level of risk is not quantified.

When setting up a monitoring programme into failure risks of systems, thorough investigation of the possible failure routes should be made in order to be able to construct a so-called failure tree. This is basically the conceptual model of the system; similar recommendations are made for aspects of the water cycle (e.g. stormwater, wastewater, aquatic ecosystems); any monitoring programme should be based on a sound conceptual model of the system. Without this, it is likely that some parameters that are irrelevant will be measured, and more importantly, some key parameters will be missed.

Water monitoring programmes focusing on public health are essentially designed to ensure that any variations in contaminant levels (usually a pathogen) are within expected or manageable bounds. From a drinking water perspective they form part of a water safety plan. The key components of a monitoring programme would include (WHO, 2004b):

- parameters to be monitored
- sampling location and frequency
- sampling methods and equipment
- sampling schedules
- quality assurance programme
- methods for interpretation
- staffing qualifications and responsibilities
- data storage and documentation
- reporting and communication requirements.

Many water-related outbreaks are preceded by extreme or unusual weather (Hrudey and Hrudey, 2004) which overwhelm treatment plants, so monitoring programmes that firstly define 'normal' variation are required, so that unusual or extreme weather can be identified. Typically treatment plants will cope with a certain amount of variation in loading but then struggle to perform adequately once loads exceed design limitations.

Monitoring water for human health considerations also needs to consider the type of system, and its points of risk. For alternative systems with multiple (often local) sources of water, and many potential points of cross-contamination, the monitoring will need to be applied at this local scale.

21.3.2 Selection of variables to monitor

As a general approach to setting up a monitoring programme, the question of which parameters to monitor is best answered by undertaking a hazard survey of the catchment of concern. If significant hazards are identified, then monitoring for that parameter in some way is prudent. For example, the Australian Drinking Water Guidelines (NHMRC and NRMMC, 2004) list over one hundred pesticides of concern in drinking water but recommend only monitoring for those that are in use in the catchment under surveillance. In alternative systems where cross-connections of wastewater are possible, indicators of faecal contamination will be of high priority.

As the focus for most public health issues associated with water are microbiological, it makes sense to sample microbiological parameters more frequently than other parameters. Typically an indicator organism (usually *Escherichia coli*) is tested for, rather than testing for pathogenic organisms directly. This is done for reasons of expense and speed. Testing for pathogens is both expensive and usually slow, and it is also often inaccurate. The most commonly used indicator organism in drinking water is *E. coli*, which is not usually pathogenic but indicates the presence of faecal contamination of water (WHO, 2004b). There are several strains of *E. coli* which are pathogenic, and these have been responsible for several waterborne outbreaks, notably the Walkerton outbreak in May 2000 (O'Connor, 2002), but generally the micro-organism is regarded as an indicator. *E. coli* is the dominant bacterial species in the intestine of man and other warm-blooded animals (AWWA, 1999).

While *E. coli* is a good indicator for bacteria, it is less impressive as an indicator for viruses and protozoa (e.g. *Giardia* and *Cryptosporidium*) which can be significant pathogens. Surrogates such as *Clostridium perfringens* may be considered, or direct sampling for protozoa may also be considered (WHO, 2004b). Other possible pathogenic bacteria indicators include *Enteroccoci*. *Enteroccoci* may be more appropriate, particularly from a recreational exposure viewpoint as it is longer lived than *E. coli*.

Monitoring of water for human health-related physical and chemical parameters are likely to be catchment specific (e.g. arsenic). However, if water is to be consumed then routine monitoring of parameters such as turbidity and colour, which are cheap to monitor and have significant impact both on the palatability of water and the impact of disinfection, would also be recommended. Monitoring of nutrients such as phosphorus and nitrogen may also be prudent if algal activity is an issue in reservoirs, wetlands or streams.

Open water bodies, such as reservoirs and wetlands, should be sampled for algae and zooplankton. Algal toxins from species such as *Anabaena circinalis* are more difficult to monitor than algae numbers, so monitoring for the algae to a certain level and then performing toxin analysis may be more effective. Monitoring of water body temperature and dissolved oxygen state will provide insight into the behaviour of the water body in question and may be used as a predictor of adverse algal behaviour.

Where household or precinct water recycling and rainwater harvesting techniques are in place, monitoring of pressure variations and failure reports of key components (return values, pumps, leakage contamination at the tap) can be used to assess the risk of cross-contamination. Unfortunately, there have not yet been any quantitative relationships or guidelines which describe the link between these parameters and contamination risks. Attention should be given to collecting the necessary data to quantify such relationships.

Local guidelines and protocols for human health monitoring should be considered in making decisions about specific monitoring programmes.

21.3.3 Sampling schedules: location and frequency

Clearly monitoring in the streams, reservoirs and offtakes of water supply systems is prudent. Monitoring above and below sewerage discharge and drainage system points will allow the impact of discharges from these points to be enumerated. Monitoring in reservoirs and wetlands will also be required to be taken at various locations and depths for the water body. Monitoring should be undertaken not just for water supply systems, but also for any system where there is actual or potential human contact (e.g. recreational water bodies, etc.). Monitoring should also be undertaken within the reticulation network, at points as close as possible to the ultimate consumers. For systems involving local recycling and rainwater harvesting, monitoring will typically need to occur at the households themselves (based on a representative sample).

Typically water sampling is undertaken on a routine basis (e.g. samples are taken on, say, every Monday at midday), the theory being that over time enough samples will be taken to statistically describe the variation of various parameters at that site. However, it is now known that large numbers of pathogens are mobilized by storm events (McDonald and Kay, 1981; Crabill et al., 1999) so monitoring of storm events is often warranted. This means that the sampling frequency and timing would have to be tailored to the catchment, as some catchments will have few storms and may have little variation between events, while others will have enormous variation.

Essentially the amount of variation of a parameter and its importance will define the frequency required of monitoring. For critical parameters that vary considerably over short periods of time, such as pathogens, frequency should be high. While others such as algal levels, which may take weeks to change, can often be monitored quite adequately on a monthly basis for much of the year. If loads are required, then the sampling regime will often be different to that used for determining peak concentrations.

Often it is not feasible or economic to test all parameters at all sites at the same frequency. A well-designed sampling schedule will allow a more efficient use of personnel and funds. As is described for other components of the urban water cycle, such decisions should be made *based on direct reference to the objectives of the monitoring programme* (see Chapter 3).

21.3.4 Sampling methods and equipment

Typically sampling is done using grab or dip samples taken using a clean bottle. Microbiological sampling will require refrigeration to prevent either growth or decay of micro-organisms. Sampling for some metals will require special bottles. Notably the volume of each sample is not large, typically one litre, except if *Giardia* or *Cryptosporidium* are to be analysed for, in which case 10 L to 20 L samples will be required.

Storm sampling is often undertaken using automatically tripped samplers, which take a sample at a given river level or volume, or on a set time after an event has triggered them (see for example Roser et al., 2002).

The process of sampling is often regarded as the most likely area for error and uncertainty in microbiological testing. The very first line in the book by Nollet (2000) states: 'First and usually the most important step in any analytical process is sampling itself'. There are possibilities of the sample being contaminated and also that as only a small volume of water is actually sampled that any results from the sample may not reflect the rest of the unsampled water (sampling representativeness). As the laboratory techniques become more refined and controlled, the sampling step becomes the major area for determining the quality of the outcome of the test (Nollet, 2000).

21.3.5 Data handling, interpretation and reporting

The ongoing storage and management of the results from a sampling programme needs to be considered (see also Chapter 9). Often there is a requirement for publication of microbiological results in reports. Even if this is not required, the storage of data is important as often long-term trends will use historically recorded data. It is always important to know the method of testing, the level of accuracy, date and time the sample was taken and other descriptive information, as well as the result. In other words, as has been emphasized throughout this book, *metadata are critical for the effective interpretation and use of data.* An excellent example of this is the recording of the limit of determination of the analysis, so that results below the limit are recorded as 'below detection' or something similar, rather than set at zero. Over time the limit will tend to go down with improved techniques, so that knowing that the results were at the historical limit will stop inappropriate trend analysis being undertaken.

Not all parameters can be analysed using the same approach. In microbiological testing many tests give a presence or absence result only. Chemical parameters will often be tested to varying concentrations depending on the method used. Determining a statistically appropriate method for examining the results will inform the end user of when the results are noteworthy or when they are simply within the normal variation for that parameter at that location.

21.3.6 Quality control

Testing laboratories should undertake quality assurance programmes to ensure appropriate levels of accuracy and confidence in the results. Engaging staff that are adequately trained to both sample the water and interpret the results is critical to a successful programme. As previously stated, sampling is the principal source of error, so using appropriate sampling staff is the first step in ensuring representative results.

21.3.7 Integration with other urban water monitoring programmes

Monitoring for human health within the urban water cycle should be integrated with (a) other human health monitoring programmes and (b) other water cycle monitoring activities. For example, data on the occurrence of other (non water-related) diseases may be very important. To interpret human health events and trends, it is clearly also necessary to understand the interactions with other parts of the water cycle. An obvious example is to have sampling programmes linked to rainfall data collection programmes (either both data collected by the same agency, or an effective data-sharing arrangement between agencies). For example, water samplers for human health-related monitoring should be triggered by rainfall gauges to monitor storm events. Other programmes referred to as 'operational monitoring' (WHO, 2004b) with implications for human health would include examination of how well treatment plants are performing (chlorine residual levels, turbidity levels, etc.), structural status of wetlands (which will aid in understanding and interpreting the results of the monitoring programme), and the number and quality of combined sewer overflows, for example. There are numerous other examples. The critical requirement is to ensure that a thoroughly considered conceptual model of all possible interactions is constructed, and the monitoring programme is designed to quantify these interactions.

REFERENCES

Ashley, R.A., Clemens, F.H.L.R. and Veldkamp, R.G. 2004. The environmental engineer, a step too far? Paper presented at the NOVATECH Conference, June 2004, Lyon, France.

AWWA. 1999. *Waterborne Pathogens*. Denver, Colorado, American Water Works Association. (AWWA Manual M48)

Clemens, F.H.L.R. 2006. Integrated urban water management modelling: challenges and developments. Paper presented at the Seventh Urban Drainage Modelling and Fourth Water Sensitive Urban Design Conference, 2–7 April 2006, Melbourne, Australia.

Crabill, C., Donald, R., Snelling, J., Foust, R. and Southam, G. 1999. The impact of sediment fecal coliform reservoirs on seasonal water quality in Oak Creek, Arizona. *Water Research*, Vol. 33, No. 9, pp. 2163–71.

Hrudey, S. and Hrudey, E. 2004. *Safe Drinking Water, Lessons from Recent Outbreaks in Affluent Nations*. London, IWA Publishing.

McDonald, A. and Kay, D. 1981. Enteric bacterial concentrations in reservoir feeder streams: baseflow characteristics and response to hydrograph events. *Water Research*, Vol. 15, No. 8, pp. 961–68.

NHMRC and NRMMC. 2004. *Australian Drinking Water Guidelines*. Canberra, National Health and Medical Research Council and National Resource Management Ministerial Council, Australian Government. Available at: www.nhmrc.gov.au (Accessed 02 July 2007.)

Nollet, L. (ed). 2000. *Handbook of Water Analysis*. New York, Marcel Dekker Publishers.

O'Connor, D. 2002. *Report of the Walkerton Inquiry. Part 1: The Events of May 2000 and Related Issues*. Toronto, Ontario Ministry of the Attorney General. Available at: www.attorneygeneral.jus.gov.on.ca/english/about/pubs/walkerton (Accessed 02 July 2007.)

Roser, D., Skinner, J., LeMaitre, C., Marshall, L., Baldwin, J., Billington, K., Kotz, S., Clarkson, K. and Ashbolt, N.J. 2002. Automated event sampling for microbiological and related analytes in remote sites: a comprehensive system. *Water Science and Technology*, Vol. 2, No. 3, pp.123–30.

Schaart, N. 2003. *Duurzaamheid in de waterketen, een eerste kwantificering van gezondheidsrisico's* [Sustainability in the water chain, a first quantification of health risks]. M.Sc. thesis, Delft University of Technology, Delft, The Netherlands. (In Dutch.)

Snow, J. 1849. *On the Mode of Communication of Cholera*. London, John Churchill.

WHO. 2004a. *Water, Sanitation and Hygiene Links to Health – Facts and Figures*. Geneva, World Health Organization.

WHO. 2004b. *Guidelines for Drinking-Water Quality, Recommendations*, 3rd edn, Vol. 1. Geneva, World Health Organization.

Chapter 22

Social and institutional components

R.R. Brown

School of Geography and Environmental Science, Monash University, Melbourne 3800, Australia

22.1 INTRODUCTION

This chapter provides a broad introduction to the social and institutional data collection requirements for managing the urban water environment. Like the bio-physical dimension of the urban water field, the socio-institutional dimension is highly complex and requires specialized skills and competencies. The capacity of socio-institutional systems may enable or constrain the adoption of IUWM, and therefore is an important component of the urban water system that should be investigated and monitored along with the bio-physical dimensions.

The socio-institutional dimension includes the people, stakeholder organizations and the formal and informal 'rules' across civil society, government and market sectors that have a role in shaping the management of the urban water environment. Socio-institutional systems are highly variable across local, regional and national contexts. Some of the cultural differences are evidenced by the variable levels of political interest and leadership around waterway protection and rehabilitation efforts. This often reflects differing community interests and associated community action for improving waterway health. Some of the structural differences relate to differing administrative arrangements for urban water management, often reflected by the variable assignment of urban water management roles and responsibilities amongst organizations. Levels of centralization may also vary.

The socio-institutional data collection programme should be designed to assist the urban water manager with:

- identifying current social and administrative priorities and activities in relation to managing the urban water environment
- understanding the relationships between the social, organizational and bio-physical components of the urban water environment
- designing implementation strategies that will strengthen social and institutional receptivity for IUWM.

An important point to highlight is that the development of an IUWM plan and associated interventions needs to be adapted and matched to the current social and organizational receptivity to the IUWM approach. Experience has shown that when these are not well matched, there is the risk that the status quo will be further reinforced within the broader sector, through community and organizational resistance. Environmental plans and policies are more likely to be successfully implemented if they are well matched to the needs of the local social context, rather than being overlooked because of a lack of strategic understanding of the values held by different community groups and networks.

22.2 INTERACTIONS WITH OTHER URBAN WATER CYCLE COMPONENTS

Chapters 13 to 21 describe how various parts within the bio-physical urban water cycle interact. For example, Chapter 18 describes how combined sewers may influence groundwater, aquatic ecosystems and even the water supply. This chapter focuses at the broad socio-institutional level, which interconnects with each of the bio-physical water cycle components presented in this book. Each part of the water cycle typically falls into a policy and regulatory arena that influences the public administration of the land and water environment. This broadly involves legislation, governments, stakeholder organizations, professionals and communities, which are all elements of the socio-institutional dimension.

Since the early to middle 1800s, modern urban water systems (including supply, waste-water and drainage infrastructure) have been designed to serve social goals, including water supply security, public health, flood protection and social amenity. As society has evolved with the rise of modern environmentalism over the last 40 years, there has been increasing social advocacy for protecting and rehabilitating the health of urban

Table 22.1 Potential interaction of social and institutional elements and other urban water system components

Urban water system component	Possible interactions with social and institutional aspects
Water supply	Community household water supply expectations. Domestic and business demand management and conservation programmes. Water-trading entitlements and regulatory regimes.
Wastewater	Institutional realignment of wastewater to become part of water supply. Community receptivity to water recycling and re-use.
Stormwater	Community land-use based polluting behaviours. Water sensitive gardening, open space and landscaping policies. Professional knowledge integration processes: engineers, urban designers, architects, social scientists.
Wastewater and combined sewers	Recycling and re-use targets, and associated policy settings. Industrial-ecology land-use zoning principle for co-generation.
Groundwater	Distribution of organizational accountabilities for conjunctive water-use. Natural resource data management sharing agreements.
Aquatic ecosystems and urban streams	Social amenity and recreation. Indigenous rights, history and connection. Community values of waterways.
Human health	Community engagement and education programmes associated with wastewater recycling and re-use. Political/media reporting systems on waterway conditions for human contact.

waterways. Therefore, 'environmental protection' has been institutionalized as an additional social goal to the existing well established suite of goals for urban water management. This has stimulated a new field of work in the bio-physical area, such as the development of stormwater quality treatment technologies (e.g. porous pavements, constructed wetlands and biofiltration technologies) and rehabilitation programmes. Table 22.1 briefly lists some examples of how the socio-institutional dimension can interact with other parts of the urban water system.

22.3 THE IMPORTANCE OF SOCIAL AND INSTITUTIONAL DATA

Having a detailed understanding of the socio-institutional context is essential to advancing the IUWM approach, because moving from the traditional urban water management approach is highly dependent on achieving mutually reinforcing change across both the bio-physical and socio-institutional realms (Brown et al., 2006b). For example, in urban places where there are sophisticated water quality management programmes and treatment technologies, there is typically a high level of organizational leadership and competency around waterway protection and rehabilitation. However, when investigated more closely, this status is likely to be the result of a history of strong local community concern and associated action, sometimes called social capital, underpinning the necessary political platform for leadership supporting waterway health protection and providing the associated external resources. The capacity of socio-institutional systems can therefore enable or constrain the adoption of the IUWM approach.

In particular, the social profile can inform the urban water manager of current receptivity to the IUWM approach. This is important information as it is essential that the IUWM plan and associated interventions can be adapted and matched to the current social receptivity to the IUWM approach. There are a number of experiences across the world, particularly with wastewater re-use schemes, where community resistance has inhibited or halted innovative projects. A number of social scientists would argue that the project proponents not only did not conduct adequate social profiling, but also failed to engage effectively the community. It is therefore important that social capacity building actions are skilfully designed so that they enable a shift from the *status quo* towards IUWM, *but are not beyond the receptivity limits of the community*.

Given reports in some of the organizational change literature that around 70% of major change management processes fail, it is important that the urban water manager understands the current organizational profile and key IUWM organizational capacity deficits. Otherwise the process of organizational change to foster more sustainable urban water management is likely to be highly tentative. Therefore actions within the final IUWM plan need to be skilfully matched to the existing capacity of the implementing organizations. If not, there is a strong likelihood of well-intentioned IUWM plans not being effectively implemented. It is also important that the IUWM initiatives proposed, such as new development assessment and approvals procedures and/or professional development training, enable a shift from the *status quo* towards IUWM, *but are not beyond the reach of the organization*. Therefore the projects built into IUWM plans need to be realistically matched to the organization's capacity to deliver in the short, medium and long term.

Outcomes of the social and organizational profiling provide good information to identify the necessary capacity building strategies that foster social and organizational change, to facilitate IUWM implementation. Overall, at a catchment and/or regional scale, social and organizational profiling can be employed as a benchmarking tool for identifying opportunities and 'weakest links' within the socio-institutional system for advancing IUWM.

22.4 CONTEXT MAPPING APPROACH TO COLLECTING SOCIAL AND INSTITUTIONAL DATA

To date there has been limited research or available guidance on how to assess and determine the quality of socio-institutional data for IUWM across a particular location and/or region. There has been a similar lack of practical tools such as templates, benchmarks or set of institutional indicators developed to assist with measuring social and institutional capacity for IUWM. While there are a number of research programmes that are now attempting to address this important knowledge gap, there is much more work to be done in this area. However, given that urban water sustainability issues are increasing in both significance and scope, lack of available practical guidance is an insufficient rationale for not assessing and improving institutional capacity for IUWM. Therefore, the guidance provided in this chapter is based on the author's knowledge gained through research and practice in this area.

To determine the quality of the socio-institutional capacity across a region or catchment, it is recommended that a *practical and participatory approach be employed*. Such approaches should be critically informed by systematically collecting contextual data that integrates the social, organizational and bio-physical attributes of the urban water system under consideration. 'Context mapping' (Figure 22.1) involves an integrated

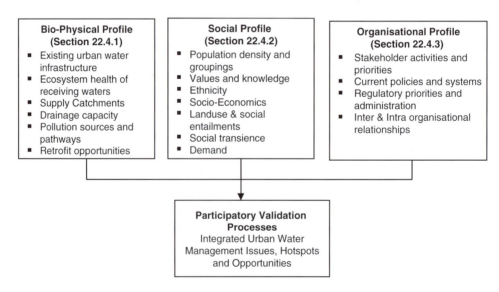

Figure 22.1 Context mapping for assessing the social, organizational and bio-physical dimensions of urban water systems

assessment of the social, organizational and bio-physical attributes of a catchment. Once each of the three profiles are developed they can be analysed together in a participatory context mapping exercise to identify relationships, drivers and barriers across the multiple domains (i.e. bio-physical, social and organizational) that define current urban water issues, practices and opportunities. It is essential that this appraisal is conducted with a good knowledge of the current level of on-the-ground implementation of IUWM.

Taking a participatory approach has a number of benefits. It recognizes the value of the experiential knowledge of frontline implementers, decision makers and community leaders, as well as allowing for collective problem framing, learning and diagnosis of the key capacity development issues at that particular time. Experience suggests that it is important to maintain focus on how to improve the overall 'socio-institutional capacity' of urban water management, rather than permitting focus to drift to technical aspects that participants may feel more confident discussing, such as the design of local water treatment facilities.

22.4.1 The bio-physical profile

The bio-physical profile of the urban water system includes an assessment of each of the urban water components described in this book. The profile is interlinked with the social and organizational profiles in a number of ways, for example:

- Water supply and associated treatment parameters are determined by estimated social (including agriculture, industry etc.) water supply demand profiles and public health expectations
- Water distribution, sewer and drainage services both constrain and are constrained by social values and land-use decisions
- The quality of stormwater runoff is influenced by human behaviour in the catchment area, especially as related to pollution hot spots and urban construction activities.

The bio-physical profiling, in conjunction with the organizational and social profiling, assists in identifying IUWM management issues and, when comprehensively drawn together in a visual reporting format, can be a powerful tool for focusing deliberations amongst stakeholders.

22.4.2 The social profile

The social profile of the target catchment area is an essential component of IUWM planning and management activities. Significantly, the social profile should assist urban water managers in determining the level of community receptivity to new and existing IUWM initiatives for addressing local/regional urban water problems. The process of profiling should not only reveal local urban water knowledge and values, but also provide a rigorous basis for designing future community communication, engagement and participation strategies (previously agreed with the community) and inform the successful implementation of future IUWM interventions.

A social profile contains information about the residential, commercial and industrial land-use values and activities within a given catchment area as it relates to the urban water environment. This information is typically derived from review of local and regional social assessment and policy information (when available). Furthermore, the social profiling approach needs to be based on an integrated qualitative and quantitative research design and data collection and analysis processes. For more information on research design see Creswell (2003) and Maxwell (2005). For further information on qualitative and quantitative social research methods see Bryman (2004), Neuman (2005) and Patton (2002).

The quantitative approach provides a broad statistical picture of the catchment population demography (e.g. age profile, socio-economic status etc.), range of dwelling and transportation stock, and levels of community receptivity to IUWM (i.e. community attitudes, preferences, activities and behaviours) within the catchment. The qualitative approach provides an in-depth understanding of a range of factors, including how community values the water environment in relation to other catchment values, the range of community understandings of the urban water issues and the quality of local and regional social networks and knowledge-sharing pathways.

While social profiling appears to be a relatively under-used practice in the water sector, it has a long history within the social sciences and is well applied in other policy fields, such as community health. The qualitative dimension is often considered particularly important because once social groupings, such as local sporting, religious and environment groupings, and their networks have been identified, they are often relied upon for gaining in-depth insights into the dominant local community values, behaviours and land-use patterns. This qualitative context provides a valuable opportunity for urban water managers to improve their knowledge of the current status of urban water catchment values, water use and polluting activities and to more reliably identify and design strategies for improving community receptivity, and preferably ownership, of future IUWM initiatives.

Social profiling has still not been incorporated as accepted practice in urban water management in many organizations. Therefore it will be more difficult for some urban water professionals to attract the necessary time and resources to develop a social profile. Given this context, a staged social profiling framework is suggested (as presented in Table 22.2) to cater to the different levels of organizational interest and resources available for this task. The framework includes three different qualities or 'tiers' of social profile assessment with increasing levels of complexity, time and cost. Table 22.2 also provides a list of possible techniques related to each of the three tiers of social profiling. The list is not exhaustive, but provides a useful snapshot of possible approaches.

Table 22.2 Three levels of intensity in social profiling

	Data collection techniques	
Tier 1	*Tier 2*	*Tier 3*
Social plan	Social plan	Social plan
Land-use and census data	Land-use and census data	Land-use and census data
	Professional staff interviews	Professional staff interviews
	Community leader interviews	Community leader interviews
		Resident/business survey(s)
		Resident/business focus groups and interviews

Based on past experience, Tier 1 can typically be executed within a couple of weeks; Tier 2 could take from one to two months, and Tier 3 is likely to take up to six months. The following sections provide a brief overview of the three tiers.

While this chapter is intended to be an introduction to social profiling, it must be emphasized that social profiling requires specialist skills. There are a number of complexities that the social scientist must consider, and sometimes control, due to the dynamic nature of communities and social phenomena more broadly. While these issues are beyond the scope of this book, they typically include, but are not limited to, the following:

- changes in social parameters over time, for example, residential turnover, age distribution, household composition, and land-use patterns
- the representativeness of groups and survey respondents in comparison to the total population within the study area, which significantly affects the ability to generalize social survey data
- the impact of external and broader drivers and initiatives, such as national and state policies and regulations on local and regional social knowledge and expectations
- the need for data collection to be updated and conducted ethically to ensure anonymity of participants, where appropriate, and to manage effectively politically sensitive issues.

For further information on these types of social research issues see Neuman (2005), Seale, (2004) and Simeon (2004).

22.4.2.1 Tier 1 social profiling components

Tier 1 is directed at developing a quick and general overview of the community demography and range of community groups and associations that have interests related to the water environment. It provides indicative insight into land-use practices and may reveal some possible issues related to urban water management. It largely relies on existing information available within government agencies, but not always located in organizational areas directly responsible for urban water management. Tier 1 includes a review of the local and regional social plans (where available), analysis of census data and municipal land-use data. For more information on conducting secondary data analyses, including documentation content analysis techniques, see Stewart and Kamis (1993).

Social plans

Social plans are usually accessible through government agency websites and libraries. The social plan typically details the priority issues for the community and the government's actions to address them. Specific theme areas often include housing and infrastructure, environment, transport, safety, living conditions for marginalized sections of the community, recreation and art and culture.

Statements of the local/regional social vision and priorities can provide some insight into community values and the level of environment-related concern. Frequently, a demographic profile and an overview of social groupings and associations are presented within the plan and while the data may not be specific to the catchment boundary, it can offer a readily available and useful snapshot, indicating points of entry for exploring IUWM opportunities.

The strategies listed within the social plan can be considered for aligning priority community development activities that are either in development or already underway, with future urban water initiatives identified in the IUWM planning process.

Census data: residential composition

National census data can be used to develop a statistical social overview of the catchment. The data can often be closely matched to the study or catchment boundary by identifying the data collection districts falling within the specific catchment area that may also cross political boundaries. This can provide a more accurate reflection of the residential character within the catchment if the area of interest does not align with local municipal jurisdictional boundaries.

Census data can be used to analyse the social, economic, dwelling and transportation characteristics of the catchment. Residential information that can typically be extracted includes population density and groupings, ethnicity, socio-economic levels, and rates of residential turnover. These statistics can provide the basis of the 'water balance' calculation by estimation of potable water consumed and wastewater discharged. Population data, when combined with land-use and dwelling data, also enables more specific water balance calculations, including on a sub-catchment and household basis. Residency status (or owner occupier status) can also provide insight into the strategies needed for catchment-wide domestic retrofitting, such as rainwater tanks and stormwater treatment systems within the private property boundary.

The range of socio-cultural backgrounds in the area can also be gauged through census data on age distribution, gender, education, place of birth and language spoken at home. This information can highlight possible cultural sensitivities, as well as the most appropriate communication and education strategies for IUWM within the catchment. It also allows the development of more tailored incentives and disincentives to encourage community acceptance and eventual ownership of sustainable urban water values and practices. As a result, the urban water manager can be more confident in targeting significant social groupings within the catchment.

While this is highly context dependent, there is some evidence to suggest that review of the education status and qualifications combined with the distribution of age statistics can provide an indication of the likely levels of familiarity with environmental matters within the catchment. This information, combined with information on resident occupation and dwelling type within the catchment, can improve the design of outreach and education initiatives.

Relying only on census data for a social profile, cannot provide a complete sense of local environmental and water knowledge, values and attitudes. However, census data themes can often be selected for their potential to inform a range of specific urban water management issues, including water usage, conservation, community education and residential retrofitting.

Land-use data

Knowledge of public and private land-use patterns is essential to accurately measure water use and demand. The proportion of residential, open space and industry water users can aid decision making when setting targets for IUWM interventions. Comprehensive residential data can be retrieved through national and regional census data, while local government records provide considerable information on local business and industry operations.

The number and type of commercial and industrial premises, and open space areas are typically available from local municipal zoning and other land use records. Industry data collected could include the location of commercial and industrial precincts, the type of production, and the density and size of all premises, while open space areas will include both private and public lands such as ovals, parks, golf courses, bushland and other reserves.

The relative size of commercial, industrial and open-space land use is important when planning for IUWM as it directly influences the form, design and number of interventions required. Matched with water demand, wastewater and pollution data, such data help identify major decision makers in business that have the greatest potential to impact on IUWM planning. Key business representatives can be targeted for establishing partnerships, sponsoring projects and education campaigns.

22.4.2.2 Tier 2 social profiling components

Tier 2 builds on the activities of Tier 1 and is focused on developing a more reliable insight into the social profile by applying a qualitative approach to data collection. This will include interview with key community leaders and professional staff from across multiple areas of government agencies. This data collection process is based on the assumption that local community leaders and key professional staff can collectively provide a reasonably reliable insight into local community values, issues and behaviours. However, it must be noted that if social capital is relatively low within the study area (a low range of community connections, relationships and activity within the catchment), it is likely that community leaders and professional governmental staff will have a much more limited perspective to offer. The Tier 2 approach also risks being perceived as a top-down government initiative that may exclude many in the local community.

Professional staff and community leader interviews

This data collection technique requires significant planning and preparation involving:

- setting objectives for the interview process
- designing the interview format, including deciding on issues such as will the interview be exploratory, semi-structured or structured?
- identifying and approaching the key interviewees and determining the importance of their representativeness and whether interviewees should be offered anonymity
- executing the interview, considering data collection and storage methods and whether interviews should be conducted by an independent party (i.e. local university researcher or consultant).

There are numerous social research textbooks that provide detailed advice on how to design and execute successful interviews (see for example Gubrium and Holstein, 2001; Holstein and Gubrium, 2003).

When identifying professional staff to interview it is recommended to first identify all stakeholder organizations related to the study catchment. It is important to identify professionals that work across all the major sectoral areas, such as strategy and policy, engineering, planning, construction and maintenance, and community. The interviews with professional staff can help identify relevant documents (i.e. organizational management plans, water audits, stormwater and drainage plans) and key stakeholders (internal and external) and build relationships with colleagues across many areas and levels. This information can help ascertain internal organizational resources and interest and willingness to commit, promote, plan, implement and manage IUWM in practice.

Community leaders (such as religious leaders, sports club presidents, local school principals, environmental leaders) can provide insights on the local community's activities and relationships towards water use, environmental values, and levels of social receptivity to changing practices and technologies. In addition, community leaders are likely to provide insights into the level of local knowledge regarding hotspots and issues relating to IUWM within the catchment. Information generated from community interviews can also be used to verify interpretations made from the census and business data about the local community. Interviewing community leaders may also assist in identifying individuals who may be effective for championing local IUWM initiatives and providing catalysts for future social change.

22.4.2.3 Tier 3 social profiling components

Tier 3 is the most resource intensive and reliable method of social profiling, and builds on the activities associated with Tiers 1 and 2. It includes detailed residential and business surveys and focus groups to gain insight into community water attitudes, behaviours and receptivity to more sustainable urban water practices. It can also potentially serve as a social capacity building intervention through raising local awareness and interest in urban water issues, providing subsequent opportunities for actively educating and influencing communities and strengthening community relationships and trust through collaborative planning. The Tier 3 approach provides the most reliable and representative account of the catchment and acts as a basis for evaluating the effectiveness of future IUWM initiatives on social change and learning over time. This level of profiling is relatively resource intensive, requiring expertise and months instead of weeks to undertake.

Residential and business survey

The design and implementation of residential or business water surveys needs to include key questions to establish community knowledge, attitudes and behaviours around water management. This information is essential to ensure that the urban water manager is in step with broad community expectations and desires, understands existing levels of support, and targets areas of greatest need. This data collection technique requires significant planning and preparation involving:

- setting the objectives of the survey
- designing the survey, for example, deciding what type of survey instrument will be most effective (i.e. postal, face-to-face or online)
- determining survey sample size

- executing the survey and data analysis, including decisions on, for example, what statistical programmes will be needed to support the data analysis.

There are numerous social research textbooks that provide detailed advice on how to design successful surveys that can assist the urban water manager (see for example Corbetta, 2003 and Neuman 2005). There are also numerous texts that can assist the urban water manager with understanding the statistical analysis and modelling systems for social data (see for example Cramer and Howitt, 2004; Hardy and Bryman, 2004 and Field, 2005).

Experience suggests that the survey should typically consist of both quantitative and qualitative questions in two main categories, namely, knowledge, attitudes and behaviour (including receptivity to change), and demographic characteristics.

Urban water knowledge, attitudes, and behaviour. Specifically, these can include questions to determine community:

- *knowledge of urban water systems*, that is, macro-system knowledge, such as drainage, local knowledge of household/business premise plumbing, volumes used, household/business consumption patterns and water pollution hotspots
- *attitudes toward water and the environment*, that is, priority and concern for the environment and water conitions and the means and responsibility for addressing these concerns
- *behaviour*, that is, current water-use practices and receptivity to sustainable urban water management practices, such as rainwater use and recycled greywater use.

Demographic factors. To ascertain the representativeness of the survey responses, questions about respondents' demographic variables should be included for comparison with similar data for the same area. Categories include gender, age, length of residency in the area, housing type, family type, tenure, employment, income, origin, language and education.

The analysis should statistically assess response frequencies, representativeness, demographic relationships and other correlations that match the hypotheses developed during the survey design. Such analyses help to clarify significant trends and verify key relationships to aid interpretation and explanation of results. This data can then inform IUWM planning processes and local community and business engagement strategies.

Residential and business focus groups and interviews

In depth exploration of key issues identified from the survey results can be provided through community-based focus groups and follow-up interviews. These involve a facilitator directing specific questions to participants in a structured or semi-structured way, depending on the research aims and group dynamics. Using focus groups and interviews allows for wider interpretation of the survey data, and more detailed investigation of research questions. Attitudes and perceptions can develop through group interaction, which ensures the findings are grounded in the social context, reducing potentially erroneous findings.

Focus groups and group interviews are usually designed to include a similar demographic group or sector such as environment, education, community services, business, and sport and recreation. This is likely to improve communication and encourage involvement of all group members, thereby increasing the likelihood of rich discussion. Participants may be recruited through existing contacts at the local municipal level, such as membership of committees or local organizations, or through advertising more widely in media and public access areas.

Information from both the community leader interview groups and resident focus group discussions can be gathered through note-taking or audio-recording for later transcription, depending on resources available. Numerous social research textbooks providing detailed advice on how to conduct successful focus groups are available to assist the urban water manager (see, for example, Putcha and Potter, 2004). Analysis of the interviews and focus groups should aim to establish and compare common themes or unusual or varying responses to surveys and any other data collection. This process requires a systematic process of data coding and analysis with consideration to a range of social research principles such as data triangulation and testing [see Corbetta (2003) and Neuman (2005) for more detail]. Final reporting should include these responses to provide improved understanding of the community composition in terms of urban water and environmental values, knowledge, behaviours and receptivity to IUWM technologies.

22.4.3 The organizational profile

The pursuit of IUWM typically involves many organizations, particularly water utilities, central government agencies and municipalities, as key stakeholders in the management of urban water systems. Integrated urban water management also requires collaboration and coordination across many more disciplinary and sectoral areas (and hence organizations) in comparison to the traditional urban water management approach. Therefore, as part of the IUWM planning and data collection processes, it is important to identify the range of stakeholder organizations within the study catchment, and their existing organizational characteristics and systems that reinforce or inhibit the IUWM approach.

It is broadly recognized that organizations that are successful with advancing more sustainable management practices typically have strong corporate commitment to sustainability, foster a continuous internal learning culture, value inter-departmental interaction, reward organizational champions for sustainability initiatives, and seek to implement cutting-edge management concepts. However, these types of organizational indicators need to be refined for the IUWM field, and can be difficult for urban water managers to reliably measure without well developed benchmarking systems and organizational analysis competencies. While this area of research within the IUWM field is still in its infancy, it is recommended that a practical and participatory approach be undertaken.

The organizational profiling task proposed here is, therefore, designed to assist in identifying the scope of organizational aspects that can provide insight into advancing the IUWM approach. This essentially involves a stakeholder analysis, and case studies of the key implementing organizations in relation to IUWM activities. It is important to note that different organizations will have differing levels of interest and expertise for urban water management. Key implementing organizations typically include the local water utility and municipality. Organizational profiling can be focused by reviewing four capacity building areas (see Brown et al., 2006b), namely:

(1) *external institutional rules and incentives*, such as the regulations, policies and incentive schemes that work to encourage or inhibit IUWM in a given region.
(2) *inter-organizational capacity*, including the agreements, relationships and consultative networks that exist between organizations to allow them to cooperatively promote IUWM.
(3) *intra-organizational capacity*, such as the key processes, systems, cultures and resources within organizations to promote IUWM.
(4) *human resources*, including the technical and 'people' knowledge, skills and expertise available within a region to promote IUWM.

Further information on organizational profiling is provided in Chapter 5 of Australian Runoff Quality (Brown et al., 2006b).

The data collection effort will need to involve a multistage approach, with the first stage focused on mapping the institutional setting that governs the urban water system. This involves a detailed examination of legislation and policy across all areas of land–water interaction. A review of stakeholder organizational literature relating to roles and responsibilities, organizational systems, human resources, as well as strategic political and organizational commitments affecting the urban water environment, should also be undertaken. Typical, relevant organizational documentation may include:

- organizational management plans, including urban water budgetary expenditure, recurrent funding of capital works programmes and urban water services
- local and regional environment plans and development control plans (e.g. zoning regulations)
- water related management policies, such as on-site detention, riparian management,
- IUWM plans, where available
- environmental management policies
- water conservation plans
- government agency websites
- local professional association and non-government organization (NGO) newsletters
- information on external grant funding of projects, addressing what type of urban water projects or programmes get funded by higher tiers of government, how are they received across the relevant organizations and what percent of external funding contributes to the overall project.

For more information on conducting secondary data analysis, including documentation content analysis techniques, see Stewart and Kamis (1993) and Neuman (2005).

The next stage would involve conducting a number of professional staff interviews with representatives across key stakeholder organizations and sectors related to land and water management. This is an important step to clarify gaps and contradictions that are likely to be identified in the documentation analysis.

It is strongly recommended in the last data collection stage of the organizational profiling activity that a participatory stakeholder workshop is facilitated. Key urban water professionals, managers, policy makers and other related implementers are invited to discuss current IUWM organizational implementation impediments and opportunities for change within the catchment and/or region. This is likely to provide a valuable opportunity for validating, completing and seeking agreement on the organizational profile for the catchment. Depending on levels of commitment and available resources, this could be a small informal process, or a large professionally facilitated region-wide process, involving several workshops dedicated to deliberating each of the four capacity areas listed above.

The following sections briefly outline the range of data types that could be collected and synthesized to support the development of the catchment's organizational profile.

22.4.3.1 External institutional rules and incentives data

External institutional rules and incentives provide the overarching governance framework within which urban water management takes place. External institutional rules typically involve formal regulative initiatives requiring the preparation of management plans and the adoption of development assessment and approvals procedures which ensure compliance with IUWM objectives by government and other stakeholders.

External institutional incentives, such as market-based instruments that use financial incentives and disincentives to achieve desired outcomes, are increasingly being advocated as efficient resource management strategies. For example, trading schemes are now emerging in places such as Australia that allow stormwater managers on highly constrained development sites to purchase 'stormwater treatment credits' offsite, so that the desired environmental outcome can be achieved at minimum cost. Such systems could potentially allow financial resources to be channelled from urban to rural parts of the catchment, when greater water quality benefit could be derived for the same cost.

Data collection on the external institutional rules and incentives essentially involves a process that a policy analyst would typically undertake. This is an analysis of all the formal and informal external institutional rules and incentives that may either enable or inhibit the IUWM approach. The following is a suggested list of 'attributes' to be reviewed to provide insight into the external institutional rules and incentives that currently govern urban water management practices across a catchment and/or region. These include:

- land- and water-related policy statements, regulations and standards
- formal design objectives and technical guidelines
- distribution of urban water management accountabilities
- scope of existing IUWM related enforcement strategies
- existing funding mechanisms, available financial resources and incentive structures
- application of market-based instruments for IUWM related practices
- organizational and intergovernmental reward and incentives systems
- support for cross-sectoral stakeholder networks and stakeholder participation
- auditing and performance reporting mechanisms, such as annual 'report card' systems for waterway health.

While the above list is not exhaustive, it does reflect the significant breadth of possible external institutional rules and incentive systems that influence the current and future application of IUWM.

22.4.3.2 Inter- and intra-organizational capacity data

Data on inter- and intra-organizational capacity involves an assessment of organizational relationships, management structures, processes and procedures, and how they influence the practice of IUWM within the catchment or region. This assessment should not only focus at the individual organizational level, but also between water management organizations and other relevant stakeholder organizations. However, it is important to note that specific types of organizations, such as local catchment management organizations, are more dependent on other organizations, because they require support and incentives from state and/or national governments. In a given region, there will be some organizations that are quite advanced with the IUWM approach and associated interventions, while a significant proportion of organizations will be developing sufficient organizational capacity for implementing IUWM.

Inter-organizational capacity data

Recognizing that IUWM involves multiple organizations with a wide range of potentially differing objectives, it is important that there are well established mechanisms for clear communication and cooperation between stakeholder organizations. Data collection efforts need to be aware of, and sensitive to, the tensions which often occur between different levels of government organizations involved in land and water management decision-making and administration. More advanced inter-organizational capacity within a given catchment and/or region typically reflect a well documented and shared understanding of the roles, needs and operational context of each tier of government, organizations and other stakeholders.

Mapping the capacity of the inter-organizational context and its influence on the realization of IUWM is still an underdeveloped area of data collection analysis. However, below is a suggested list that could be considered as a tentative list of benchmark 'indicators' of inter-organizational strategies, mechanisms and activities that could significantly improve the implementation of the IUWM approach across catchments. The following list is not exhaustive, but provides a useful introduction to a number of important inter-organizational capacities for advancing IUWM:

- existence of IUWM inter-agency steering committees, technical advisory groups and project teams
- jointly developed and high level management strategies that include a clear IUWM vision statement or policy, objectives, key actions, responsibilities, time frames and monitoring and reporting mechanisms to ensure accountability of responsible agencies
- memoranda of understanding between agencies to clarify roles and core issues such as funding and to articulate a clear vision with key actions budgeted
- existence of inter-organizational team-building workshops and IUWM events
- resolution procedures for conflicts at the individual and agency levels
- incentives for agency staff to cooperate
- trial and/or pilot IUWM projects with objectives associated building relationships between organizations and departments and sharing knowledge and recognition
- staff exchange programmes to improve human resource competencies and knowledge of other organizational contexts in relation to IUWM

- transparent and readily accessible information management systems which share IUWM information across all relevant IUWM organizations (see also Chapters 9 and 10).

Intra-organizational capacity data

Unlike other assessment techniques, such as rainfall-runoff modelling, a widely verified procedure for intra-organizational capacity assessment of IUWM has not been developed. Recent research into local government capacity by Brown (*in press*) provides a suggested template that could be treated as a formative assessment technique. While this area of research and practice evolves, organizations should recognize the importance of using such tools to advance their own capacity for IUWM while contributing to the ongoing knowledge base and refinement of such techniques.

It is recommended that only the organizations that conduct the bulk of the urban water management activity be subject to intra-organizational analysis, as this is a task that requires sophisticated organizational evaluation skills. Importantly, the level of corporate commitment, organizational policies and systems and human resource competency seem to be some of the most significant organizational variables reflecting capacity for IUWM practices.

In addition to reviewing the organizational documentation described above, there is a need to conduct a series of interviews with internal organizational staff and extended stakeholders. These interviews should be with both senior organizational staff and others more involved in the day-to-day activities associated with urban water management. There are numerous social research textbooks that provide detailed advice on how to design and execute successful interviews that can assist the urban water manager (see for example Gubrium and Holstein, 2001; Holstein and Gubrium, 2003). The types of information sought from organizational staff would include:

- departmental budget allocation to water and related activities, which will gauge the current and possibly the historical importance of various programmes and the department in which they are situated
- knowledge of the functional location of IUWM responsibilities within the organization, in areas such as flooding, water quality, design, construction, maintenance, planning, regulation monitoring and enforcement, and funding
- key positions for IUWM in the organization, that is, who is responsible for what, and where this responsibility rests within the organizational hierarchy
- nature of the internal and external relationships between staff with urban water management responsibilities
- organizational communication channels for IUWM, formal and informal
- the range of trial and externally funded IUWM projects.

22.4.3.3 Human resource data collection

It is important to recognize that the ability of organizations to apply IUWM knowledge and skills is largely dependant on the level of professional IUWM competencies (i.e. technical skills and knowledge) within the organization. In addition, there is also a suite of 'people skills' that are being increasingly recognized as important to ensure effective engagement with professionals from other disciplines, organizations and communities. Such skills are thought to include consultation and participation techniques, group facilitation, leadership, change-management facilitation, relationship-building, networking and principled negotiation.

The purpose of collecting human resource data is to assess the quality of the human resource competencies within the area of focus. This can be thought of as similar to conducting a 'needs analysis'. In an ideal scenario, each organization and/or region would have conducted a knowledge-and-skills-gap analysis as part of assessing the quality of their organizational capacity to critically inform their specific human resource development needs. This would be part of an ongoing, coordinated research and development programme to supply water management practitioners with relevant and up-to-date knowledge. It would typically need to be coordinated by all key stakeholders with responsibility for urban water management to answer critical research questions and fill specific gaps in the knowledge of local stakeholders. This would ideally include peer review processes as a quality control mechanism.

These human resource competencies will vary from region to region, due to different cultural and bio-physical conditions, as well as regional specific technologies and modelling programmes. It is important for the urban water manager to be abreast of what these competencies should be, recognizing that the technical knowledge and skills of professionals working in urban water management need to be frequently updated, given the pace at which new knowledge is being generated, particularly in relation to the design of new water management measures. IUWM professionals also need to keep up to date with the regulatory and policy environment that typically undergoes bursts of rapid change. Required knowledge and skills may include those listed below.

Core knowledge and skills of professionals dealing with water in the region

Projects that involve IUWM will typically require professionals with skills in hydrology, civil engineering, geography, ecology, town planning, geology/hydrogeology, landscape architecture, policy science and social science. These skills are generally acquired through tertiary education. Increasingly, more sustainable approaches to urban water management are also being taught at universities to supplement core skills.

Knowledge of natural resources in the region

Sound water management decisions, particularly during the design phase of a large urban development, are often limited by available knowledge of local water resources. Information is needed on aspects such as sustainable yields, necessary environmental water provisions, meaningful receiving water quality objectives, local relationships between catchment imperviousness and ecological health, the required hydrological regime for receiving wetlands, and the value of ecosystem services. Findings of technical studies assessing the ecological health of receiving water bodies in urban areas, or studies determining sustainable water yields from an urban catchment and/or aquifer, can fill such knowledge gaps.

Ongoing, proactive and well coordinated government-funded natural resource management programmes are necessary to build this type of knowledge over time. It is essential that natural resource management scientists in government agencies have a clear understanding of what knowledge and information is needed by local stakeholders in the water management industry and in what form it needs to be available to support decision-making.

Knowledge of and skill in assessing the performance and cost of water management measures

Capacity for the adoption of IUWM requires knowledge of the performance and cost of locally applicable water management measures. Strategies for water conservation and re-use, minimization of wastewater discharges, improved management of ground-water quality and quantity, as well as management of stormwater quality and quantity, all depend on water managers' understanding of IUWM applications. These include applications such as constructed wetlands, biofiltration systems, aquifer storage and recovery systems, water-efficient appliances, greywater re-use systems and other IUWM technologies. This knowledge provides a sound framework for selecting the optimum combination of water management 'best management practices' at a variety of scales.

Knowledge and skills of technical assessment tools to support water management decisions

A range of tools can be used to assess alternative best management practices to manage water. They include a wide array of computer models, cost-benefit analysis incorporating externalities, and multi-criteria analysis such as triple bottom-line assessment method-ologies. These technologies and assessment frameworks appear to vary across the world with some more region-specific technologies available in some places. The development of these tools is a significant focus within the IUWM research field, and it is important that the urban water manager stays abreast of available methods.

Knowledge of social acceptance and expectations of urban water management practices

Increasingly, water managers are assessing technologies against a 'triple bottom line' (i.e. environmental, social and economic outcomes). That is, the social pros and cons of vari-ous options are being evaluated with reference to their cost and technical-environmental performance. With many water management strategies, such as household water con-servation, xeriscaping and treated effluent re-use for toilet flushing, now being targeted at householders and communities, social receptivity become increasingly important. This includes both qualitative and quantitative receptivity to issues such as using treated wastewater to irrigate local parks and playgrounds or buying a property with a sepa-rate ('third-pipe') non-potable water supply system.

Knowledge of water governance issues and research

High-quality research on the effectiveness of strategies to change people's water-related behaviour is needed at all levels, from the individual to the organizational scale. Across the spectrum of water management activities such research is often absent or under-developed compared with bio-physical research. Some research is available on specific non-structural measures for water management, such as water pricing, education, and how best to build the capacity of people, organizations and organizational networks for more sustainable urban water management. Useful information can also be sought from findings of research into specific institutional capacity-building practices, such as the effectiveness of leadership programmes.

REFERENCES

Brown, R., Mouritz, M. and Taylor, A. 2006a. Chapter 5: Institutional capacity. T.H.F. Wong (ed.), *Australian Runoff Quality: A Guide to Water Sensitive Urban Design*, Canberra, Engineers Australia. pp. 5.1–5.22.

Brown, R.R., Sharp, L. and Ashley, R.M. 2006b. Implementation impediments to institutionalizing the practice of sustainable urban water management. *Water Science and Technology*, Vol. 54, No. 6–7, pp. 415–22.

Brown, R.R. *in press*. Local institutional development and organizational change for advancing sustainable urban water futures. *Environmental Management*.

Bryman, A. 2004. *Social Research Methods*. Oxford, Oxford University Press.

Corbetta, P. 2003. *Social Research: Theory, Methods and Techniques*. London, Sage.

Cramer, D. and Howitt, D.L. 2004. *The Sage Dictionary of Statistics: A Practical Resource for Students in the Social Sciences*. London, Sage.

Creswell, J.W. 2003. *Research Design: Qualitative, Quantitative and Mixed Methods Approaches*. Thousand Oaks, CA, Sage.

Field, A. 2005. *Discovering Statistics Using SPSS*. 2nd edn. Thousand Oaks, CA, Sage Publications.

Gubrium, J.F. and Holstein, J.A. 2001. *Handbook of Interview Research: Context and Method*. Thousand Oaks, CA, Sage.

Hardy, M.A. and Bryman, A.E. 2004. *Handbook of Data Analysis*. Thousand Oaks, CA, Sage.

Holstein, J.A and Gubrium, J.F. (eds) 2003. *Inside Interviewing: New Lenses, New Concerns*. Thousand Oaks, CA, Sage.

Maxwell, J. 2005. *Qualitative Research Design: an Interactive Approach*, 2nd edn. London, Sage.

Neuman, W.L. 2005. *Social Research Methods: Quantitative and Qualitative Approaches*, 6th edn. Boston, Allyn and Bacon.

Patton, M.Q. 2002. *Qualitative Research and Evaluation Methods*, 3rd edn. Thousand Oaks, CA, Sage.

Putcha, C. and Potter, J. 2004. *Focus Group Practice*. London, Sage.

Seale, C. 2004. *Social Research Methods: A Reader*. London, Routledge.

Simeon, J.Y. 2004. *Social Science Research*. London, Sage.

Stewart, D. and Kamis, M. 1993. *Secondary Research: Information Sources and Methods*. Thousand Oaks, CA, Sage.

Wong T.H.F. (ed.). 2006. *Australian Runoff Quality: A Guide to Water Sensitive Urban Design*. Canberra, Engineers Australia.

Part III

Case studies

INTRODUCTION

The following chapters (Chapter 23 and 24) describe two case studies which illustrate many of the principles discussed in this book. They provide tangible examples of how advances in the collection, validation, storage and use of data for Integrated Urban Water Management can be employed. It is important to be aware, however, that neither of these case studies illustrates a completely integrated approach to managing the entire urban water cycle. Indeed, both were designed to meet specific objectives. There are as yet no known cases in which a genuinely integrated monitoring programme across the entire urban water cycle has been undertaken. These case studies therefore provide an illustration of what is possible and allow the reader to imagine, using the principles outlined in this book, how they could be further improved.

The first case study derives from a project of the *Observatoire de Terrain en Hydrologie Urbaine* (OTHU), or Field Observatory for Urban Hydrology, in Lyon, France. It describes an integrated programme for monitoring stormwater runoff and infiltration, as well as subsequent impacts on the level, quality and ecology of groundwater. The OTHU Project was developed with a truly multi-disciplinary approach, involving hydrology, hydraulics, chemistry, biology, ecology and soil science. The result is a monitoring programme that solves many of the problems of integrated monitoring, because issues such as measurement of compatible variables at appropriate spatial and temporal scales were addressed by all scientific disciplines at the start of the project. Processes of data validation and assessment of uncertainty are also very well elaborated in the OTHU Project.

It is possible to imagine an expansion of the OTHU Project to include water supply, wastewater, ecosystem health of surface waters, and the role of society in affecting these aspects. This is particularly the case because the OTHU Project has already achieved one of the key prerequisites of such expansion: the collaboration between many relevant agencies – universities, engineering schools, the municipality and the water agency – with each being involved in the formulation of project objectives and the sharing of results and data.

The second case study describes the application of a wireless sensor network (WSN) to large-scale water and wastewater infrastructure in Boston, USA. In this project, the focus is on the application of technology to allow integrated monitoring of hydraulics, water quality and combined sewer overflows, as well as acoustic leak detection. The network showcases the use of low-cost, reliable sensors capable of collecting extremely high resolution (both spatial and temporal) data even in environments traditionally too harsh for permanent sensors. The sensor network uses technology with very low power consumption, thus enabling a battery life of several years and facilitating remote deployment. Collected data are relayed to a central server for validation, analysis and use. Future expansion will see the network developed into an integrated monitoring and real-time control system for water supply and wastewater networks.

Again, one could imagine further developments of such a sensor network to include data collection on surface waters (e.g. streams, lakes, and water supply reservoirs), groundwater and stormwater. This would provide water managers with high resolution, time-synchronized data, making the task of understanding interactions between the components a much more readily achievable task.

Chapter 23

The OTHU Case study: integrated monitoring of stormwater in Lyon, France

J.-L. Bertrand-Krajewski[1], S. Barraud[1], J. Gibert[2], F. Malard[2], T. Winiarski[3] and C. Delolme[3]

[1]Laboratoire LGCIE, INSA-Lyon, 34 avenue des Arts, F-69621 Villeurbanne CEDEX, France
[2]Laboratoire d'Ecologie des Hydrosystèmes Fluviaux (ESA/CNRS 5023), Université Claude Bernard Lyon I, Bât 403, 43 bd du 11 Novembre, F-69622 Villeurbanne CEDEX, France
[3]Laboratoire des Sciences de l'Environnement, ENTPE, rue Maurice Audin, F-69518 Vaux-en-Velin CEDEX, France

23.1 INTRODUCTION

Greater Lyon is a French urban community of 55 municipalities, 50,000 ha and 1.25 million inhabitants. Two main rivers cross the greater Lyon territory: the Saône River from north to south and the Rhône River from northeast to south. The confluence is located in the municipality of Lyon. The sewer and stormwater drainage systems are combined in the hilly western suburbs and in the centre between the two rivers and separate in the eastern suburbs on the left bank of the Rhône River (Figure 23.1).

In the eastern suburbs there is no surface watercourse and thus for centuries surface runoff has been infiltrated into the ground where it reaches the underlying aquifer at 2 m to 15 m below the surface. The aquifer has very high water quality and needs to be effectively protected. Indeed, it is the single, alternative source of drinking water in case of pollution or an accident in the Rhône River, the normal source of drinking water, initially filtered through natural soil layers, before being pumped, slightly chlorinated and distributed.

Until the late 1960s, surface runoff in the eastern suburbs of Lyon was not significantly polluted and there was no great risk of contamination of the underlying eastern aquifer. Since the rapidly growing urbanization in the 1970s, stormwater has been collected in separate sewers and infiltrated at the detention-settling and infiltration facilities. This is the case in the municipality of Chassieu, where the Django Reinhardt stormwater infiltration facility was built in 1975, rehabilitated in 1985 and in 2002 and retrofitted in 2004 (Bardin and Barraud, 2004) (Figure 23.1).

Stormwater infiltration is increasingly applied in France and especially in Lyon. However, given the characteristics of urban surfaces, and especially the loads of various pollutants contained in stormwater, it is important to assess the impact of stormwater infiltration systems on soil and groundwater. In order to answer these questions, field monitoring and observations are absolutely necessary. The main difficulty in such research lies in the complexity of the observed system and the need for multidisciplinary approaches including hydrology, ecology, biology, chemistry and soil

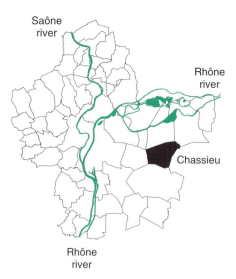

Figure 23.1 The Greater Lyon Metropolitan Area, highlighting the Chassieu municipality (see also colour Plate 23)

sciences. Another difficulty is that measurements have to be carried out *in situ*, in an uncontrolled environment which experiences highly variable phenomena and interferences. Thus very long-term monitoring is required to ensure the representativeness of the results.

Improving the scientific knowledge of the operation and management of the stormwater infiltration facilities and their impacts is one of the key objectives of the OTHU launched in Lyon in 1999 (Bertrand-Krajewski et al., 2000a, 2000b). In 2007, the OTHU research group includes thirteen research laboratories from eight universities, engineering schools and other research institutions in Lyon. Four-year contracts are signed with the Greater Lyon Urban Community, a key partner of the Project both as an end-user of the results and contributor to the definition of the scientific programme. The Rhône-Méditerranée and Corse Water Agency is also associated with the OTHU. More information about the OTHU is available at www.othu.org (in French) (Accessed 02 July 2007).

This chapter describes the Django Reinhardt experimental site in Chassieu, the objectives of the monitoring and how the monitoring system has been designed according to input from various disciplines (biology, ecology, hydrology, chemistry and soil sciences) with an over-arching goal of assessing all of the uncertainties in the measurements.

23.2 DESCRIPTION OF THE DJANGO REINHARDT FACILITY AND ITS ENVIRONMENT

23.2.1 Catchment and drainage system

The catchment is a 185 ha in the industrial area of Chassieu with a flat topography (mean slope of 0.4%) and an imperviousness of about 75% (Figure 23.2). The aquifer is located approximately 13 m below the base of the infiltration basin. The unsaturated zone is thus very thick and plays a major role in retention and processing of stormwater pollutants before they reach the underlying aquifer (Gautier, 1998).

The catchment is drained by a separate stormwater system. Its outlet is the Django Reinhardt facility, which extends over 2 ha and is composed of two compartments: (i) a detention and settling basin and (ii) an infiltration basin, each of about 1 ha. The volumes of the two compartments are respectively 32,000 m^3 and 61,000 m^3 (Figure 23.3). Stormwater flows successively through: (i) the detention and settling basin, (ii) a flow

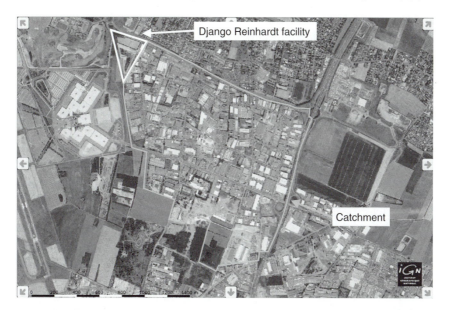

Figure 23.2 Aerial view of the Chassieu catchment and Django Reinhardt stormwater infiltration facility (see also colour Plate 24)

Figure 23.3 The two basins of the Django Reinhardt stormwater infiltration facility (see also colour Plate 25)

Source: OTHU.

control device, (iii) a 60 cm circular connection pipe and (iv) the infiltration basin. The detention and settling basin is also equipped with an overflow structure in case of exceptional storm events (Figure 23.4).

23.2.2 Soil and alluvial aquifer characteristics

The Django Reinhardt infiltration basin is located over quaternary fluvial and glacial deposits of the St Laurent de Mure-Decines-Chassieu corridor (direction SE to NW). The aquifer material overlying an impervious substratum of crystalline formations (tertiary mollassic sands), has a local thickness of approximately 30 m to 35 m, a permeability of about 7 to 9×10^{-3} m/s (Burgéap, 1995) and a transmissivity of 0.0075 to 0.075 m^2/s. The groundwater natural flow averages 110,000 m^3/d. The aquifer material is composed mainly of coarse material: 30% pebbles (diameter $d > 20$ mm), 45% gravel (20 mm $> d > 2$ mm), 20% coarse sand (2 mm $> d > 0.2$ mm) and 5% fine sand (0.20 mm $> d > 0.08$ mm). Samples taken every metre show that the proportions of each category are about the same all along the depth gradient until a depth of 26 m (Figure 23.5a). The grain-size characteristics (deciles d_{10}, d_{30} and d_{60}) do not show any particular structure in relation to the depth (Figure 23.5b), indicating a high degree of homogeneity.

The alluvial deposits, coarse and very permeable, contain highly mineralized groundwater (electrical conductivity of about 950 µS/cm) with relatively high concentrations of sulphate (130 mg/L), nitrate (30 mg/L) and chloride (60 mg/L). There are no significant amounts of organic or metallic micropollutants. The main physical and chemical soil characteristics are: pH (8.2), humidity (5.4%), field capacity (22.6%), saturation capacity (27.3%), organic matter content (1.75%), organic carbon (4.66 mg C/g of dry soil), carbonate (22.7%), TKN (0.59 mg N/g of dry soil), and cation exchange capacity (CEC) (2.75×10^{-5} mol/g).

The global quality of the groundwater is not significantly degraded in this area. Due to the great thickness of the unsaturated zone, self-purification of the infiltrated stormwater can be expected if travel time through the profile is sufficiently long (Ronen et al., 1987; Gibert, 1990), and especially in the first metres of the unsaturated zone (Bouwer, 1991; Roberston et al., 1991).

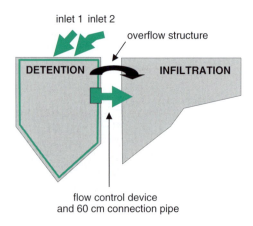

Figure 23.4 Schematic of the Django Reinhardt facility

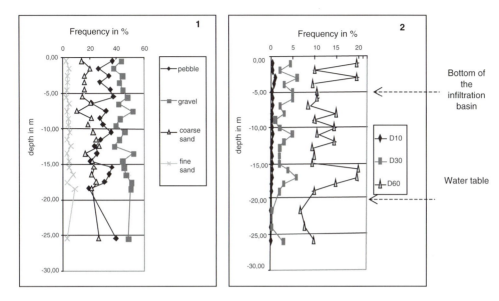

Figure 23.5 Grain size distribution of alluvial aquifer material by depth: Left percentage of different fractions in the 26 samples analysed: Right grain size indices for the 26 samples analysed (see also colour Plate 26)

23.3 MONITORING SYSTEM

The monitoring of the whole system includes the measurement of (i) climatic conditions, (ii) inflow to the detention and settling basin, (iii) settling process and efficiency, (iv) inflow to the infiltration basin, (v) transfers of water and pollutants through the unsaturated soil layers and into the groundwater and (vi) impacts of these transfers on groundwater quality. Two types of impacts are monitored: short-term impacts (infiltration impacts on groundwater consecutively to individual storm events) and long-term (annual and multi-annual) impacts.

23.3.1 Monitoring climate

The site is equipped with a weighing rain gauge, an evaporometer and a thermometer for air temperature monitoring. All devices work continuously and are connected to a data logger.

23.3.2 Monitoring of inlet discharges and pollutant loads in both basins

The objective is to measure inlet discharges and pollutant loads in both basins of the Django Reinhardt facility. Two inlets are considered:

- Inlet 2 of the detention and settling basin (Inlet 1 functions only under very exceptional conditions and is not yet equipped with monitoring apparatus)
- the connection pipe, which is the inlet of the infiltration basin as well as the outlet of the detention and settling basin.

Discharge is measured in the 1.6 m pipe of Inlet 2 and in the 0.6 m connection pipe (Figure 23.6). Each point is equipped with two water-depth sensors (one piezoresistive sensor and one ultrasonic sensor) and with two velocity meters (two different Doppler sensors, one for mean velocity, and one for maximum velocity). As discharge measurements are critical for the project, redundant sensors have been installed (i) to reduce the probability of simultaneous failures, (ii) to achieve a better assessment of uncertainties and (iii) to achieve a better reliability of data through a rigorous data validation methodology (see Chapter 8 for a detailed description of the data validation method applied for the OTHU Project).

Pollutant concentrations and loads are measured through continuous monitoring of pH, electrical conductivity, turbidity and temperature. As turbidity provides critical continuous information about pollutant concentrations, because of its correlation with total suspended solids (TSS) and/or chemical oxygen demand (COD) concentrations (Bertrand-Krajewski, 2004), redundant nephelometric sensors are used. Since 2005/2006, UV-visible spectrometers have been tested to evaluate online TSS and COD equivalent concentrations (Gruber et al., 2006; Figure 23.6).

Pollutant loads are also evaluated from samples taken with a refrigerated automatic sampler. Two modes of sampling may be used: instantaneous sampling to determine pollutographs, or mean (composite) samples to evaluate event mean concentrations. The following pollutants are analysed: suspended solids, organic descriptors (COD, TOC, BOD_5, DOC), nutrients (NO^{3-}, TKN, TN, PO_4^{3-}, TP, etc.), heavy metals (Zn, Pb, Cr, Cu, Cd, etc.), hydrocarbons, and pesticides. Not all are measured systematically, but, as much as possible, the same pollutants are measured at all steps from the surface runoff to the groundwater.

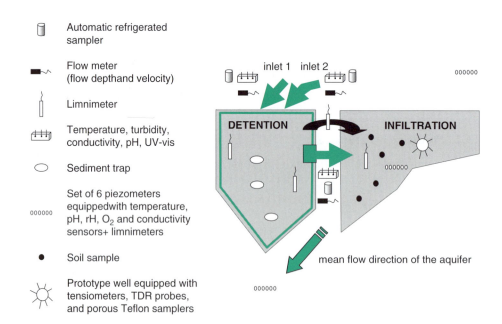

Figure 23.6 Schematic of discharge and pollutants monitoring equipment installed in the Django Reinhardt facility

All sensors (except flow meters) are located in a shelter where a transit flume is continuously supplied with effluent by means of a 1 L/s and 1 m/s peristaltic pump (Figure 23.7). Each sensor is connected to a central data logger which stores all values at a two-minute time step. All data are sent by modem each night to the laboratory. Each sensor is initially and then regularly calibrated to ensure data quality and reliability. Specific procedures for calibration, verification and maintenance have been written (Mourad, 1999; Bertrand-Krajewski et al., 2000a). A specific methodology for data validation has also been implemented (Mourad and Bertrand-Krajewski, 2002, 2003).

23.3.3 Monitoring of settling processes

In order to assess the settling efficiency of the detention and settling basin and of the infiltration basin, two types of measurement are carried out.

- In the detention and settling basin, twelve sediment traps laid at the base of the basin collect sediments which settle during storm events. Using hydrodynamic modelling, the location of the twelve traps was chosen according to recirculation zones, flow velocities and the observed sediment accumulation zones in the tank. The traps are numbered from 1 to 12 according to their altitude (Figure 23.8). Mixture (water and sediment) samples are transported as quickly as possible to the laboratory, where grain size, pollutant loads and settling velocity distributions are measured. In addition, some sediment samples are analysed immediately after storm events, while others analysed after a 'holding period' to evaluate their bio-chemical transformations.
- In the infiltration basin, surface soil samples are taken and analysed each year.

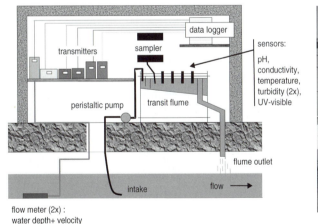

Figure 23.7 Schematic (left) of the monitoring shelter installed on each inlet, with photo (right) of the peristaltic pump and of the transit flume inside the shelter (see also colour Plate 27)

Source: Mourad, 1999; photo by J-L Bertrand-Krajewski.

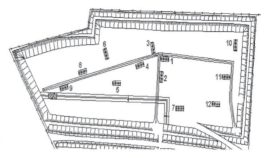

Figure 23.8 **Honeycomb structured sediment traps (left) and their location in the settling basin (right) (see also colour Plate 28)**

Source: photo by A. Torres.

23.3.4 Monitoring the non-saturated zone

This component aims to investigate the role of the carbonated soil in heavy metal retention in unsaturated conditions. Previous laboratory results (Plassard et al., 2000; Fevrier and Winiarski, 1999) show that the retention potential of this carbonated soil does not guarantee the groundwater quality. The temporal stability of these mechanisms, given the environmental conditions, will be investigated.

Two types of measurements are needed to evaluate the fate of pollutants in field conditions, namely:

• physical measurements using tensiometers for estimating the interstitial pressure, and time domain reflectometry (TDR) probes for assessing the soil water content, allowing the evaluation of flows and of soil hydraulic properties in the unsaturated zone
• chemical measurements of the interstitial water, obtained by both drainage tubes and porous Teflon samplers.

The monitoring system shown in Figure 23.9 is composed of a 2.2 m diameter well, sunk in the base of the infiltration basin. This well is presently 2 m deep (with a planned extension to 6 m if the 2 m prototype gives satisfactory results), whereas the groundwater level is about 13 metres under the basin bottom. The lining of the well is made of concrete. The measurement equipment is installed inside at several depths: 0.50 m, 1 m, 1.50 m, 2 m (and proposed for 2.5 m, 3 m, 4 m, 5 m in the future). Horizontal pipes for tensiometers, TDR probes, porous Teflon samplers and drainage tubes are 2 m long.

23.4 GROUNDWATER MONITORING

The objective is to evaluate the infiltration hydrological regime and seasonal variation in the pollution of the alluvial aquifer, in order to assess the biological and chemical impacts of the infiltration basin (recharge) on groundwater quality. To achieve this, it is necessary to (i) analyse environmental gradients in the water table zone and then identify

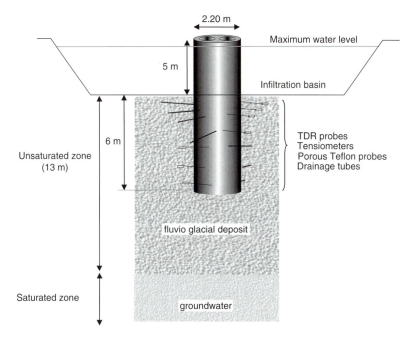

Figure 23.9 Diagram of the prototype of the experimental well for monitoring unsaturated soil (see also colour Plate 29)

the transit of water for various storm events, (ii) determine the ecological processes that create these gradients, and (iii) study the strategies of hypogean living species in response to environmental gradients, and identify the resulting biodiversity patterns. Biodiversity and ecological processes in the upper part of the water table are expected to react to changes of the infiltration regime. The links between the hydrological regime of the infiltration basin, the groundwater quality, the biocenosis and the self-purification efficiency need to be elucidated (Notenboom et al., 1994; Mauclaire et al., 2000).

In order to achieve the above objectives, three sets of six piezometers in three locations have been built: one set is located in the recharge zone of the infiltration basin, while the two other sets are located upstream and downstream of the infiltration basin according to the mean flow direction of the aquifer (Figure 23.6). One piezometer (labelled PZ1 in Figure 23.6) is sunk in the unsaturated zone extending to a depth of about 20 m (measured from the level of the natural soil, which is higher than the base of the infiltration basin) and then in the saturated zone down to a depth of 5 m below the water table. It is screened along its entire length in the saturated zone. The other five piezometers have a screen length of only 0.5 m and respectively intersect the saturated zone at 0, −1 m, −2 m, −3 m and −5 m below the water table (Figure 23.10).

Water depth profiles and physical and chemical parameters are recorded in the piezometers in the screen-region, once per hour, at five depths below the water table. The measured physical and chemical parameters are: temperature, electrical conductivity, pH, redox potential and dissolved oxygen. Measurements are made with a YSI 600 XLM water quality logger.

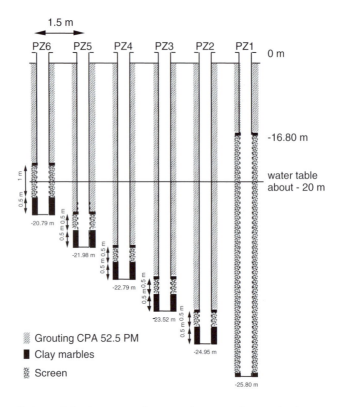

Figure 23.10 Installed set of six piezometers for groundwater quality monitoring

Moreover, for specific storm events, the groundwater is sampled at the five depths below the water table, by means of an inertial pump or a submerged pressure pump. Analyses of groundwater quality include pollutant analyses, potential conservative tracers, electron acceptors for aerobic and anaerobic respiration or end products of the metabolism and substrates for respiration (carbon). Abundance and activity of bacteria are measured and groundwater invertebrates are also sampled.

23.5 CONCLUSION

Like any other natural science, urban hydrology, defined as the science of the production, transfer and impacts of water and pollutants on the urban aquatic environment, is based on field observations and measurements. The observation of such systems is complex because many phenomena interact and involve many related disciplines, including hydraulics, chemistry, biology, ecology, and soil science. Additional difficulty stems from the fact that all these disciplines are often partitioned and do not use the same variables and descriptors, which inhibits global and integrated understanding of the phenomena. Finally, the evolution of the systems (both urban systems and natural hydrological systems) requires long-term monitoring and investigation to enable evaluation of the sustainability of stormwater drainage strategies. This is the reason why the integrated experimentation described in this chapter was developed. Its originality lies in the common thinking and coordinated approach by partners of various disciplines and various points of view (operators and researchers). It lies also in the installation of a durable and coherent monitoring system (the expected duration of measurement is at least 10 years). Integrating several redundancies in the monitoring system is necessary in order to minimize errors and to evaluate better the uncertainties in all measurements.

REFERENCES

Bardin, J.-P. and Barraud, S. 2004. *Aide au Diagnostic et à la Restructuration du Bassin de Rétention de Chassieu*. INSA de Lyon, URGC Hydrologie Urbaine, Villeurbanne, France. (Rapport pour la Direction de l'Eau du Grand Lyon).

Bertrand-Krajewski, J.-L. 2004. TSS concentration in sewers estimated from turbidity measurements by means of linear regression accounting for uncertainties in both variables. *Water Science and Technology*, Vol. 50, No. 11, pp. 81–88.

Bertrand-Krajewski, J.-L., Barraud S. and Chocat, B. 2000a. La mesure de l'impact environnemental des systèmes d'assainissement: exemple de l'observatoire de terrain en hydrologie urbaine (OTHU). *Actes du Third Congrès Universitaire de Génie Civil, 27–28 June 2000, Lyon, France*, pp. 35–42.

Bertrand-Krajewski, J.-L., Barraud, S. and Chocat, B. 2000b. Need for improved methodologies and measurements for sustainable management of urban water systems. *Environmental Impact Assessment Review*, Vol. 20, No. 3, pp. 323–31.

Bertrand-Krajewski, J.-L., Laplace, D., Joannis, C. and Chebbo, G. 2000. *Mesures en Hydrologie Urbaine et Assainissement*. Paris, Editions Tec et Doc.

Bouwer, H. 1991. Groundwater recharge with sewage effluent. *Water Science and Technology*, Vol. 23, No. 10–12, pp. 2099–108.

Burgéap. 1995. *Etude de la Nappe de l'Est Lyonnais*. Lyon, France, Ministère de l'environnement, Direction Départementale de l'Agriculture et de la Forêt du Rhône.

Fevrier, L. and Winiarski, T. 1999. Caractérisation de l'écoulement dans un dépôt fluvio-glaciaire de l'est lyonnais. *Bulletin du GFHN*, Vol. 45, pp. 34–39.

Gautier, A. 1998. Contribution a la connaissance du fonctionnement d'ouvrages d'infiltration d'eau de ruissellement pluvial urbain. PhD thesis, Institut National des Sciences Appliquées (INSA), Lyon, France.

Gibert, J. 1990. Behaviour of aquifers concerning contaminants: differential permeability and importance of the different purification processes. *Water Science and Technology*, Vol. 22, No. 6, pp. 101–08.

Gruber, G., Bertrand-Krajewski, J.-L., de Bénédittis, J., Hochedlinger, M. and Lettl, W. 2006. Practical aspects, experiences and strategies by using UV/VIS sensors for long-term sewer monitoring. *Water Practice and Technology*. Vol. 1, No. 1 Available at: paper doi10.2166/wpt.2006.020 (Accessed 02 July 2007).

Mauclaire, L., Gibert, J. and Claret, C. 2000. Do bacteria and nutrients control faunal assemblages in alluvial aquifers? *Archiv für Hydrobiologie*, Vol. 148, No. 1, pp. 85–98.

Mourad, M. 1999. *OTHU: Etalonnage des Appareils de Mesure et Premier Regard sur le Traitement, la Critique et la Validation de Données*. Villeurbanne, France, Institut National des Sciences Appliquées (INSA) de Lyon, URGC Hydrologie Urbaine.

Mourad, M. and Bertrand-Krajewski, J.-L. 2002. A method for automatic validation of long time series of data in urban hydrology. *Water Science and Technology*, Vol. 45, No. 4–5, pp. 263–70.

Mourad, M. and Bertrand-Krajewski, J.-L. 2003. Pré-validation automatique de données environnementales en hydrologie urbaine. *Actes du colloque A&E 2001 Automatique et Environnement*. Saint-Etienne, France, Ecole Nationale Supérieure des Mines de Saint-Etienne.

Notenboom, J., Plénet, S. and Turquin, M.J. 1994. Groundwater contamination and its impact on groundwater animals and ecosystems. J. Gibert, D. Danielopol and J.A. Stanford (eds), *Groundwater ecology*. San Diego, Academic Press, pp. 477–504.

Plassard F., Winiarski, T. and Petit-Ramel, M. 2000. Retention and distribution of three heavy metals in a carbonated soil: comparison between batch and unsaturated columns studies. *Journal of Contaminant Hydrology*, Vol. 42, No. 2–4, pp. 99–111.

Roberston, W.D., Cherry, J.A. and Sudicky, E.A. 1991. Groundwater contamination from two small septic systems on sand aquifers. *Ground Water*, Vol. 29, No. 1, pp. 82–92.

Ronen, D., Magaritz, M., Almon, E. and Amiel, J. 1987. Anthropogenic anoxification (eutrophication) of the water table region of a deep phreatic aquifer. *Water Resources Research*, Vol. 23, No. 8, pp. 1554–60.

Chapter 24

Wireless sensor network for monitoring a large-scale water infrastructure system, Boston, USA

I. Stoianov[1], C. Maksimović[1], L. Nachman[2], A. Whittle[3], S. Madden[4] and R. Kling[2]

[1]Department of Civil and Environmental Engineering, Imperial College, London SW7 2AZ, UK
[2]Sensor Network Operations, Intel Research, Santa Clara, CA, USA
[3]Department of Civil and Environmental Engineering, MIT, Cambridge, MA, USA
[4]Department of Electrical Engineering and Computer Science, MIT, Cambridge, MA, USA

24.1 INTRODUCTION

This chapter outlines the application of wireless sensor networks for monitoring large scale engineering infrastructure such as water supply and sewer systems. The proposed monitoring solutions are based on the results of research carried out at the Massachusetts Institute of Technology (MIT), Boston (USA) and the Imperial College, London (UK). A trial application of the technology was carried out in the city of Boston over a two-year period (December 2004 – December 2006). While specific aspects are described, the implemented solution is generic and may be readily applied to a wide range of urban water cycle components.

24.1.1 Overview of Wireless Sensor Networks

A Wireless Sensor Network (WSN) is a collection of autonomous nodes which offers distributed services. Each node is capable of sensing, acquiring and processing the data *in situ* under remote control; all nodes are connected as an ad hoc wireless network to process information in a distributed manner. WSNs link the digital world of computers to the physical world by monitoring variables such as temperature, discharge, pressure or acceleration. WSNs have great potential for the monitoring of large-scale infrastructure.

Over the last two years the Infrastructure Sensing Lab and UWRG (Urban Water Research Group) in cooperation with the Computer Science and Artificial Intelligence Lab (CSAIL) at MIT and Intel Research, developed a generic monitoring system for urban water systems, which bridges advances in wireless sensor networks with advances in hydraulic, water quality and hydrological modelling, to develop a range of end-to-end solutions. These solutions encompass several key areas such as: hydraulic steady-state and transient analyses and water quality monitoring in water supply systems, monitoring and control of combined sewer outflows and neighbouring water bodies, monitoring and management of ground water in urban areas, and urban flood prediction and control.

24.1.2 Application potential

The monitoring system that has been developed together by Imperial College, MIT, and Intel researchers provides unique functionalities in terms of sampling rates (up to 2 kHz), time synchronization, bi-directional communication, adaptive sampling, and capabilities to form self-healing ad hoc networks. Within the water industry, the developed system can be used in several key applications:

- near real-time monitoring of water supply systems, including hydraulic and water quality monitoring
- remote acoustic leak detection, for permanent installation in high consequence areas, such as central London or downtown Boston
- urban flood forecasting, control and mitigation of damage
- monitoring and managing groundwater in urban areas, where the problems arise either from groundwater rising (causing flooding of underground assets and structural instability) or lowering (causing soil subsidence and building collapse).

Many of the outlined applications require capturing high frequency data, processing data locally to minimize communication, and automatically varying sampling regimes and rates based on the observed data values. The required spatial and temporal density can only be achieved, however, if the cost of deploying and maintaining these systems is relatively low.

Unfortunately, current monitoring solutions have significant limitations in terms of local intelligence, sampling regimes and cost which precludes their mass adoption. This chapter provides a brief overview of current telemetry and remote data acquisition options.

24.1.3 Conventional data acquisition systems

Currently water utilities are using a variety of telemetry solutions. Frequently these solutions are integrated into SCADA (Supervision, Control, Data Acquisition and Data Analysis) systems (Figure 24.1), which have four major components, interconnected via a network: (a) remote telemetry and automation devices, such as outstations, data-loggers and programmable logic controllers; (b) data gatherers, which acquire and manage the telemetry data; (c) a data server providing telemetry data for users and other applications; and (d) workstations, which provide a user interface.

The outstations are connected to the data gatherers via a range of different systems including telephone lines (PSTN lines, cellular phone modem), leased lines, radio, private networks, fieldbuses (e.g. Profibus), and satellites. The workstations communicate with the data gatherers via local and wide area networks as appropriate. Communications interfaces between workstations, data gatherers and corporate systems are provided through industry standard protocols such as TCP/IP or OSI standards. The data gatherers (DGs) provide the data collection service at the heart of the system by scheduling and executing telemetry polling, managing and distributing the real-time database, and serving the workstations. The outstations are grouped into sets (clusters) and each set is interfaced to redundant DGs: a primary and a secondary, to minimize the risk of failure. During normal operation, the primary DG polls the

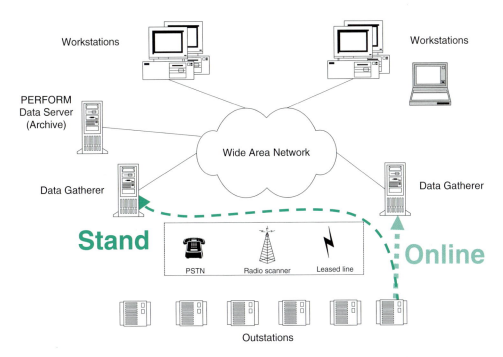

Figure 24.1 Schematic of a conventional SCADA system (Supervision, Control, Data Acquisition and Data Analysis system)

set's outstations and collects the corresponding data. This data is forwarded to the secondary DG where a further copy of the set's database is maintained.

Current SCADA systems are expensive and their deployment within the water industry is limited to critical sites. Many SCADA protocols are proprietary as the legacy of the early low-bandwidth protocols. In general, SCADA systems serve low data-rate applications (e.g. collecting data once every 15 minutes) and provide little flexibility in terms of adaptive sampling regimes, local processing and remote software updates. Many SCADA protocols now contain extensions to operate over TCP/IP, although many utilities prefer not to connect SCADA systems to the internet for security reasons.

24.1.4 Innovative solutions based on Wireless Network System

There is a growing need for monitoring solutions that can be deployed more quickly, and at much lower cost, using in-house expertise, while providing much higher spatial and temporal resolution. These novel monitoring solutions are expected to complement traditional telemetry and SCADA systems while generating a high level of monitoring redundancy. As an illustration of this trend, many sensor vendors have started to offer embedded cellular connectivity in their products.

For example, ABB (www.abb.com (Accessed 02 July 2007)) provides GSM/SMS connectivity to its FieldIT AquaMaster water meters, enabling information to be remotely collected via SMS messaging. The AquaMaster flowmeters can be remotely configured by sending an SMS message. Data are being recorded at predefined intervals of 15 minutes with an option for a high resolution one-minute sampling rate. The data are transmitted once every 24 hours by battery operated units with a projected battery life of 5 years. Data can also be downloaded on demand by dialling an individual sensor to retrieve sensor data (flow rate, pressure, total water consumption, alarms) and status information such as battery level. The battery life is reduced to two months for data collection and communication once every 5 min; and to approximately 6 months for data collection and communication every 15 min. Using commercially available SMS Gateway solutions, the automated SMS meter readings can then be received, decoded and exported to an existing billing application or database to provide near real-time usage information via internet. Wireless communication is also supported via packet-switched or circuit-switched cellular (e.g. CDMA).

The wide-scale adoption of these communication solutions illustrates that the water industry is actively looking for novel low-cost monitoring solutions. Bandwidth, however, remains limited, as the data are primarily used for billing with limited use for near real-time monitoring. High-frequency data collection is nearly exclusively done manually.

24.2 PROJECT RATIONALE AND OBJECTIVES

24.2.1 Rationale

The applications that were selected for the development and evaluation of a novel monitoring solution based on advances in wireless sensor networks include (i) hydraulic and water quality monitoring of water transmission and distribution systems (this also includes capturing transient pressure events); (ii) remote acoustic leak detection including remote cross-correlation; and, (iii) monitoring the water level in sewer collectors and combined sewer outflows. This approach was selected for the following reasons:

- The monitored infrastructure, water transmission, distribution and sewer pipes, share the same spatial distribution. Many cities worldwide face the same challenges in using streets as conduits for utilities to transport water, gas, electricity and telecommunication services.
- Water distribution and sewer pipes are frequently located within close proximity. Under certain hydraulic (pressure) conditions, leaks can become entry points for the intrusion of contaminants which might introduce significant public health hazards.
- Pipeline infrastructure is generally operated by a single company (water utility) and the opportunity to address all these applications within a generic monitoring system enables an integrated modelling and management approach while keeping the cost for the monitoring solution low.
- In previous applications, a number of problems had been identified in attempting to use custom-built data loggers for time synchronized data collection of hydraulic transients (Stoianov et al., 2003b). Extensive research was carried out to address these problems. The practical implementation of proposed solutions, however, was

hindered by costly manual data collection and the technological limitations of current telemetry solutions.

24.2.2 Objectives

The overall goal was to develop a complete solution which includes sensors, wireless data collection system, middleware and back-end applications for data analysis. As this was a proof-of-concept project, the following objectives were defined:

- *outline the requirements for a wireless data collection system*, with applications considered the most demanding within the industry in terms of sampling rates, bandwidth and operational environment in order to demonstrate the system's suitability for a wide range of water-related purposes
- *evaluate the cost of deployment, maintenance and ownership*
- *assess the reliability and robustness of sensors and wireless sensor nodes* under extreme environmental conditions (e.g. in sewer collectors)
- *evaluate the performance of the deployed platform* in terms of reliable data transfer, processor speed and network bandwidth in dense urban environments
- *learn from operating the data collection network* over an extended period of time (24 months) in collaboration with Boston Water and Sewer Commission (BWSC), under real-life conditions.

24.3 COMPONENTS OF THE BOSTON WIRELESS NETWORK SYSTEM

24.3.1 Hydraulic and water quality monitoring

Near real-time hydraulic and water quality monitoring in water supply and transmission systems are essential for detecting failures such as leaks and bursts, optimizing operational control, pump scheduling, chlorination and implementing an early warning system for contaminant intrusion. The monitoring process is highly dependent upon the resolution of the measurements and the accuracy of the simulation model. The hydraulic model which solves a system of non-linear equations approximates the network behaviour by calculating pipe flows, velocities, head-losses, pressures and heads, reservoir levels and reservoir inflows. State estimation techniques are well suited for the purpose of online monitoring as they allow tracking of time, varying flows and pressures. These techniques are frequently used in the electrical and gas industries, but the limited number of monitoring points precludes their use in the water industry. State estimation is defined as the computation of the minimum set of values necessary to completely describe all pertinent variables in a given system from some measurement data. The state estimator algorithm maps the available new information from measurements into a state-space using an over-determined set of equations. This is typically formulated as a projection resulting in a minimization problem.

The principles of hydraulic modelling and state estimation can be extended for modelling water quality parameters. Figure 24.2 outlines a monitoring system that is designed to use near real-time data coupled with accurate hydraulic and water quality models for detecting and tracing a contamination event. In this example, simulations were

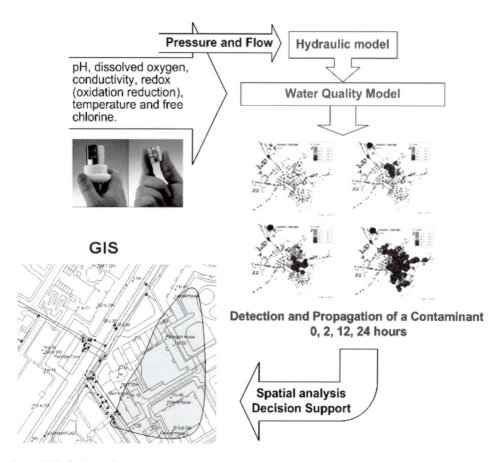

Figure 24.2 Outline of monitoring system using near real-time data coupled with accurate hydraulic and water quality models for detecting and tracing a contamination event (See also colour Plate 30)

carried out to demonstrate how hydraulic data from pressure sensors and flow meters can be combined with water quality data, such as pH, dissolved oxygen, electrical conductivity and free chlorine, obtained from multi-parameter water quality sensors. The developed model simulated the spatial spread of an introduced contaminant at time 0, +2 h, +4 h and +24 h. The data are then projected over the GIS (Geographic Information System) and can be used to minimize the effects of contamination.

The sampling regimes are split into continuous (periodic) mode and burst mode. The sampling regimes and rates defined for this application are summarized as follows:

Continuous mode: Collect for A seconds (e.g. 5 s, 10 s, 15 s, etc., specified remotely by the user) every B minutes (e.g. 1 min, 5 min, 15 min, etc., specified remotely by the user) with a sampling rate (SR) of C samples per second (S/s) (e.g. 1 S/s, 10 S/s, 100 S/s, 1000 S/s, 2000 S/s etc., specified remotely by the user). The outputs include average, minimum, maximum, and standard deviation. The acquired data are communicated to the gateway once every D minutes (e.g. as collected, 1 min, 5 min, 15 min, 30 min, 60 min, 12 h, 24 h or when a threshold is exceeded, options specified remotely by the user and can be changed in near real time). Complete remote control (bi-directional)

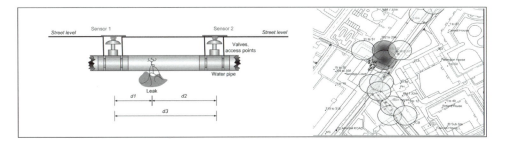

Figure 24.3 Acoustic leak detection with fixed-point remote monitoring (See also colour Plate 31)

enables researchers to change the collection regime, sampling rate and communication frequency. Adaptive sampling is employed for specified parameters. That is, if the data exceeds a pre-determined threshold, the sampling rate is increased while the communication intervals are decreased. For example, if pH goes above 9.5, then collect data every E minutes (e.g. 1 min) and communicate data to data gatherer every F minutes (e.g. every 5 min).

Burst Mode: Employ a sampling rate of 1000 S/s burst mode over a period of 5 min. This will be performed under burst demand request from the server at a pre-defined start time. A minimum of 15 min will be allowed for the server to send the start time for the burst data collection mode to the sensor nodes. The acquired data are used for a sophisticated analysis and modelling of hydraulic transients, detecting large bursts and failures in air valves and fine tuning of control valves in large diameter transmission pipelines (Stoianov et al., 2003a). The acquired high-frequency data are compressed in near real-time using lossless data compression algorithms to reduce communication time and power consumption as the sensor nodes are battery operated. The acquired burst mode data will be time synchronized between sensor nodes located in separate clusters to 1 ms, and remote sensor nodes re-programmed remotely (e.g. a software or firmware update).

24.3.2 Remote acoustic leak detection

Acoustic emission and vibration signals have been widely used as a non-destructive method for detecting and locating leaks in pipes. Generally, a leak generates noise due to the rapid release of energy, which results in a transient elastic wave. To perform leak detection, vibration or acoustic signals are manually acquired at two access points using sensors such as accelerometers or hydrophones on either side of the suspected leak (Figure 24.3).

If a leak exists, a distinct peak may be found in the cross-correlation of the two signals $s_1(t)$ and $s_2(t)$. This gives the time delay τ_{peak} that corresponds to the difference in arrival times between the signals at each sensor. The location of the leak relative to one of the measurement points, d_1, can be calculated using a relationship between the time delay τ_{peak}, the distance d between the access points, and the propagation wavespeed c in the buried pipe:

$$d_1 \frac{d - ct_{peak}}{2}.$$

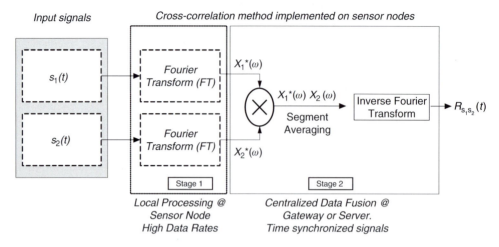

Figure 24.4 Procedure for calculating the cross-correlation function

If $s_1(t)$ and $s_2(t)$ are two stationary random signals with zero mean, the cross-correlation function (Figure 24.4) is defined by (Gao et al., 2006; Oppenheim et al., 1986):

$$R_{s_1 s_2}(\tau) = E[s_1(t)s_2(t + \tau)]$$

where τ is the time lag and E is the expectation operator. The value of τ that maximizes the equation provides an estimate τ_{peak} of the time delay. A procedure to calculate the cross-correlation function using sampled data is illustrated in Figure 24.4. The cross-correlation estimator can be obtained from the inverse Fourier transform of $X^*_1(f)X_2(f)$ and scaled appropriately for normalization, $X_1(f)$ and $X_2(f)$ are the Fourier transforms of $s_1(t)$ and $s_2(t)$ (* denotes complex conjugation).

Commercial products such as MLOG [available at: http://www.flowmetrix.com/ (Accessed 02 July 2007.)] and Phocus2 [available at: www.primayer.co.uk (Accessed 02 July 2007.)] provide functionalities for remote and drive-by data acquisition. Various processing algorithms are used to locally analyse the noise characteristics to provide status information which is defined as leak, possible leak, and no-leak. While these products facilitate unattended, night-time data collection (during hours of low background noise) and approximate identification of leaking areas, they still require manual intervention for accurately pinpointing leaks (the cross-correlation).

The data collection system developed for the Boston project can provide functionalities that go beyond the listed commercial systems, by enabling both local processing of status information and centralized pair-wise data processing of high-frequency time-synchronized data. The developed data collection and processing system can provide significant benefits for monitoring high-consequence pipelines and areas.

The sampling regime required for the remote acoustic leak detection can be summarized as follows:

- *Burst data* collection is carried out **G** times per 24 hours (e.g. **G** = 4 as specified remotely by user)
- Collect data for **H** minutes (e.g. **H** = 5 as specified by user) with a SR of 1000 S/s or 2000 S/s
- Process data to identify a status (*Leak, Possible Leak, No Leak, Do Not Know*). The local data analysis includes time-frequency algorithms together with a classification algorithm. The processing algorithms perform real-time processing on the sensor node (mote). This requires the development of middleware for plugging computational routines that can be remotely queried and updated (over-the-air software update)
- Time-synchronize (time stamp) acquired data with accuracy of 1 ms
- Communicate status information. If status information differs from *No Leak* status, then transfer high-frequency data to a central server and carry out pair-wise cross-correlation.

24.3.3 Monitoring combined sewer overflows

Combined sewer systems are sewers that are designed to collect rainwater runoff, domestic sewage, and industrial wastewater in the same pipe. Most of the time, combined sewer systems transport all of their wastewater to a sewage treatment plant, where it is treated and discharged to a water body. During periods of heavy rainfall, however, the wastewater volume in a combined sewer system can exceed the capacity of the sewer system or treatment plant. For this reason, combined sewer systems are generally designed to overflow and discharge excess wastewater directly to nearby streams, rivers, or other water bodies. These overflows, called combined sewer overflows (CSOs), are among the major sources for water quality degradation of receiving waters, as the discharge contains not only stormwater, but also untreated human and industrial waste, toxic materials, and debris. They are a major water pollution concern for 772 large cities in the US that have combined sewer systems (EPA 2006).

Combined sewer systems would greatly benefit from real-time control (RTC). While RTC systems were first used in the mid-1960s (EPA 1974), recent developments in wireless sensor networks, telecommunication, instrumentation, and automation are turning RTC into a viable solution. RTC management provides a cost-effective solution in comparison to construction projects designed to separate combined sewers in urban areas. RTC systems are designed to assist managers to perform a variety of management functions in a given sewerage system, such as routing flows to a treatment plant or other designated points; controling flooding, overflows, or surcharges; maximizing storage space; optimizing treatment plant capacity; preventing operational problems and protecting receiving waters. Field et al. (2000) defines the basic components of RTC systems as sensors, automated gates, and strategies. The RTC equipment includes measurement devices for water level, flow, rainfall intensity and sometimes pollutant concentration, and regulators for pumps, gates and weirs. The reliability of the RTC equipment, calibration and maintenance present significant challenges as the monitoring equipment is subjected to extreme fouling and corrosion and is frequently placed in difficult-to-access locations. Furthermore, the equipment needs to be intrinsically safe as in the sewer atmosphere it could potentially cause the ignition of gases.

In this study, reliable measurement of water levels in sewer collectors is demonstrated. The equipment is generic to allow the interface of additional sensors. The key requirements of the sampling regime for monitoring combined sewer outflows are as follows:

- *Multiple sensors*: Use multiple sensors to create hardware redundancy for reliable monitoring and sensor fault identification
- *Periodic mode of data collection*: Collect data for I seconds (e.g. $I = 10$ s as specified remotely by user) every J minutes (e.g. $J = 5$ min as specified by user) with SR of 1 S/s
- *Regular data communication*: Communicate acquired data (outputs include average, minimum, maximum, and standard deviation) to gateway every K minutes (e.g. $K = 15$ min as specified by user)
- *Adaptive sampling*: If the collected data exceeds a user-specified threshold, then start collecting data once every minute and communicate the data at 5-minute intervals until the level drops below the threshold
- *Radar input*: Use radar-measured precipitation and/or data from rain gauges to modify the sampling regime parameters
- *Remote specification*: All data collection and communication parameters are specified remotely by the user.

24.4 WIRELESS MONITORING SYSTEM ARCHITECTURE

The major challenge in developing the wireless monitoring system is how to balance the conflict between long-distance communication, bandwidth, local data processing and the constraints of low-power consumption (since many sites do not have mains power available, and must rely on battery operation).

To better address these challenges, a prototype hierarchical wireless monitoring system was developed. The schematic of the system with its main components (sensors, communication, middleware and back-end) is presented in Figure 24.5. The system consists of a three-tier (three subsystems) communication structure which utilizes a cluster-based power management protocol, and a reliable bulk transport. In this way, the subsystems work together to coordinate periodic and burst data collection across a large number of sensing points while maximizing sleep time (to save power). The first tier contains energy-constrained sensor nodes with low transmission ranges, which form clusters. The data from the sensor nodes are transmitted to local data gatherers which comprise the second tier. The data gatherers are not energy constrained as these can be installed at street lights, illuminated street signs and bollards, or equipped with solar panels. This setup eliminates the need for digging up the pavements and maintaining power cables, both costly and risky, particularly in dense urban environments.

The data gatherers combine cluster head nodes which control the sleep schedule of each sensor node and a gateway (an industrial single board computer, SBC). The gateway initiates and controls the long-range communication to a central server via

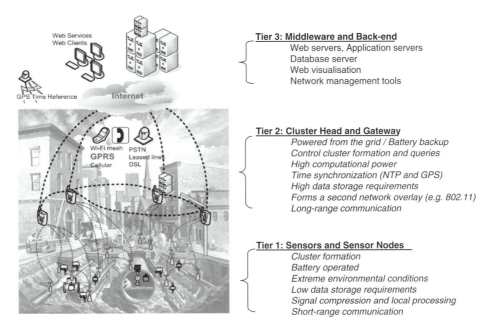

Tier 3: Middleware and Back-end
  Web servers, Application servers
  Database server
  Web visualisation
  Network management tools

Tier 2: Cluster Head and Gateway
  Powered from the grid / Battery backup
  Control cluster formation and queries
  High computational power
  Time synchronization (NTP and GPS)
  High data storage requirements
  Forms a second network overlay (e.g. 802.11)
  Long-range communication

Tier 1: Sensors and Sensor Nodes
  Cluster formation
  Battery operated
  Extreme environmental conditions
  Low data storage requirements
  Signal compression and local processing
  Short-range communication

Figure 24.5 Schematic of the urban water system architecture for WMS, Boston (See also colour Plate 32)

TCP/IP over GPRS (General Packet Radio Service). Secure shell network protocol, SSH-2 (Barrett et al., 2005), is used to establish a secure tunnel between the gateway and the server for bi-directional communication. The SSH protocol guarantees confidentiality and integrity of the data exchanged between the gateway and the server using public-key cryptography and message- authentication codes.

A data control centre on the back-end stores and processes the data on a server at MIT, then displays the acquired data via a web browser [http: //db.csail.mit.edu/dcnui/ (Accessed 02 July 2007.)]. The application is built on open-source web technology, deploying a Linux/Apache/PostgreSQL/PHP stack in a client-server model. Google Maps Programmer's toolkit (API) was used to build geospatial viewing tools for the deployed monitoring locations (Figure 24.6). This open source framework facilitates a rapid application development at low cost. As the user interface is just a common browser window, it runs on any computer and on a hand-held device enabling quick data interrogation and validation by office and field engineers. A basic set of additional functionalities were added to the control centre such as charting near real-time data and status information, alarm notification via email and SMS messages based on thresholds, executing pre-defined queries on historical data, account management for authorized users to view acquired data, and interface between PostgreSQL database and Matlab.

The following sub-sections provide a brief overview of the system components and operation of the first two tiers of the data collection network.

24.4.1 Tier I: sensor nodes

In the proposed two-tier monitoring setup the transmission range requirement of the sensor units is within 10 m to 100 m as they communicate with the data gatherer. At this

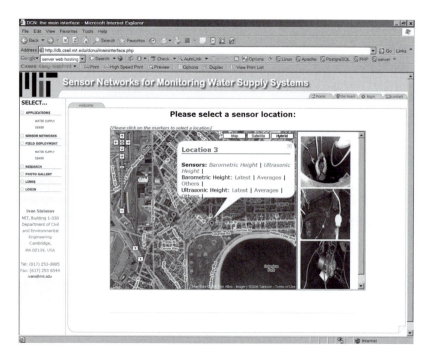

Figure 24.6 Web interface, including Google MAP API, for geospatial viewing of sensor data (See also colour Plate 33)

initial stage of development it was decided to use Bluetooth for short-range communication within a cluster, due to its robustness, high data-transfer rate (1 Mbps) and low cost. The cluster formation includes a small number of nodes with one to two hops exchanging periodic or burst data over relatively short periods of time. The choice of the radio is particularly important as it impacts not only energy consumption, range and reliability of data transmission, but also the software design (e.g. network self-assembly, multi-hop routing and time synchronization).

The applications under consideration relied on the implementation of computationally intensive real-time processing of high-frequency data, which required more advanced microprocessor architecture for the sensor nodes while maintaining low power consumption. These requirements were addressed by a novel sensor node platform developed by Intel Research (Kling, 2003). The first version of the Intel Mote deployed in the Boston Water trial, was built on a 3×3 cm circuit board that integrates a wireless microcontroller module, 64 kB of Ram, 512 kB of FLASH, a CMOS Bluetooth radio, and various digital I/O options using stackable connectors. The radio range was up to 100 m. TinyOS [http://www.tinyos.net/ (Accessed 02 July 2007.)] was used as an open-source operating system for the Intel Mote. Another advantage of the Intel Mote is its modularity, which allows custom sensor boards, interface boards and debug boards to be attached to the system in a flexible manner.

Key components in the tiered communication structure are the cluster-based power management protocol and the reliable bulk transport of high-bandwidth data. These elements work together to coordinate periodic data collection across the nodes within a cluster, while minimizing power consumption and utilizing Bluetooth master/slave and piconet/scatternet operation. Subsequently, the network is self-organizing on start-up employing a distributed node discovery and connection procedure (Nachman et al., 2005). After establishing the basic network, routing information is exchanged between the nodes to permit automatic network repair in the event of node or link failures, while a low-power mode maintains network connectivity. The nodes in a cluster wake up based on a *Wake_UP* parameter communicated by the cluster head at the end of a previous period. Once the nodes are awake, the cluster head initiates routing (Yarvis et al., 2002) to allow all nodes to find a path to the cluster head. Next, each node sends periodic *Trace Route* packets to the cluster head, allowing the cluster head to discover the nodes in its clusters. The cluster head waits a predefined period to allow all nodes to report. Once discovery is complete, the cluster head sends a data capture and transfer request to each node. The resulting data is transferred using the bulk transfer protocol (Nachman et al., 2005). Once data collection is complete, the cluster head sends beacons indicating a start time and duration of the sleep phase to the nodes.

24.4.2 Tier II: data gatherers and gateway

The second tier acts as a cluster head, data gatherer and a gateway, which manages the cluster, controls the long-range communication with the remote server and sends time beacons for time synchronization using a pulse-per-second signal provided by an embedded GPS. These functions generally could be performed by many single board computers. For our field trial a research platform developed by Intel called Stargate [http://platformx.sourceforge.net/ (Accessed 02 July 2007.)] was used, running Linux OS Kernel 2.4.19. General Packer Radio Service (GPRS), available worldwide via GSM cellular networks, was used for long-range communication. For this purpose, a Sierra Wireless A750 GPRS modem was interfaced with the Stargate platform. Furthermore, 802.11b (WiFi) connectivity was added in order to access the gateway locally for drive-by data collection and software upgrade.

24.5 DEPLOYMENT OF THE SYSTEM

In December 2004, in collaboration with Boston Water and Sewer Commission, three monitoring clusters were deployed as a proof-of-concept. The website [http://db.csail.mit.edu/dcnui/PhotoAlbum/index.html (Accessed 02 July 2007.)] presents a series of photos detailing the installation. The trial, which is still running, aims to address a wide range of technical and economic issues, as outlined in Section 24.2, such as ease of deployment, cost of installation and maintenance, reliability of data communication, and the reliability of sensors and their packaging.

24.5.1 Installation

The three clusters selected focused on three monitoring applications, namely, (a) hydraulic and water quality monitoring, (b) acoustic leak detection, and (c) monitoring combined sewer overflows. For all three clusters, the gateways were installed at neighbouring lamp posts which provide direct access to power. In addition, the gateways have back-up battery with re-charging circuitry which allows one week operation in the case of a power failure. A number of additional design challenges had to be overcome, including packaging, temperature control via heat sinks and water proof

Figure 24.7 Cluster 1: Installation of pH probe (top) with antenna embedded in the road surface (below, as visible in circle) (See also colour Plate 34)

ventilation (temperature measured in the gateway enclosure on a hot, sunny day reached 60°C), antenna design and its installation in the road surface.

Cluster 1 (Figure 24.7) includes monitoring pressure and pH in a 300 mm cast-iron pipe which supplies potable water. Data are collected at 5 minutes interval for a period of 30 seconds. The pH probe is warmed up (powered) for a period of 15 s, and then readings are taken once per second for 15 s. The data are communicated to the data gatherer and the server every 5 min. A pH glass electrode with silver/silver chloride (Ag/AgCl) reference cell was used, for which an immersion apparatus was developed to lower the probe in the pipe through a 25 mm access point. Signal conditioning circuitry was also developed to condition the output signal to 0.5 Vdc to 4.5 Vdc, which corresponds to a pH range of 3 to 11. The signal conditioning circuitry for the pH probe consumed less than 10 mW of power. Significant amounts of time and effort were spent on the selection and modification of the pressure sensor. A low-cost sensor (less than US$200) with good accuracy (+/−0.3% FS) and long-term stability was

Figure 24.8 Cluster 2: Installation (right) of the pressure sensor node (left) (See also colour Plate 35)

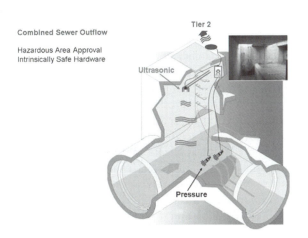

Figure 24.9 Schematic of Cluster 3 for monitoring of combined sewer overflows (CSOs) (See also colour Plate 36)

needed. The most critical parameters, however, were the start-up time, the dynamic response for capturing pressure transients and the sensor performance under aggressive power cycling. In order to address this challenge, an OEM piezoresistive silicon sensor was used, for which an advanced ASIC compensation technology was developed to achieve accuracy better than $+/-0.2\%$ FS, including effects of non-linearity, hysteresis and repeatability; start-up time of less than 20 ms; fast dynamic response and power consumption of less than 10 mW. Pressure data are collected at 5 minutes intervals for a period of 30 seconds with a SR of 600 S/s. The raw data are communicated to the data gatherer every 5 min where the data are compressed and sent to the server.

Cluster 2 includes monitoring pressure in a 225 mm cast-iron pipe (Figure 24.8). Data are collected in a similar manner as for Cluster 1.

Cluster 3 (Figure 24.9) includes monitoring the water level in a combined sewer out-flow collector. As this is an aggressive environment, we decided to use hardware redundancy and to implement a voting algorithm that identifies sensor failures or drifts. This information will optimize maintenance and increase the reliability of data. For this purpose, three sensors were installed: two pressure transducers at the bottom of the collector and an ultrasonic sensor on the top. The pressure sensors are low-power devices consuming less than 10 mW while the ultrasonic sensor is a high-power device consuming around 500 mW to 600 mW. Therefore, pressure sensors were used for continuous (periodic) monitoring while the ultrasonic sensor was only used to ver-ify the readings from the pressure sensors when their difference exceeded a threshold or when the water level exceeded the weir height. Data from the pressure sensors are collected at 5 min intervals for a period of 30 s. Sensors are powered for 10 s before readings are taken with a sampling rate of 1 S/s. Both raw data and average data are transmitted to the data gatherer after every data collection.

24.5.2 Performance

The performance of the data collection network is being evaluated against four criteria: (i) the ability to collect and deliver data to the gateway, (ii) the ability to transfer the data from the gateway to the server via the GPRS link, (iii) the ability to recover from loss or errors and (iv) the long-term performance and durability of the deployed sensors.

During the initial stage (December 2004 to July 2005) a series of problems was observed with the gateways, ranging from strange GPRS modem power modes to cor-ruption of the Linux kernel. Detailed analysis of these problems identified design faults with voltage regulators and the watchdog timer on the Stargate platform. An external watchdog and automated reset feature were added to the gateway nodes to monitor gateway performance. The gateway is rebooted if the application software halts. In addi-tion, the external watchdog timer reboots the gateway once every 24 hours. Adding these features eliminated the observed problems and reduced the risk of unforeseen problems in the gateway software that would require manual intervention by an operator.

The modified gateways were installed in July 2005 and they have been operating for one year of continuous trouble-free operation. The Intel Motes have been operating successfully without hardware failures so far.

The communication performance is variable. The packet reception rate for the clus-ter ranges from 65% to 85%. The GPRS packet reception rate is within 78% to 90%, however, all collected data are transmitted from the gateway to the server as the gate-way archives the data if a connection to the server cannot be established. The sensor node, however, does not currently have the functionalities to separate data acquisition from communication. Thus, if a connection to the cluster head cannot be established, then data are not acquired. A newer version of the hardware (Intel Mote v2) has already addressed this limitation. We are also in the process of logging weather conditions to correlate humidity and rainfall to the packet reception rate.

Battery (6 V 12 Ah) life has been consistent with a duration of around 50 days to 62 days. The Intel Mote consumes 2 mA in sleep mode; 16 mA for Intel Mote plus pres-sure sensor and A/D board; around 30 mA for Intel Mote plus radio, sensor and A/D board. This short battery life is due to the very aggressive data acquisition and communication cycles. Separating the acquisition from communication and adopting

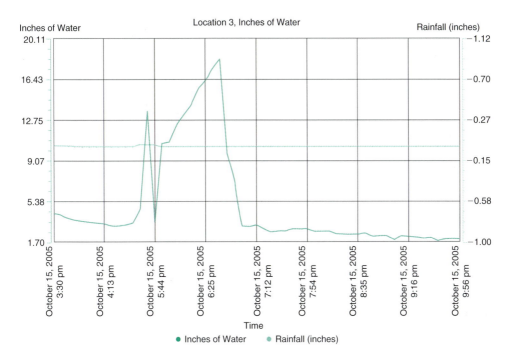

Figure 24.10 Hydrographic record of sewage release, Deer Island Sewage Treatment Plant, Boston, MA, 15 October, 2005

communication intervals of 15 min with adaptive data acquisition and storage will increase the battery life beyond one year.

The performance of the pressure sensors exceeded expectations. The sensors have operated since December 2004 under extreme environmental conditions. The pH sensor, however, has required frequent maintenance and replacement. The replacement of the pH sensor is under consideration with a micro non-glass ISFET probe.

The monitoring system successfully captured several critical events such as the emergency failure of the power supply at the Deer Island Sewage Treatment Plant in Boston in October 2005 (Figure 24.10), when approximately 95 ML of untreated sewage were released into Quincy Bay. The availability of a larger number of monitoring stations such as the one deployed in this case study could have provided near real-time information for utilizing the spare buffer capacity of the system thus significantly reducing the discharge volume.

The system has operated through the snowiest winter on record showing excellent results in terms of reliability, cost of ownership and value of acquired information. Scaling up of the trial is now in progress to monitor all combined sewer outflows in Boston. The ultimate aim is to build an integrated monitoring and control system which reduces the effect of sewer spillages in Boston Bay. The data will also be available on the internet so that users can be notified as soon as a pollution event occurs. This will have significant social benefits for the local community and will assist the water utility in managing such events.

24.6 CONCLUSIONS

This case study has been demonstrated how advances in wireless sensor networks, communication and sensing technologies, could provide much needed increases in spatial and temporal resolution of hydraulic and water quality data. Such advances will deliver better understanding and monitoring of large-scale water supply and sewer systems. The prototype described in this chapter enables us to remotely acquire, view and process both high and low-frequency time-synchronized data from large-scale water supply systems. The field trial with Boston Water and Sewer Commission has provided invaluable information about the performance of sensors, sensor nodes, data collection networks, radio, hardware and software tools.

Finally, this project has demonstrated the use of a sensor network to meet almost one year of continuous operation with a minimum of technical support, even under extreme physical conditions. Several techniques, including careful protocol design, external watchdogs and periodic resetting of the system state, enabled sufficient reliability for completely unattended operation. The data collection in this trial was primarily focused on the proof of concept *vis-à-vis* the communication protocol, reliability of hardware and sensor operations. Future work will involve extending the trial to acquire data of sufficient quality and quantity to support substantially enhanced analytical and operational models.

REFERENCES

Barrett, D.J., Silverman, R.E., and Byrnes, R.G. 2005. *SSH, the Secure Shell: The Definitive Guide*. Cambridge, Mass., O'Reilly Media.
EPA. 1974. *Computer Management of a Combined Sewer System (Seattle, WA)*. Cincinnati, OH, Municipal Environmental Research Laboratory. (Report no. EPA-670/2-74-022)
EPA. 2005a. *Drinking Water Infrastructure Needs Survey and Assessment: Third Report to Congress*. Washington DC, US Environmental Protection Agency. (EPA 816-R-05-001)
EPA. 2005b. *FACTOIDS: Drinking Water and Ground Water Statistics for 2004*. Washington DC, US Environmental Protection Agency. (EPA 816-K-05-001)
EPA. 2006. *Combined Sewer Overflows Demographics*. Available at: *http://cfpub1.epa.gov/npdes/cso/demo.cfm?program_id=5* (Accessed 02 July 2007.).
Field, R., Villeneuve, E., Stinson, M.K., Jolicoeur, N., Pleau, M. and Lavallee, P. 2000. Implementing real-time control schemes offers combined sewer overflow control for complex urban collection systems. *Water Environment and Technology*, Vol. 12, No. 4, pp. 64–68.
FlowMetrix, 2005. *MLOG: Comprehensive Pipeline Integrity Management for Water Distribution Systems*. Available at: *www.flowmetrix.com/doc/MLOG_White_Paper_2005.pdf* (Accessed 02 July 2007.).
Gao, Y., Brennan, M.J. and Joseph, P.F. 2006. A comparison of time delay estimators for the detection of leak noise signals in plastic water distribution pipes. *Journal of Sound and Vibration*, Vol. 292, No. 3–5, pp. 552–70.
Kling, R. 2003. Intel Mote: An Enhanced Sensor Network Node. Paper presented at the International Workshop on Advanced Sensors, Structural Health Monitoring, and Smart Structures, 10–11 November 2003, Keio University, Yokohama, Japan.
Nachman, L., Kling, R., Adler, R., Huang, J. and Hummel, V. 2005. The Intel Mote platform: a bluetooth-based sensor network for industrial monitoring. Paper presented at Fourth International Symposium on Information Processing in Sensor Networks (IPSN 2005), April 25–27, 2005, Los Angeles, CA, US.

Oppenheim, A.V., Schafer, R.W. and Shaffer, R. 1986. *Digital Signal Processing*. Englewood Cliffs, NJ, Prentice-Hall.

Primayer. 2006. Phocus 2: intelligent acoustic leak detection system. *http://www.primayer. co.uk/downloads/phocus/PHOC2.pdf* (Accessed 02 July 2007.).

Stoianov, I., Maksimovic, C. and Graham, N.J.D. 2003a. Designing a Continuous Monitoring System for Transmission Pipelines. Paper presented at CCWI 2003 Advances in Water Supply Management Conference, 15–17 September 2003, London, UK.

———.2003b. Field validation of the application of hydraulic transients for leak detection in transmission pipelines. Paper presented at CCWI 2003 Advances in Water Supply Management Conference, 15–17 September 2003, London, UK.

Yarvis, M.D., Conner, W.S., Krishnamurthy, L., Chhabra, J., Elliott, B. and Mainwaring, A. 2002. Real-world experiences with an interactive ad hoc sensor network. *Proceedings of the International Workshop on Ad Hoc Networking* (IWAHN 2002), Vancouver, pp. 143–51.

Index

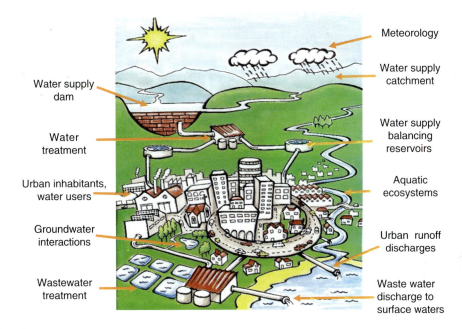

Plate 1

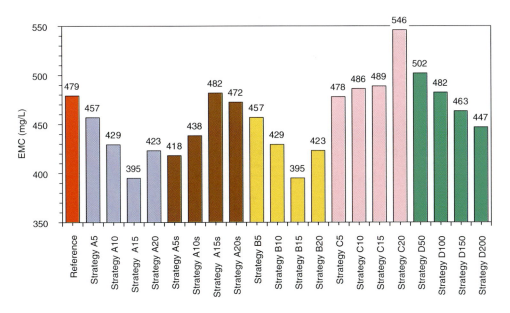

Plate 2

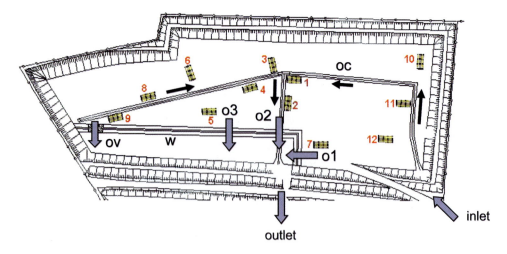

Plate 3

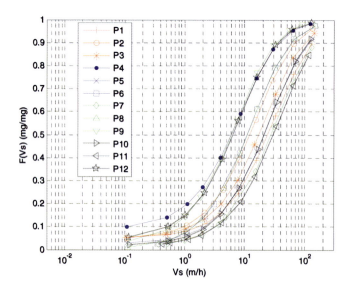

Plate 4

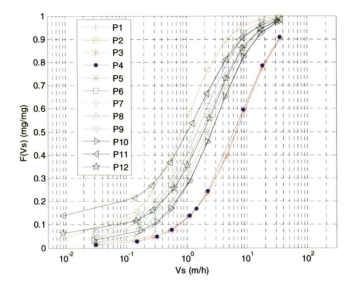

Plate 5

Plate 6

Plate 7

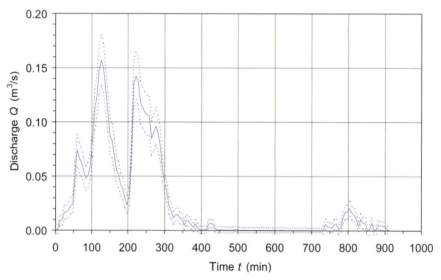

Plate 8

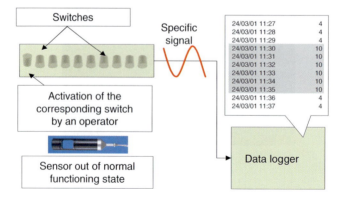

Plate 9

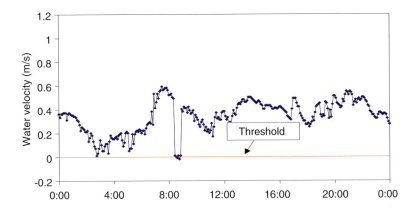

Plate 10

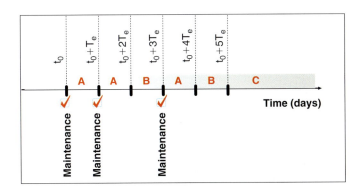

Plate 11

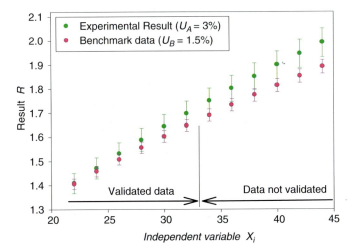

Plate 12

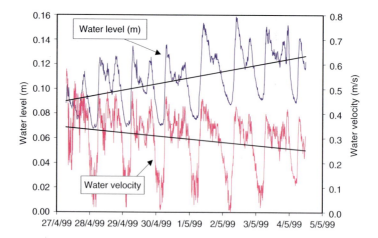

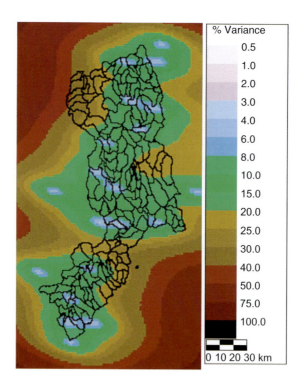

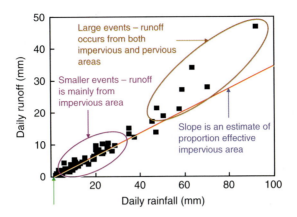

Intercept of rainfall axis is an estimate of
initial loss (or impervious store capacity)

Plate 15

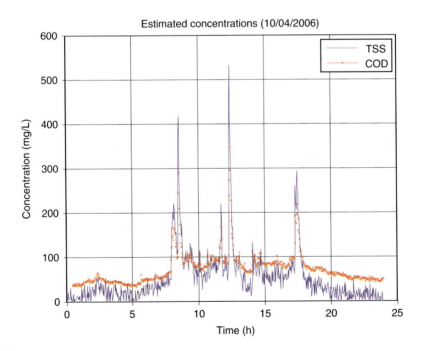

Plate 16

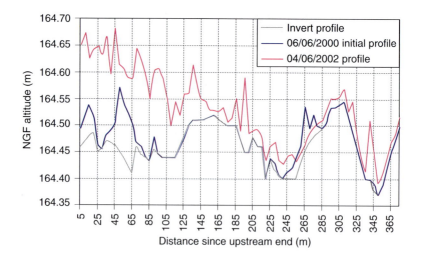

Plate 17

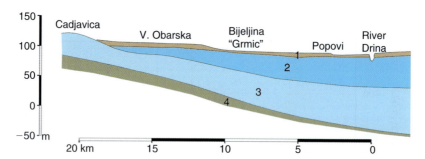

Plate 18

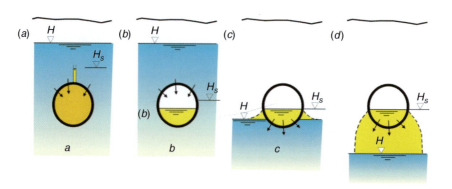

Plate 19

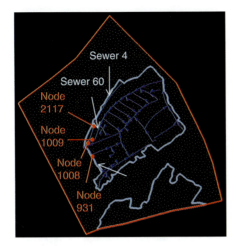

Plate 20

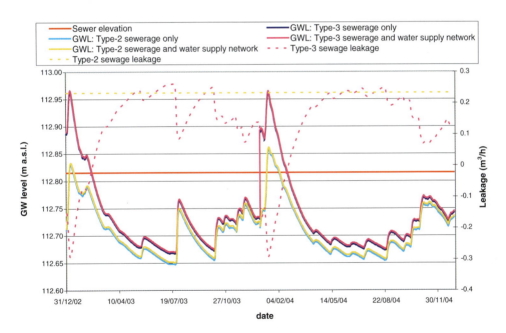

Plate 21

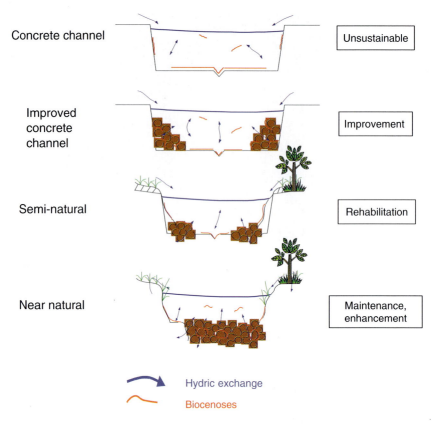

Concrete channel — Unsustainable

Improved concrete channel — Improvement

Semi-natural — Rehabilitation

Near natural — Maintenance, enhancement

→ Hydric exchange

〜 Biocenoses

Plate 22

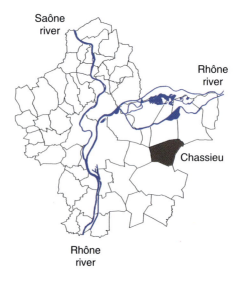

Saône river

Rhône river

Chassieu

Rhône river

Plate 23

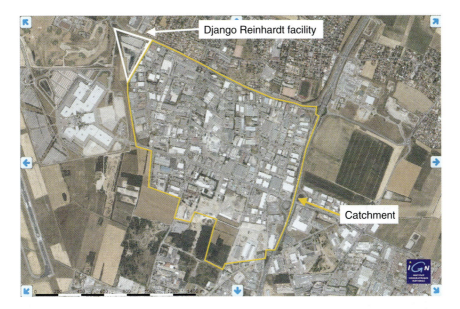

Plate 24

Plate 25

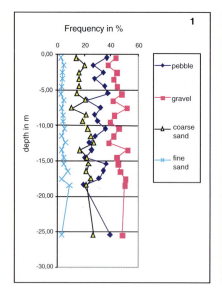

Frequency in % — **1**

Frequency in % — **2**

Bottom of the infiltration basin

Water table

Plate 26

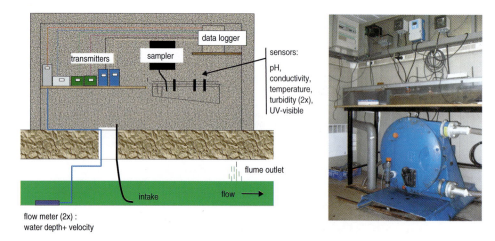

Plate 27

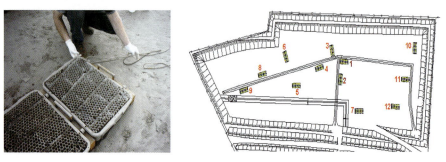

Plate 28

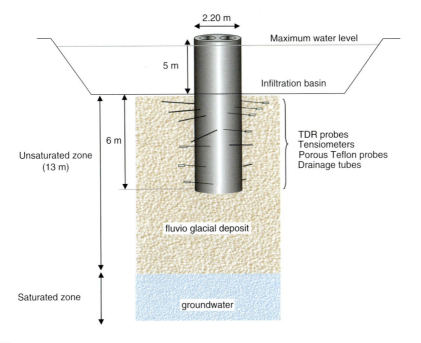

2.20 m

Maximum water level

5 m

Infiltration basin

6 m

Unsaturated zone
(13 m)

TDR probes
Tensiometers
Porous Teflon probes
Drainage tubes

fluvio glacial deposit

Saturated zone

groundwater

Plate 29

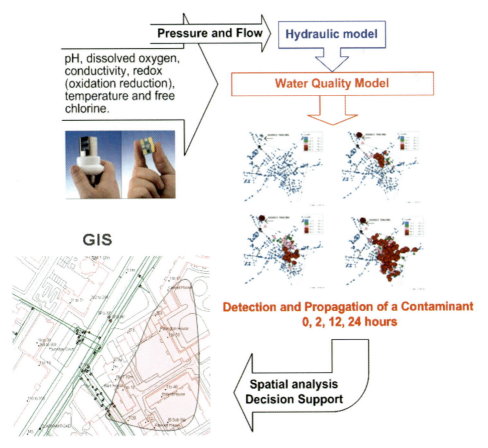

Pressure and Flow   **Hydraulic model**

pH, dissolved oxygen,
conductivity, redox
(oxidation reduction),
temperature and free
chlorine.

Water Quality Model

GIS

**Detection and Propagation of a Contaminant
0, 2, 12, 24 hours**

**Spatial analysis
Decision Support**

Plate 30

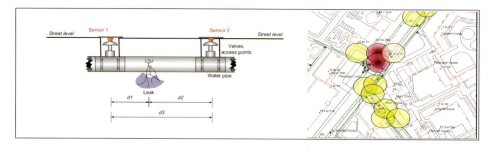

Plate 31

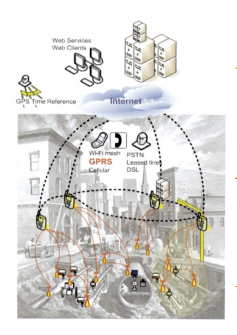

Tier 3: Middleware and Back-end
    Web servers, Application servers
    Database server
    Web visualisation
    Network management tools

Tier 2: Cluster Head and Gateway
    Powered from the grid / Battery backup
    Control cluster formation and queries
    High computational power
    Time synchronization (NTP and GPS)
    High data storage requirements
    Forms a second network overlay (e.g. 802.11)
    Long-range communication

Tier 1: Sensors and Sensor Nodes
    Cluster formation
    Battery operated
    Extreme environmental conditions
    Low data storage requirements
    Signal compression and local processing
    Short-range communication

Plate 32

Plate 33

Plate 34

Plate 35

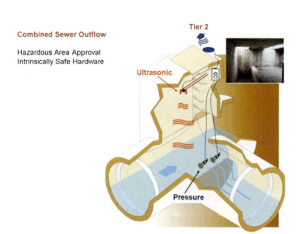

Combined Sewer Outflow

Hazardous Area Approval
Intrinsically Safe Hardware

Tier 2

Ultrasonic

Pressure

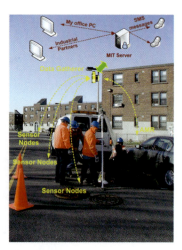

Plate 36